The
Genetics of
Behavior

The Genetics of Behavior

LEE EHRMAN
State University of New York, Purchase

PETER A. PARSONS
LaTrobe University, Bundoora, Australia

SINAUER ASSOCIATES, INC · PUBLISHERS
Sunderland, Massachusetts

TO OUR FAVORITE PHYSICIAN AND DENTIST,
LOUISE AND RICHIE.

THE GENETICS OF BEHAVIOR

First Printing

Printed in U.S.A.

Library of Congress Catalog Card Number: 75-30152

ISBN: 0-87893-173-2

THE COVER

Mutants at *Drosophila melanogaster* shown as mosaics of female (black) and male
(brown) parts. (Courtesy of K. Ikeda and W. D. Kaplan; see pages 165–167.)

Contents

Preface

"The time seems ripe for a modern statement of the division of knowledge we have called 'behavior genetics' . . . not presented as a definitive work, because that would be impossible in a field of study which is in a dynamic stage of growth." The time was May 1960 and the writers, John L. Fuller and W. Robert Thompson, co-authors of the very first book devoted to this hybrid subject (*Behavior Genetics*, J. Wiley, New York). Indeed, behavior genetics may be said to have begun in 1869 with the publication of another book, F. Galton's *Hereditary Genius* (followed by his *English Men of Science. Their Nature and Nurture* [1874] and *Inquiry into Human Faculty* [1883], Macmillan, London).

To us and those (some listed below) who advise us, the time now seems ripe, one and a half decades later, for the appearance of an advanced textbook devoted to behavior genetics, although the field continues in its dynamic stage of growth. This being so, the present text is again " . . . not presented as a definitive work," a task which would even now, still be impossible. Our textbook is directed to those undergraduate — or graduate — level students who already possess a basic background in general genetics. These could be students in biology or psychology or in what now often appears in colleges and universities as programs and majors in psychobiology, itself an increasingly fertile hybrid as can be seen by the rapid development of areas of studies in the behavioral sciences.

Furthermore, since behavior genetics taught as a formal course is a recent innovation, it is our hope that this book will be useful for those already trained in a variety of ways for a multitude of careers. Professional geneticists, animal biologists, and psychologists come to mind first, but in addition aspects of behavior genetics increasingly infringe upon the work of physicians, veterinarians, animal breeders, sociologists, and educators generally, as should become clear from the examples and the organisms discussed. Then too thoughts about some major political controversies of our day may be clarified by an understanding of behavior genetics, not merely that of man but other organisms as well.

We shall consider as behavior any and all of the activities performed by the holistic entity, the organism, in relation to this organism's surroundings, its environment. We do so according to the recommendation of Ethel Tobach (1972), but we confine ourselves to those aspects of an organism's muscular, glandular and neural responses which have been demonstrated, albeit with varying degrees of firmness, to have an underlying hereditary basis — one transmitted via germinal tissues from generation to generation.

The examples we have chosen are necessarily selective, meant to illustrate various aspects of behavior genetics. Omission of some excellent studies is inevitable, just as inclusion of some studies has occurred by virtue of our familiarity with them. Even so, because of the need to be selective, it is our hope that we have managed to provide a relatively advanced and comprehensive text in behavior genetics. We apologize to those who may feel that their work has been neglected, and we would be grateful if our readers draw our attention to matters of this sort that they feel strongly about. Indeed, any comments will be most welcome.

Besides our patient spouses to whom this book is dedicated, we were aided in manifold ways by our students and staff. At the State University of New York at Purchase, they were Goeffrey Ahern, Roslyn Black, Luba Burrows, Dan Cannizzo, Lila Ehrenbard, Toni Faucher, Alena Leff, Max Kirsch, Eileen O'Hara, Dr. Anita Pruzan, Jodi Rucquoi, and Gary Rosenfeld. Bertha Inocencio bore an especially heavy burden. Beside reading the entire manuscript, she typed parts of it and tended flies, phones, et cetera, while we wrote and rewrote. Bless her!

At LaTrobe University in Australia, they were Jeff Cummins, Dr. David Hay, Michele Jones, Lon McCauley, Glenda Wilson, and Cheryl Wynd.

A most special mention must be made of Dr. Nikki Erlenmeyer-Kimling, who improved the entire manuscript with her kind and perceptive criticism.

LEE EHRMAN
PETER A. PARSONS

Introduction

The literature of the earlier part of this century shows clearly that the study of behavior and the study of genetics proceeded independently of each other, with few exceptions. The geneticist, preoccupied with the study of easily defined, mainly morphological or anatomical genetic types, tended to ignore possible genetic components of behavioral traits. No doubt one reason for this was the greater difficulty of measuring behavioral traits as compared with morphological ones; a second reason was that few geneticists had any training in psychology. And when one looks at the psychological literature of the period, it is apparent that experimental and certainly chemical psychologists took little note of genetic components of behavior. Beach (1950), in a rather lighthearted but scientifically serious article, "The Snark Was a Boojum," discussed why genetic variability was largely ignored by psychologists. Nonhuman behavioral work was and still is largely conducted on the Norway rat, *Rattus norvegicus*. A relatively constant genetic type was and is assumed, and this "one" type is then surveyed for a series of behaviors, so that the behaviors themselves are the variables of study. A geneticist, on the other hand, primarily manipulates genetic types, genotypes, in order to see how traits vary according to the genetic type.

Although this was the major trend earlier in the century, the leads necessary for combining the two approaches have been embedded in the literature for a long time. For example, in *Drosophila melanogaster*, the vinegar fly commonly used in genetic experiments, differences in male sexual vigor in different strains were reported by Sturtevant as early as 1915. This is all the more remarkable since research activities on this species began only 5 years or so earlier (by T. H. Morgan and his colleagues in the famous Columbia University Drosophila room). These early experiments in Drosophila behavior, however, were mainly by-products of genetic or evolutionary investigations having other objectives, although the 1940s did see a number of pertinent investigations, principally by Dobzhansky and his colleagues, into sexual isolation between many of the then newly discovered races and species of *Drosophila* (for references see Parsons, 1973). Similarly, during this early period there were reports of behavioral differences among different genetic types in some rodents, principally in house mice and to a lesser extent in rats. These are ably summarized in Fuller and Thompson's (1960) classic, *Behavior Genetics*, a thorough account of the behavior-genetics literature up to the end of the 1950s.

In man, despite a few early reports about twins (e.g., Newman, Freeman, and Holzinger, 1937), the development of a recognizable behavior-genetics approach is recent. Studies carried out by psychologists deal mainly with traits of social significance in which measurements are difficult, as are precise genetic interpretations.

What then are the factors that differentiate behavior from other traits, such as morphological ones, that a geneticist may use? While this question cannot be answered in any absolute sense, the study of behavior genetics has emphases differing from those of other areas of genetics. As such it must be regarded as a true discipline, but one certainly interacting with other subdivisions of genetics such as developmental, population, and evolutionary genetics, and with other subdivisions of behavioral studies. Three main factors suggest themselves as being of greater concern to the behavior geneticist than to other geneticists; the third is essentially unique to behavior genetics:

• Difficulty of environmental control. In an organism like *D. melanogaster*, the environment can relatively easily be precisely and accurately controlled. This also means that the effects of environmental variations can be assessed and quantified successfully, given appropriate experimental designs. In rodents this is also possible. However, here complications begin to appear, since, as we will soon see, variations in early experiences may affect later behavior, an observation perhaps even true for Drosophila. For example, whether mice are brought up together or separately may influence fighting behavior within a given strain. Often these environmental influences on behavior (by no means restricted to work with rodents) are difficult to assess, or worse, may occur without our being aware of them, and differences in results between laboratories could be due to factors of this sort. With man we are dealing with a species in which we have no hope of defining early experiences or of using controlled environments. This stress on the need for environmental control and its study was not always considered important by classical geneticists, but it is no less than imperative for the behavior geneticist.

- Difficulty of objective measurement. For an accurate assessment of the relative importance of genetic influences, environmental influences, and interactions between them, a trait must by definition be measured completely objectively, i.e., without any bias from the person carrying out the measurements. Clearly, in Drosophila objectivity is normally possible for traits such as mating speed (the time elapsed from meeting to mating), duration of copulation, or phototaxis as measured in a maze. In rodents, objective measurement may be somewhat more difficult. However, for traits such as activity measured by using automatic counters in activity wheels or photoelectric cells that count the number of times the animal passes a certain defined spot, high objectivity is possible. Objective measurements of mating rituals, social behavior, and territoriality present greater difficulties, though such measurements have indeed been achieved in well-designed experiments. In man, except for relatively simple sensory perception traits such as colorblindness, objective measurement is a problem of great difficulty. For traits such as intelligence and personality, which are so frequently assessed, it is difficult to avoid the conclusion that some subjectivity is likely to occur in measurement. The problem is that once an element of subjectivity appears, it becomes inaccurate if not impossible to assess the relative importance of heredity and environment, and in our own species we cope with the greatest difficulties of all. This element of subjectivity, which should be minimal for biochemical, physiological, or morphological traits, is therefore a factor that partly differentiates the work of the behavior geneticist from the work of other geneticists.

- Learning and reasoning. Behavior geneticists are concerned with learning and reasoning, while other geneticists generally are not. This concern should be regarded as essentially unique to behavior genetics, when viewed as a branch of genetics. Learning is probably of minor significance in Drosophila, since most behaviors surveyed are innate (i.e., a direct property of the nervous system). However, learned behavior in Drosophila may occur in species-recognition patterns (see Parsons, 1973, for review), but very few other reports of learning exist, and those that do need further substantiation. In rodents, there is ample evidence that early experience affects later behavior patterns (see Erlenmeyer-Kimling, 1972, for review). Patterns and rates of learning also are found to vary among different strains. Therefore, we have both heredity and environment involved in learning, and interactions between heredity and environment. In man, in whom learning and reasoning are developed to the highest level, we have little hope of environmental control, and generally we do not have known behavioral genotypes.

These are three reasons why we believe that a text in behavior genetics should be written. For the geneticist, there are emphases that differ from, or do not occur in, other subdivisions of genetics. For the animal behaviorist or psychologist, the need to appreciate the extent to which the behavioral variability within and among species is under genetic as well as environmental control is a major reason for the need to understand behavior genetics.

But these are not the only reasons why a text on behavior genetics should be written: it is important to consider the role of behavior in evolution. This can be fully accomplished only through an understanding of the genetics of behavior. It is therefore reasonable that the book should conclude with considerations on behavior and evolution, treating topics such as behavior as a component of fitness, behavioral factors in habitat selection, population dynamics in rodents, genetic and sociocultural differences in man, and the evolution of behavior in man. As will be shown in many organisms, such studies are hindered by a lack of knowledge of behavior genetics. However, it is hoped that this book will assist both the advance of behavior genetics and the realized importance of behavioral studies in evolutionary biology.

Our book is organized into four main sections:

- Chapters 2 to 5 provide an introduction to genetics as applied to behavior, proceeding from behaviors under the control of single genes and chromosomes, to those controlled by many genes. Chapter 2 is a brief introduction to genetics using behavioral examples. Its object is to show that the principles of genetics can be studied using behavioral examples. A person without a knowledge of genetics should read it in conjunction with an elementary genetics text, some of which are listed at the end of that chapter.

- Chapters 6 and 7 provide the theoretical bases of analyses of traits controlled by many genes in experimental animals and in man.

- Chapters 8 to 12 look at behavior phylogenetically, considering Drosophila, rodents, man, and various other animals on which behavior-genetics studies have been or may be undertaken.

- Chapter 13 discusses the role of behavior in evolution and in this sense begins the integration of the material of the preceding chapters. Chapter 14, the concluding chapter, presents a final discussion of the place of behavior genetics in evolutionary biology. Some specific areas where behavior genetics has been uniquely successful are discussed, with comments on likely future trends. The point is made that the behavior geneticist of the future must look beyond questions of how heredity and environment, considered discretely, control behavior. To this incipient behavior geneticist, we offer our apologies: we have tried to be more selective than comprehensive, with goals both instructive and inspirational. This text should be worthy of its new vigorously hybrid subject, but textbooks, like organisms, evolve. With this in mind, we formally request suggestions for improvement.

All the references cited in the body of the text are collected together at the end of the book. In addition, selected books and reviews are listed at the end of some chapters under the heading General Readings. They are of relevance to major areas of that chapter and should provide a guide to additional reading.

At the end of this chapter a list of general readings in behavior genetics is given, and at the end of Chapter 2 is a list of some of the elementary genetics texts.

General readings

Ehrman, L., G. S. Omenn, and E. Caspari (eds.). 1972. *Genetics, Environment and Behavior: Implications for Educational Policy.* New York: Academic Press. The proceedings of a research workshop on the genetics of behavior, human and animal, at molecular, cellular, individual, population, and evolutionary levels, with the aim of seeking possible applications in research of interest to education.

Fuller, J. L. and W. R. Thompson. 1960. *Behavior Genetics.* New York: Wiley. The classic text in the field ably summarizing it to the end of the 1950s.

Hirsch, J. (ed.). 1967. *Behavior-Genetic Analysis.* New York: McGraw-Hill. An overview of much of behavior genetics that developed in the early 1960s.

Manosevitz, M., G. Lindzey, and D. D. Thiessen. 1969. *Behavioral Genetics: Method and Research.* New York: Appleton. A comprehensive collection of important original articles contributing to the development of the field.

McClearn, G. E. and J. C. DeFries. 1973. *Introduction to Behavior Genetics.* San Francisco: Freeman. A recent representative account of the field at a relatively elementary level and assuming no previous knowledge of genetics.

Parsons, P. A. 1967. *The Genetic Analysis of Behaviour.* London: Methuen. An account of how behavior can be analyzed genetically, with specific reference to Drosophila, rodents, and man. A discussion of evolutionary implications is included.

Spuhler, J. N. (ed.). 1967. *Genetic Diversity and Human Behavior.* Chicago: Aldine. The proceedings of a conference on the behavioral consequences of genetic differences in man.

Thiessen, D. D. 1972. *Gene Organization and Behavior.* New York: Random House. A brief account of behavior genetics with some stress on evolutionary aspects.

Requisite Genetics

The object of this chapter is to review such basic principles of genetics as are necessary for an understanding of the chapters to follow. A list of appropriate reference texts on general genetics appears at the end of this chapter for those needing further detail, as space is insufficient to discuss the principles of genetics fully.

2.1 Mendelian genetics

If we observe differences in eye or hair color, and note that these differences tend to run in families, it is clearly insufficient to say that such traits are inherited. We wish to find out *how* traits are inherited, and this is one of the main objects of research in the science of genetics. Thus, we must turn to the transmission of observed traits from one generation to the next to see what rules can be formulated.

The appearance of an organism is referred to as its *phenotype,* and this depends on the sum of the genes it possesses (its *genotype*) and on any effects of the environment in which the organism occurs. As will rapidly become apparent, the environment is of particular significance in the study of behavior. Thus behavior may alter according to the environment. A good example in man is phenylketonuria, a genetically controlled disorder arising from an upset in phenylalanine metabolism. As we shall see, phenylalanine is an essential amino acid. One of the effects of this metabolic upset is a lowered intelligence quotient (IQ), a commonly used assessor of intelligence. However, if the error in metabolism is corrected by a special diet, IQ is improved. This shows that the phenotype depends on both the genotype and the environment.

The major theme of the brief survey contained in this chapter is the nature of the genotype. The effect of variations in environment is not greatly considered; however, this issue is taken up in later chapters of the book (especially Chapters 6, 7, and 12). This is because it is important to understand genetic principles as such before introducing complications due to the environment. The units of inheritance are *genes,* and they are carried on *chromosomes* located in cell nuclei. Chromosomes can be observed during cell division. In man there are 46 chromosomes arranged in 23 pairs: in females all 23 pairs are homologous; in males there are 22 homologous pairs plus a nonhomologous pair consisting of an X and a Y chromosome (Fig. 1). The chromosome pairs differ in length and appearance so that some of these pairs can be recognized. During sperm and ovum formation, referred to collectively as *gamete* formation, the number of chromosomes is halved. All the chromosomes in a single gamete differ, i.e., each gamete has *one* member of each chromosome pair. At fertilization, two gametes unite, each containing a set of 23 chromosomes to form the fertilized cell (*zygote*) with 23 pairs or 46 chromosomes once again. This process is shown diagrammatically in Figure 2. The chromosome number of 23 for gametes is referred to as the *haploid* number, and that for zygotes ($2 \times 23 = 46$) is the *diploid* number. More generally, we can write n as the haploid number and $2n$ as the diploid number.

Genes occupy various positions on chromosomes referred to as *loci*. In the mouse, for example, there is a gene on chromosome V called *fidget,* which when present on both members of chromosome V leads to a behavioral phenotype in which the mouse's head moves from side to side continuously. The gene can be written *fi* for short, and such a mouse is said to have a genotype *fifi*. In most mice the fidget gene is not present at this locus, but in place is its normal alternative gene, which can be written as *+*. By normal we mean the gene that is usually present at a locus. Two possibilities then occur: the mouse is genotypically either *fi+* or *++*, and in neither case is the behavioral alteration seen. The fidget phenotype is seen only if two *fi* genes are present, so gene *fi* is said to be *recessive* to *+*. On the other hand, the normal nonfidget phenotype appears if one or two *+* genes are present, so the gene *+* is said to be *dominant* to *fi*. Alternative forms of genes at a given locus (singular of loci), in this case *fi* and *+*, are referred to as *alleles*. Individu-

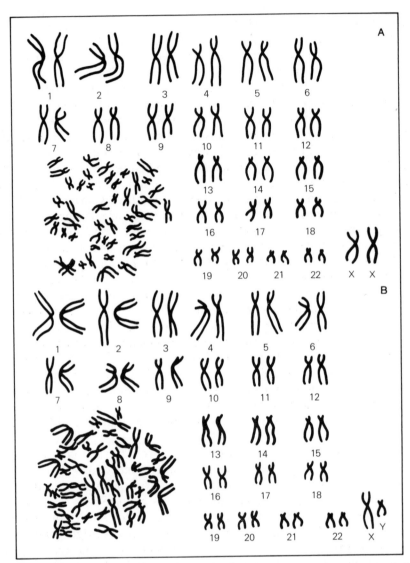

1. **Chromosomes of man. A.** Normal female cell with 46 chromosomes and the normal female karyotype (XX). **B.** Normal male cell with 46 chromosomes and the normal male karyotype (XY). (Courtesy of Professor Raymond Turpin.)

als with identical alleles at a given locus on both chromosomes (*fifi* or *++*) are referred to as *homozygotes*, and individuals with nonidentical alleles such as *fi+* are referred to as *heterozygotes* (mixed zygotes).

It must be stressed that dominance and recessivity are not necessarily complete, as heterozygotes are often distinguishable from both

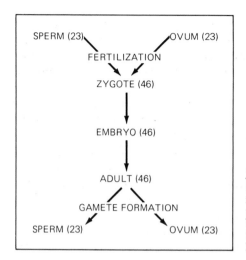

2. **Chromosome number changes in man** during gamete and zygote formation. This can be generalized for sexually reproducing organisms which will be designated in this book by *n*, where *n* is the haploid number and 2*n* the diploid number.

homozygotes. Furthermore, even if dominance at first sight appears to be complete, detailed biochemical studies or other tests may reveal differences between the heterozygote and homozygote. Untreated phenylketonuria in man is an example: to the superficial observer it is controlled by a recessive gene *p*, with phenylketonurics being *pp* and normals *p+* or *++*. However, at the biochemical level *p+* and *++* may be distinguishable, as the *p+* individual tends to have more phenylalanine in his serum. Thus dominance is incomplete, a situation called *semidominance*. Therefore, depending on the level of phenotypic observation, it is possible to obtain somewhat differing conclusions about levels of dominance according to the component of the phenotype being measured.

Suppose a male mouse, *fifi*, is crossed with a female *fi+* mouse. From the *fifi* mouse only *fi* gametes are possible, while from the *fi+* mouse gametes carrying *fi* or *+* are produced. In other words, there is *segregation* into gametes carrying one gene or the other but not both. By chance from *fi+* mice we should get about half the gametes containing the *fi* gene and half with the *+* gene. Diagrammatically the expected gametes and zygotes are:

	♀	½ *+*	gametes	½ *fi*
♂ gametes *fi*		½ *+fi*	zygotes	½ *fifi*

Thus in the offspring we expect ½ *fi+*:½ *fifi* or ½ normal:½ fidget. In other words, the offspring show the effects of segregation during gamete formation. (Often, slightly fewer than $\frac{1}{2}$ fidget mice occur in breeding data because fidget mice are less likely to survive than normal mice.) The principle of *segregation* was first demonstrated by Mendel (1865) in his classic studies on peas and is in fact frequently referred to as Mendel's first law.

Mendel also studied the segregation of two pairs of alleles at two loci simultaneously. If at one locus there are two alternative alleles A and a, and at a second locus B and b, and a *double heterozygote AaBb* is crossed with a *double homozygote aabb*, what should be expected? (Note: we assume A and B to be dominant to a and b, respectively; this is an alternative but equally used terminology.) From the double recessive homozygote, we expect gametes ab. From the double heterozygote, considering each locus separately, there should be $\frac{1}{2}A:\frac{1}{2}a$ and $\frac{1}{2}B:\frac{1}{2}b$. When considering simultaneous segregation at these loci, the simplest hypothesis is that the segregation of the two pairs of alleles occurs independently of each other. This implies that the chromosomes carrying the alleles assort independently during gamete formation. On this hypothesis, the double heterozygote is expected to give the following gametes in equal proportions, thus:

$$\frac{1}{4}AB : \frac{1}{4}Ab : \frac{1}{4}aB : \frac{1}{4}ab$$

The constitutions of these gametes are then revealed by being fertilized with ab gametes to give four discrete and recognizable phenotypic classes:

$$\frac{1}{4}AaBb : \frac{1}{4}Aabb : \frac{1}{4}aaBb : \frac{1}{4}aabb$$

Many pairs of gene loci in many organisms give ratios approximating to this. This principle of *independent assortment* is known as Mendel's second law.

However, if loci are on the *same* chromosome, assortment is not generally independent, and the closer to each other loci are along a given chromosome the more closely *linked* they are in gamete formation. This is because, during gamete formation, genes on the same chromosome may recombine with each other during a complex process referred to as *meiosis*, and the frequency of such *recombination* depends on the spatial distance between the relevant loci. From these frequencies, *chromosome maps* have been constructed for individual chromosomes. Genes located on the same chromosome are said to belong to the same *linkage group*. Thus in man we can expect 23 linkage groups. The mouse, which is discussed extensively in this book, has 40 chromosomes or 20 pairs of chromosomes and 20 linkage groups. In the vinegar fly,

Drosophila melanogaster, another organism of major importance in behavior genetics, the figures are 8 and 4, respectively. A chromosome map of *D. melanogaster* (Fig. 3) consists of 4 linkage groups, as expected. It incorporates loci having mainly behavioral effects, in addition to some other loci commonly used in experimental breeding work. (Note that the number of linkage groups corresponds to the haploid number.)

A further complication relates to sex. In the human female there are 23 pairs of chromosomes totaling 46, but the male has 46 chromosomes made up of 22 pairs plus 1 chromosome, the X chromosome, which corresponds to one of the pairs in the female, and another, the Y chromosome, which does not correspond to any chromosome in the female (Fig. 1 *part B*). Thus the female can be written as 22 + XX and the male as 22 + XY, these being the 22 *autosomal* pairs plus the *sex chromosomes*. Generally, in the organisms discussed in this book, the sex chromosomes function as the sex-determining mechanism. We refer to genes on the X chromosome as being *sex-linked*. Little genetic activity has been identified as being located on the Y chromosome. This means that in females, the principles of heterozygosity and homozygosity apply for the sex chromosomes just as for the autosomes already described. Because the X is paired with a Y in males, however, a rare sex-linked recessive gene shows up more frequently in males than females, since the recessive genes cannot be masked by the corresponding dominant genes. This follows from the general observation that loci on the X chromosome are for the most part not matched by corresponding loci on the Y. Males with loci only on the X are referred to as *hemizygous* for such loci. (Note that other modes of sex determination exist in other organisms but are of minor significance for this text.) Of behavioral interest is the fact that in man the genes for red-green color blindness and for one form of muscular dystrophy are under the control of sex-linked recessive genes, and as expected, these conditions occur in males much more frequently than in females. It is shown in subsequent chapters that various and profound behavioral changes may be associated with sex chromosome changes. Chromosome I in Figure 3 is the X chromosome of *D. melanogaster*. The appearance of the X and Y chromosomes of *D. melanogaster* is shown in Figure 4.

2.2 Quantitative genetics

So far we have discussed variation under the control of specific genes assignable to specific loci on chromosomes, but many behavioral traits are *quantitative* and do not segregate into discrete classes. Examples in man include height, weight, and IQ within a population. This does not mean that specific genes affecting these traits are not known. Indeed,

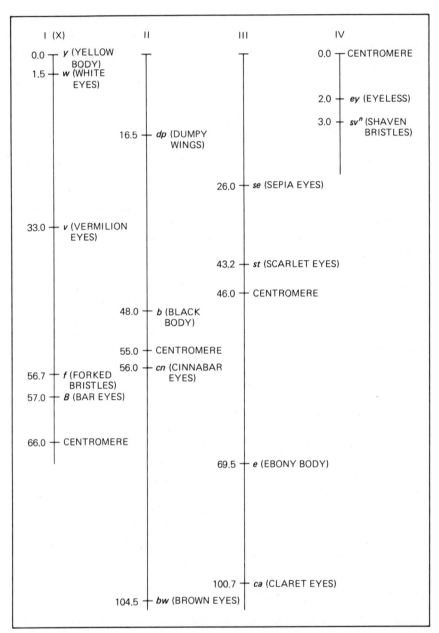

3. **Linkage map of** *D. Melanogaster.* Some of the commonly used genes are included, especially those that have been involved in behavioral work. (The **centromere** is the body to which spindle fibers attach during cell division.) (After Bridges and Brehme, 1944, and other sources.)

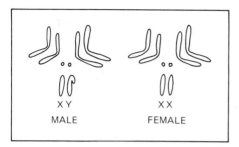

XY
MALE

XX
FEMALE

4. **Chromosomes of D. Melanogaster.** Note the X and Y chromosomes
of the male and the two X chromosomes
of the female.

the phenylketonuria gene has a discrete effect in reducing IQ, as already
seen. The frequency distribution of many quantitative traits approximates more or less closely the *normal distribution* (Fig. 5). The normal
distribution can be completely described in terms of two quantities
(*parameters*). One is the mean or average value. If x_i is an individual
observation and there are n observations, then the mean $\bar{x}$ is given by:

$$\bar{x} = \frac{\Sigma x_i}{n} \tag{2.1}$$

The other parameter is an expression of variability around the mean. In
some cases the variability around the mean is small and in other cases
large (Fig. 5). The term for the parameter measuring variability is the
variance, which is estimated as:

$$\frac{1}{n-1} \Sigma(x_i - \bar{x})^2 \tag{2.2}$$

The square root of the variance is the *standard deviation*. It can be shown
that 95 percent of the population of a normally distributed trait lies
within two standard deviations of the mean. Much of the theory of
quantitative genetics is based on the assumption of a normal distribution. If it is impossible to assume a normal distribution, it may be
possible to find a suitable algebraic transformation that will convert the
data to an approximately normal distribution.

Assuming that a continuously varying trait is partly under genetic
control, it must be asked how the intrinsically discontinuous variation
caused by genetic segregation is expressed as continuous variation.
Suppose two individuals $A/a \cdot B/b$ are crossed, where A,a and B,b are
gene pairs at two unlinked loci, and further suppose that genes A and B
act to increase the measurement of a quantitative trait by one unit, and
genes a and b act to decrease the trait by one unit. It is perhaps less
confusing to write $A/a \cdot B/b$ as $+/- \cdot +/-$, considering A and B genes as

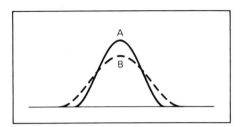

5.   **Normal distribution.** Curves **A**
and **B** have the same mean, but the variance of **B** is larger than that of **A**.

+ genes, and a and b as − genes. Counting the number of + and − genes
gives a metric or quantitative value for a genotype.

The above cross gives five genotypes distributed as in Figure 6, going
from one with four − genes to that with four + genes. The most common
genotype is +/− · +/−, having a genotypic value of zero since it has two
+ and two − genes. This is the mean *genotypic value*. The least common genotypes are the two extremes, +/+ · +/+ and −/− · −/−, with
values of +4 and −4, respectively. If there is a third locus with two
"similar" alleles, then for a cross between multiple heterozygotes the
number of genotypic classes rises to seven, and with a fourth, to nine,
and so on. The differences among the classes become progressively
smaller as the number of segregating loci rises. At the stage when the
differences among classes become about as small as the error of measurement, the distribution becomes continuous, as in Figure 5. In addition, any variation due to nongenetic causes blurs the underlying
discontinuity implied by segregation, so that the variation seen may
become continuous irrespective of the accuracy of measurement.

Therefore, many genes, each with small effects controlling a trait,
superimposed upon variability due to nongenetic or environmental
causes, lead to the expectation of a continuous distribution similar to
that given in Figure 5. Genes that contribute to a quantitative trait, but
that are not directly identifiable by classic Mendelian segregation (i.e.,
cannot be studied individually), are referred to as *polygenes*, and genes
whose effects can be studied individually are referred to as *major genes*.
There is no fundamental distinction between major genes and
polygenes. The terms are merely a matter of convenience, since the
breeding methods used to study the effects of major genes cannot, in
general, be used in the study of polygenes. Even so, it is possible under
certain circumstances to increase the effects of polygenes by statistical,
and perhaps biochemical, techniques to such an extent that, to all intents
and purposes, they appear as major genes.

Behavioral traits such as duration of copulation in Drosophila, activity
scores in mice, and IQ in man are essentially quantitative: for their
analysis an appreciation of the aims and methods of quantitative genet-

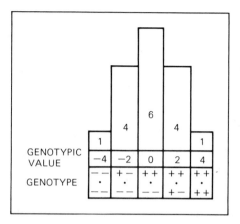

6. **Genotypic frequencies** from the cross $+/- \cdot +/- \times +/- \cdot +/-$ plotted according to the genotypic value (i.e., the relative number of $+$ and $-$ genes). The frequencies of each genotype are given in the histogram.

ics is essential. The basic aim is to divide the phenotypic value (*P*), which we measure, into genotypic (*G*) and environmental (*E*) components. This can be written most simply as:

$$P = G + E \tag{2.3}$$

As we will see later, since continuously varying traits are being considered, we need the phenotypic variance (V_P), which can be split into variance components due to genotype (V_G) and environment (V_E). Assuming no interaction between genotype and environment — which is the simplest assumption possible and which frequently does not hold for behavioral traits — the phenotypic variance is the sum of the genotypic and environmental variances, thus:

$$V_P = V_G + V_E \tag{2.4}$$

It is reasonable to compute the proportion of the total phenotypic variance that is genotypic, thus:

$$\frac{V_G}{V_G + V_E} = \frac{V_G}{V_P} \tag{2.5}$$

This ratio is referred to as the *heritability in the broad sense* or the *degree of genetic determination* of a trait; it is an extremely important entity in the study of quantitative traits including behavioral traits. This concept is further developed in Chapters 6 and 7.

Traits controlled by many genes (*multifactorial* or *polygenic inheritance*) make up many of the behavioral characters we study, especially in man. Except for rare diseases that can be traced in pedigrees and shown to be controlled by a single locus, many diseases, including many forms of mental deficiency, are interpreted as being controlled polygenically. It should be clear that a further complicating factor, again especially in man, is that such traits are often profoundly labile environmentally when compared with morphological traits such as height and weight. As already stressed in Chapter 1, this environmental lability is one of the unique difficulties of behavior genetics.

Another type of trait we must consider is the *threshold trait*. These are traits for which organisms can be phenotypically classified into those having a given trait and those not having the trait. Morphological examples with behavioral consequences in man include gross abnormalities of the nervous system such as anencephaly, hydrocephaly, and spina bifida, probably all originating within the first 8 weeks of embryonic life. Polygenic inheritance has often been invoked, even occasionally with major gene effects. In addition, environmental factors may be relevant. This is so because the normal development of complex morphological traits depends on many processes and reactions, offering many possibilities for interference positively or negatively by environmental factors. Information about the heritable component of such traits emerges from comparisons of relatives; the closer the relationship to the *index case* or the *proband*, the higher the incidence of the disorder if there is a heritable component. The same occurs for disorders such as epilepsy, schizophrenia, and manic depressive psychoses (Chapter 11). Family studies provide evidence for their genetic control, but the problem of variation due to the environment and to interactions between the genotype and the environment makes it difficult to distinguish genetic and environmental components (see Chapters 7 and 11).

2.3   Population genetics

In Section 2.1 we consider segregation in single progeny at the family level. We now extend the unit under consideration to the population, which is made up of a number of individuals with many progenies. In the absence of an elementary behavioral example, the M-N blood group system in man provides a suitable model for the segregation of a pair of alleles in human populations. The blood groups are determined by two alleles L^M and L^N. The correspondence between genotype and phenotype is direct, since $L^M L^M$ individuals are blood group M, $L^M L^N$ individuals blood group MN, and $L^N L^N$ individuals blood group N. The three phenotypes can be detected by serological tests.

Considering a whole population of people, there must be a certain number of L^M alleles and a certain number of L^N alleles. These numbers can be assessed by counting every homozygote ($L^M L^M$) as 2 L^M alleles, every heterozygote ($L^M L^N$) as $1 L^M$ and $1 L^N$ allele, and every homozygote ($L^N L^N$) as $2 L^N$ alleles.

In a sample of 100 people, let us say there are $40 L^M L^M$, $40 L^M L^N$, and $20 L^N L^N$. We can therefore count the number of L^M and L^N alleles in these individuals:

	No. of L^M alleles	No. of L^N alleles	Total
$40 L^M L^M$	80		80
$40 L^M L^N$	40	40	80
$20 L^N L^N$		40	40
Total	120	80	200

The total number of alleles, of course, adds up to 200, since every diploid individual has 2 alleles.

The ratio

$$\frac{\text{number of } L^M \text{ alleles}}{\text{total number of alleles}} = \frac{120}{200} = 0.60$$

is called the *gene (allele) frequency* of gene L^M, and the ratio

$$\frac{\text{number of } L^N \text{ alleles}}{\text{total number of alleles}} = \frac{80}{200} = 0.40$$

is called the *gene (allele) frequency* of gene L^N.

The two ratios add up to unity, as might be expected, as only 2 alleles are being considered.

Thus in the whole population in the next generation, the eggs have a gene frequency of $L^M = 0.6$ and $L^N = 0.4$, and similarly for the sperm. What happens if these gametes unite at random? We get:

Female gametes	Male gamete	
	$0.6 L^M$	$0.4 L^N$
$0.6 L^M$	$0.36 L^M L^M$	$0.24 L^M L^N$
$0.4 L^N$	$0.24 L^M L^N$	$0.16 L^N L^N$

In other words, the genotype frequencies are:

$$L^M L^M = 0.6^2 = 0.36$$
$$L^M L^N = 2 \times 0.4 \times 0.6 = 0.48$$
$$L^N L^N = 0.4^2 = 0.16$$

which add up to unity.

We then wish to find out what happens in the next generation. From L^ML^M only L^M gametes are produced; from L^ML^N, $\frac{1}{2}L^M$ and $\frac{1}{2}L^N$ gametes; and from L^NL^N, only L^N gametes. Therefore, the gene frequency of $L^M =$ 0.36 from L^ML^M genotypes $+ \frac{1}{2} \times 0.48$ from L^ML^N genotypes $= 0.6$, and the gene frequency of $L^N = \frac{1}{2} \times 0.48$ from L^ML^N genotypes $+ 0.16$ from L^NL^N genotypes $= 0.4$.

Thus after one generation of the *gametes uniting at random at fertilization the gene frequencies are unchanged.* Similarly the zygotic (genotypic) proportions remain unaltered from generation to generation.

These conclusions can be shown generally. Let the gene frequency of $L^M = p$ and $L^N = q$, such that $p + q = 1$. This gives, assuming random union of gametes, the following zygotes:

Female gametes	Male gametes	
	pL^M	qL^N
pL^M	$p^2L^ML^M$	pqL^ML^N
pL^N	pqL^ML^N	$q^2L^NL^N$

or $p^2L^ML^M + 2pqL^ML^N + q^2L^NL^N$, so that the total of the zygotic frequencies comes to $p^2 + 2pq + q^2 = (p + q)^2$.

In the next generation, the gene frequency of L^M is p^2 from $L^ML^M + \frac{1}{2}2pq$ from $L^ML^N = p^2 + pq = p(p + q) = p$, since $p + q = 1$; and the gene frequency of L^N is $\frac{1}{2}2pq$ from $L^ML^N + q^2$ from $L^NL^N = q^2 + pq = q(p + q) = q$, which is what we began with. Computing the genotypic frequencies again gives $p^2L^ML^M + 2pqL^ML^N + q^2L^NL^N$.

We have therefore established the Hardy-Weinberg law, so named after its codiscoverers. This law states that (1) gene frequencies do not change from generation to generation under random union of gametes, (2) the progeny genotypes are in the proportions $p^2{:}2pq{:}q^2$, and (3) irrespective of the initial genotypic frequencies, the Hardy-Weinberg proportions $p^2{:}2pq{:}q^2$ are established in one generation.

So far, we have discussed a situation where the gene frequency at a locus is estimated assuming the heterozygote to be distinguishable from both its corresponding homozygotes. However, this is not always the situation. For example, there is a locus with behavioral consequences which determines whether some members of a population can taste phenylthiocarbamide (PTC). Those who do taste it find it bitter and unpleasant. Tasting is controlled by a locus with two alleles, T and t, such that TT and Tt are tasters and tt are nontasters. In different populations between 50 and 95 percent of people can taste PTC. Unlike the MN blood groups, it is not possible to distinguish the heterozygotes Tt from the homozygotes TT owing to the dominance of the T allele over t. Thus gene frequencies of T and t cannot be estimated by direct allele counting as in the MN blood group example. However, if gene frequencies are p of T and q of t, there are $p^2 + 2pq$ tasters ($TT + Tt$) and q^2 nontasters (tt).

Hence $q = \sqrt{\text{proportion of nontasters}}$. For example, if in a sample of 100, there are 91 tasters and 9 nontasters (or, in proportions, 0.91 tasters and 0.09 nontasters), then $q^2 = 0.09$, so that $q = \sqrt{0.09} = 0.3$, and by subtraction $p = 0.7$, since $p + q = 1$. If the heterozygotes are recognizable, the allele counting method described for the MN blood group system must be used for estimating gene frequencies, since otherwise the information given by the heterozygotes is not taken into account.

So far we have considered *random union of gametes*, but what happens under *random mating* at the phenotypic level (also referred to as *panmixia*)? Random mating is the more important way of looking at populations, since the gametes that come together depend on whether or not mating itself is at random. Table 1 demonstrates the Hardy-Weinberg law under random mating. Clearly the genotypic frequencies, and hence gene frequencies, do not change from generation to generation as for random union of gametes.

Much theory in population genetics is based on the assumption of random mating. However, random mating does not necessarily always

TABLE 1. Demonstration of Hardy-Weinberg Law in Random Mating Population

	Under Random Mating There Are $p^2TT + 2pqTt + q^2tt$ in Each Sex:		
		Females	
Males	p^2TT	$2pqTt$	q^2tt
p^2TT	p^4	$2p^3q$	p^2q^2
$2pqTt$	$2p^3q$	$4p^2q^2$	$2pq^3$
q^2tt	p^2q^2	$2pq^3$	q^4

From This Table We Can Extract Mating Types and the Offspring They Give as Follows:

		Offspring		
Mating type	Frequency	TT	Tt	tt
$TT \times TT$	p^4	p^4		
$TT \times Tt$	$4p^3q$	$2p^3q$	$2p^3q$	
$TT \times tt$	$2p^2q^2$		$2p^2q^2$	
$Tt \times Tt$	$4p^2q^2$	p^2q^2	$2p^2q^2$	p^2q^2
$Tt \times tt$	$4pq^3$		$2pq^3$	$2pq^3$
$tt \times tt$	q^4			q^4

Frequency $TT = p^2(p^2 + 2pq + q^2) = p^2$
Frequency $Tt = 2pq(p^2 + 2pq + q^2) = 2pq$
Frequency $tt = q^2(p^2 + 2pq + q^2) = q^2$

apply. One important deviation occurs as a result of *inbreeding* (the mating of individuals related to each other by ancestry). Individuals having a common ancestor are more likely to carry replicates of one of the genes present in the ancestor, and if they mate the replicates of the ancestral genes may be passed on to their offspring. Such a process increases the number of homozygotes compared with strict random mating.

Phenotypic assortative mating is the mating of individuals based on phenotypic resemblance. *Positive* assortative mating occurs when like phenotypes mate more frequently than expected under random mating. Since like phenotypes may be under the control of like genotypes, this leads, as does inbreeding, to more homozygotes than would be expected under random mating. Positive assortative mating has been found for many traits in man — height, weight, IQ, and various other behavioral traits.

In conclusion, random mating occurs only when there is no tendency for certain kinds of males and females to pair, when considered with respect to a trait. Although random mating is almost universally assumed in theoretical considerations, behavioral evidence based on the actual observance of matings between differing genotypes shows that it is in fact a special circumstance, and in subsequent chapters some attention is given to various deviations from it.

Table 2 gives Hardy-Weinberg equilibrium genotype frequencies for various gene frequencies. One interesting class to consider is the rare recessive genetic disorder in man. If the gene frequency of gene a controlling such a character is 0.01, the expected frequency of affected individuals is $aa = 0.0001$, which is very small. However, the incidence of heterozygotes Aa (referred to as *carriers* because they carry the gene for the disorder) is 0.0198, nearly 200 times as common as the affected individuals. An example is phenylketonuria, which has a frequency of the order of 1/40,000 in some populations. Since it is controlled by a recessive gene, $q^2 = 1/40,000$ giving $q = 1/200$, and so the proportion of heterozygous carriers is:

$$2pq = 2 \times \frac{199}{200} \times \frac{1}{200} \approx \frac{1}{100}$$

which is about 400 times as common as the recessive homozygotes. Taking into account all the numerous deleterious recessives found in man, we all possess our load of deleterious genes, including those with behavioral consequences. The other point shown by Table 2 is that as the gene becomes commoner, the relative excess of carriers Aa relative to aa falls. Thus for $q = 0.10$, carrier individuals are only 18 times as common as *aa*. Many recessive behavioral disorders in man under the control of single genetic loci are rare, and so the genes controlling them occur in populations mainly in carriers.

The final situation we must consider are those genes on the X chromosome, the sex-linked genes, described at the end of Section 2.1. Since males have only one X, direct gene counting yields an estimate of gene frequency. Thus, at an equilibrium state for two alleles A and a controlling a trait, in males there should be $pA + qa$, and in females $p^2AA + 2pqAa + q^2aa$ as previously observed. For an X-linked recessive gene, therefore, the frequency of the trait in females is expected to be the square of that in males. Thus color blindness is much less common in females than in males, being expected to be 0.64 percent in a population where 8 percent of males are affected. Rare sex-linked recessive genes

TABLE 2. Hardy-Weinberg Equilibrium for $p^2AA + 2pqAa + q^2aa$ where $p + q = 1$

		Frequency		
p	q	$AA = p^2$	$Aa = 2pq$	$aa = q^2$
1.00	0.00	1.0000	0.0000	0.0000
0.98	0.02	0.9604	0.0392	0.0004
0.96	0.04	0.9216	0.0768	0.0016
0.94	0.06	0.8836	0.1128	0.0036
0.92	0.08	0.8464	0.1472	0.0064
0.90	0.10	0.8100	0.1800	0.0100
0.88	0.12	0.7744	0.2112	0.0144
0.86	0.14	0.7396	0.2408	0.0196
0.84	0.16	0.7056	0.2688	0.0256
0.82	0.18	0.6724	0.2952	0.0324
0.80	0.20	0.6400	0.3200	0.0400
0.78	0.22	0.6084	0.3432	0.0484
0.76	0.24	0.5776	0.3648	0.0576
0.74	0.26	0.5476	0.3848	0.0676
0.72	0.28	0.5184	0.4032	0.0784
0.70	0.30	0.4900	0.4200	0.0900
0.68	0.32	0.4624	0.4352	0.1024
0.66	0.34	0.4356	0.4488	0.1156
0.64	0.36	0.4096	0.4608	0.1296
0.62	0.38	0.3844	0.4712	0.1444
0.60	0.40	0.3600	0.4800	0.1600
0.58	0.42	0.3364	0.4872	0.1764
0.56	0.44	0.3136	0.4928	0.1936
0.54	0.46	0.2916	0.4968	0.2116
0.52	0.48	0.2704	0.4992	0.2304
0.50	0.50	0.2500	0.5000	0.2500

After Maynard-Smith, Penrose, and Smith (1961).

are thus almost entirely restricted in occurrence to males, as described at the end of Section 2.1.

2.4 Chi-square tests

SEGREGATION AND MATING PREFERENCE DATA

We continue the chapter on requisite genetics with a section on how to assess the meaning of experimental segregation data. Segregation data are often obtained, and hypotheses can be erected about them. We wish to know whether the data observed fit a hypothesis, since because of random variation exactly fitting ratios are unlikely. To illustrate this point we consider two crosses in mice: (1) a cross between two heterozygous agouti mice *Aa*, and (2) a cross between two heterozygous yellow mice A^Ya. A and A^Y are dominant to *a*, and all three are alleles at the same locus, providing our first example of a *multiple allelic series*. Agouti mice have dark fur with yellow tips, yellow mice have yellow fur and are fat and rather sluggish in movement, and nonagouti mice have black fur. (Some comments on associated behavioral variants of coat color mutants in mice appear in Chapter 9.)

Considering gametes from *Aa* mice in each sex, ova and sperm are expected in ratios $\frac{1}{2}A:\frac{1}{2}a$, and if these unite at random, the zygotes expected are derived from $(\frac{1}{2}A + \frac{1}{2}a)(\frac{1}{2}A + \frac{1}{2}a)$ yielding $\frac{1}{4}AA + \frac{1}{2}Aa + \frac{1}{4}aa$ genotypically, or $\frac{3}{4}$ agouti:$\frac{1}{4}$ nonagouti phenotypically.

In Table 3 some observed data from crosses between heterozygous agouti (*Aa*) mice are given. Expected numbers based on an expected 3:1 ratio are also given. The greater the difference between the observed (*O*) and expected (*E*), the greater the deviation of the data from the ratio expected. In this case $O-E$ is small for each class. However, if $O-E$ were, say 10 times greater, or 112.5, we could well wonder whether the expected ratio of 3:1 is realistic. Clearly, some deviation from the exact 3:1 is tolerable, just because of chance, but at a certain level of deviation,

TABLE 3. Data for a Cross Between Heterozygous Agouti (*Aa*) Mice

Phenotype	Genotype	No. observed (O)	No. expected (E)	O-E	(O-E)²	(O-E)²/E
Agouti	*Aa*	306	317.25	−11.25	126.5625	0.3989
Nonagouti	*aa*	117	105.75	11.25	126.5625	1.1968
Total		423	423.00			1.5957

we begin to suspect the validity of the expected ratio. The testing of deviations from an expected ratio can be done by means of a simple statistical test, which consists of computing $(O-E)^2/E$ for each phenotypic class and adding together. The resultant value is called χ^2 (chi square), and can be written:

$$\chi^2 = \Sigma \frac{(O-E)^2}{E} \qquad (2.6)$$

where Σ means the sum of, as in equation 2.1. For the above data, $\chi^2 = 1.59$, which is small. It can be shown that if $\chi^2 > 3.84$, there is a ≤ 5 percent chance of the data fitting a 3:1 ratio, and if $\chi^2 > 6.64$, there is a ≤ 1 percent chance of the data fitting a 3:1 ratio. Values of χ^2 corresponding to various probabilities are readily available in statistical tables. Table 4 provides part of a χ^2 table.

For the moment, only the top row should be considered; the remaining rows are used later. Usually it is judged by convention that if the probability comes out at ≤ 5 percent, we begin to suspect that the data do not fit the hypothesis assumed, or in statistical terminology, we say the data *differ significantly* from the hypothesis assumed, i.e., the hypothesis can be regarded as not valid. In the example, χ^2 is extremely small so the data agree with the 3:1 ratio. However, *it can never be absolutely proved that the 3:1 ratio is correct*, since if more data were collected then ultimately the value of χ^2 might increase and indicate a significant difference from the 3:1 ratio. Thus all the χ^2 test can do is *dis*prove a hypothesis at a certain level of probability.

In Table 5 are data concerning the cross between heterozygous yellow mice. Phenotypically we would expect $\frac{3}{4}$ yellow: $\frac{1}{4}$ nonagouti. Considering this simple 3:1 hypothesis, $\chi^2 = 16.46$, which is significant at the 0.1 percent level (Table 4). If P is the probability of the hypothesis being

TABLE 4.   Distribution of χ^2

Degrees of freedom (n)	Probability (P)						
	0.50	0.30	0.20	0.10	0.05	0.01	0.001
1	0.455	1.074	1.642	2.706	3.841	6.635	10.827
2	1.386	2.408	3.129	4.605	5.991	9.210	13.815
3	2.366	3.665	4.642	6.251	7.815	11.345	16.266

As n increases, χ^2 increases for a given probability, and as the probability decreases χ^2 increases for given n.
After Fisher and Yates (1967).

correct, we can write $P < 0.001$ as the chance of the hypothesis being correct, which is obviously very remote. If we look further at the data in Table 5, it is clear that there is a deficiency of yellow mice observed compared with expected. This provides a clue, because a number of situations are known where genotypes do not survive or are *lethal*, so in this case we are suggesting that $A^Y A^Y$ may be lethal. If this is so, we would then expect $\frac{2}{3}$ yellow : $\frac{1}{3}$ nonagouti, and in fact carrying out a χ^2 test on a 2:1 ratio gives $\chi^2 = 3.64$ ($P < 0.10$), which does not disagree with the hypothesis. Thus the χ^2 test is useful in determining which of several hypotheses fits best. *It is important, however, to proceed from the simplest hypothesis to the complex, since there is little point in explaining data by a complex hypothesis when a simpler one is adequate.* Biologically the 2:1 hypothesis fits, since $A^Y A^Y$ genotypes have been shown to die in utero, leaving only $A^Y a$ and aa as the two viable genotypes.

In many cases significant χ^2 values may appear in data due to poor viability of genotypes rather than direct lethality. In mice, *fidget* homozygotes tend to be nonviable; therefore, in a cross $fi+ \times fi+$, which would have an expectation of 3 nonfidget : 1 fidget, there is frequently a deficiency of fidget mice, which would, if enough progeny were obtained, lead to a significant χ^2 value. The same point would apply to a number of the neurological mutants in the mouse discussed in Chapter 9.

What happens if there are more than two classes? In the cross $AaBb \times aabb$ already discussed, the recognizable phenotypic classes are, genotypically, expected to be $\frac{1}{4}AaBb:\frac{1}{4}Aabb:\frac{1}{4}aaBb:\frac{1}{4}aabb$, if the genes a and b segregate independently of each other. The procedure adopted is to compute χ^2 over the four classes exactly as described previously, namely, $\Sigma (O-E)^2/E$. Now the larger the number of classes the greater the number of components of χ^2. This means that the value of χ^2 is expected to increase by chance as the number of classes increases, or alternatively, the value of χ^2 at which data are judged as significant at, say, the 5 percent level increases. To deal with this, the concept of the *number of degrees of freedom* (n) must be introduced, which in simple cases

TABLE 5. Data for a Cross Between Yellow Mice $A^Y a$

Phenotype	No. observed (O)	No. expected (E) at		$(O-E)^2/E$	
		3:1	2:1	3:1	2:1
Yellow	706	762	677.33	4.12	1.21
Nonagouti	310	254	338.67	12.35	2.43
Total	1016	1016	1016.00	16.46	3.64

TABLE 6. **Mating Preferences in Crosses Between Different Geographical Populations of *D. pseudoobscura***

Cross strain A × strain B	No. of matings	No. of each type of mating				χ^2_3 *for random* *mating*
		A♀ ×A♂	A♀ ×B♂	B♀ ×A♂	B♀ ×B♂	
Berkeley × Okanagan	222	60	50	72	40	10.14*
Berkeley × Austin	160	37	43	42	38	0.65
Berkeley × Hayden	28	7	7	5	9	1.14
Berkeley × Sonora	103	23	22	28	30	1.74
Okanagan × Austin	125	27	33	33	32	0.79
Okanagan × Hayden	51	14	14	10	13	0.84
Okanagan × Sonora	114	26	29	32	26	0.74
Austin × Hayden	103	21	26	30	26	1.58
Austin × Sonora	113	36	28	27	22	3.57

* Significant at 0.05 level.
After Anderson and Ehrman (1969).

is the number of phenotypic classes minus one. In Table 4, χ^2 values are given for $n = 1, 2,$ and 3. In the simple crosses analyzed in Tables 3 and 5, we are dealing with two classes for which $n = 1$, and we use the symbol χ^2_1, where the subscript indicates the number of degrees of freedom. For the above cross with four phenotypic classes, a χ^2_3 would be calculated.

In Table 6, mating preference data from five geographically distinct populations of *Drosophila pseudoobscura* are given. In each test 10 virgin females and males were used in specially designed mating chambers. The χ^2_3 tests were computed for an expected 1:1:1:1 ratio, which indicates random mating. The results indicate that in only one case, the Berkeley × Okanagan combination, is there a significant deviation that seems due to the higher mating propensity (vigor?) of the Berkeley strain male compared with the Okanagan male. Therefore, χ^2 tests are useful for any observed frequency data where they can be compared with what would be expected if a hypothesis were true.

CONTINGENCY χ^2 TEST

Occasionally data are presented in the form of 2 × 2 tables. For example, some female twin pairs are tested for smoking habit and classified according to whether they are identical twins derived from the same zygote (monozygotic twins) or nonidentical fraternal twins derived from two different zygotes (dizygotic twins). The data from Fisher (1958), tallied for concordances in smoking habit, are:

	Concordant (both smoked or both did not)	Discordant (one smoked and one did not)	Total
Identical twins (monozygotic)	44	9	53
Fraternal twins (dizygotic)	9	9	18
Total	53	18	71

The question to be answered is whether smoking habits are more alike in monozygotic twins than in dizygotic twins, since monozygotic twins have the same genotype while dizygotic twins have differing genotypes. If there is no association, we expect the concordances for smoking habit to be similar in monozygotic and dizygotic twins.

Algebraically, the 2 × 2 table can be written:

	Alike	Unlike	Total
Monozygotic	a	b	$a + b$
Dizygotic	c	d	$c + d$
Total	$a + c$	$b + d$	$a + b + c + d = N$

where a,b,c,d are observed totals corresponding to the numbers in the above table. We expect $a:b = c:d$ if there is no association. In other words, if there is no association we expect $ad = bc$ or $ad - bc = 0$. It can be proved that

$$\chi^2{}_1 = \frac{(ad - bc)^2 N}{(a+c)(b+d)(c+d)(a+b)}$$

is a test for association. Note that if $ad - bc = 0$ or $ad = bc$, $\chi^2{}_1 = 0$, and if $ad \neq bc$, $\chi^2{}_1 > 0$. The greater the inequality between ad and bc, the greater is $\chi^2{}_1$ and the greater the association.

Especially for small expected numbers, as is the case in the data under consideration, Yates' continuity correction is generally applied, as it gives a better fit to the expected χ^2 distribution. The above $\chi^2{}_1$ formula, applying Yates's correction for continuity, becomes:

$$\chi^2{}_1 = \frac{(|ad - bc| - \frac{1}{2}N)^2 N}{(a+c)(b+d)(c+d)(a+b)} \tag{2.7}$$

For the data $\chi^2{}_1 = 6.09$ $(P < 0.05)$, showing a significant association between smoking habit and twin type, such that smoking habit is more similar in monozygotic twins than in dizygotic twins, one may argue for a genotypic component in determining smoking habits.

Now, as with much human data of this type, one can argue that the environment of monozygotic twins may be more similar than that of dizygotic twins, so the above result could be environmental. The only way of dealing with this is to compare monozygotic twins separated at birth and reared apart with those reared together. The subdivision of monozygotic twins so obtained (Fisher, 1958) is:

	Concordant	Discordant	Total
Separated	23	4	27
Not separated	21	5	26
Total	44	9	53

for which the χ^2_1 for association = 0.004. In other words, the difference in upbringing has no significant effects for these limited data. (It should be noted that when expected values are less than about 3 or 4, then χ^2 tests become rather inaccurate, but we are probably in the safe range here.)

χ^2 TEST FOR RANDOM MATING

To test for random mating, one must first carefully define the population and trait being measured. The population must be as uniform as possible. Mixing populations together which do not themselves show random mating for a trait may lead to simulated random mating or panmixia. The most commonly used method of testing for random mating is to see if the distribution of phenotypes is in agreement with Hardy-Weinberg equilibrium. This involves an elementary knowledge of the χ^2 test. We compute the gene frequencies from the observed data. Thus in both groups I and II of Table 7 the gene frequency of A or $p = 0.4$ and of a

TABLE 7. χ^2_1 **Tests for Random Mating**

	No. observed			No. expected				
	AA	Aa	aa	AA	Aa	aa	χ^2_1	P
Group I	40	240	120	64	192	144	25.00	≪0.001
Group II	85	150	165	64	192	144	19.14	≪0.001
Pooled	125	390	285	128	384	288	0.20	>0.50

or $q = 0.6$. Hence the expected genotypic frequencies are ($N = 400$, the population size):

$$AA = p^2 N = 0.4^2 \times 400 = 64$$
$$Aa = 2pq\, N = 2 \times 0.4 \times 0.6 \times 400 = 192$$
$$aa = q^2 N = 0.6^2 \times 400 = 144$$

From the observed and expected frequencies so obtained, a χ^2 value can be calculated in the usual way as $\Sigma\,(O-E)^2/E$.

There is, however, one difference between these data and those considered previously: to obtain the expected frequencies, a parameter, namely the gene frequency (p), is estimated from the observed data. In these circumstances, the rule derived from statistics is that the number of degrees of freedom equals the number of phenotypic classes minus the number of independent parameters estimated from the observed data minus 1. Clearly only one independent parameter is determined from the data, since $p + q = 1$. Therefore, in this case $\chi^2{}_1$ is being calculated as a test for random mating.

Table 7 therefore shows the results of testing each of two groups as well as the two groups combined for Hardy-Weinberg equilibrium. The first group does not fit expectation because the observed number of heterozygotes exceeds the expected number, so that there is a deficiency of homozygotes for an expectation of random mating. Such a situation of an excess of heterozygotes is quite commonly observed in both experimental and natural populations, and may be due to natural selection whereby the heterozygotes are fitter than the corresponding homozygotes. The second group also gives a poor fit, with the two homozygote classes having excess numbers; this could occur as a result of positive assortative mating or inbreeding. If we ignore the differences between the two groups and pool them to test for Hardy-Weinberg equilibrium, the fit is very good. But to conclude that the combined population shows panmixia is false owing to the heterogeneity of the population.

The opposite effect can be obtained if two groups, with different gene frequencies for a particular trait, are pooled together to test for random mating. The resulting population does not necessarily show evidence of random mating, although within each homogeneous group mating may well be at random. Such a pooled population shows an excess of homozygotes over those expected, an effect first described by Wahlund (1928). An example of this may result if two ethnic groups, although geographically intermingled, continue to be partially isolated with respect to mating patterns. Some characteristics (e.g., blood groups), although not influencing mating choice, may maintain gene frequency differences in the two groups. Other forms of partial mating isolation (e.g., assortative mating for height) may not be associated with corresponding gene frequency differences of other traits (such as blood groups), so the criterion of a homogeneous population is fulfilled for these traits. A more complete discussion of the importance of homogeneity of groups when either association or independence is being sought can be found in Li (1961).

It should be emphasized that knowledge of the homogeneity of the population is imperative before tests for random mating can be meaningful, and in many cases it is assumed that no heterogeneity is present when, in fact, it is just not detected. If the homogeneity of a population can be demonstrated satisfactorily, a test for random mating can be carried out, preferably by testing mating classes rather than merely looking for Hardy-Weinberg equilibria. However, conclusions drawn, either in favor of random mating or against it, should always be assessed carefully, and they should be no stronger than the confidence one can place in the demonstration of an acceptably homogeneous population.

Finally, as a matter of completeness, it is appropriate to list causes, including those already discussed, whereby deviations from panmixia may occur in a statistical sense as described in this section:

- Selection. This possibility is considered in the discussion of Table 7, group I. Selection occurs when some genotypes leave more offspring in the next generation because of differences in overall viability or *fitness* between different genotypes. The likelihood is that behavioral parameters, especially those associated with mating, are of considerable importance in fitness differences, as discussed in later chapters.

- Mutation. Genes may change, say from A to a, at a low frequency. Over long periods of time, mutations play a role in evolutionary change, but the mutation rate is so low that over a few generations it can be ignored.

- Migration. In some senses this is related to mutation, in that new genes may be introduced into populations, but the influence on the gene pool may be much greater if there are many immigrants.

- Inbreeding. As already mentioned, inbreeding leads to an increased incidence of homozygotes. In man, inbreeding is of importance in isolated populations, where high frequencies of marriages between relatives (*consanguineous marriages*) may occur.

- Assortative mating. This has been, and will be further, discussed.

- Genetic drift. This is the term used to describe the random chance events that may occur generation by generation in the determination of gene frequencies. For example, if the population size is relatively small, it may happen just by chance that a sample of gametes from that population which provides the next generation is not representative. The consequence is a change in gene frequency in the new population merely because of chance phenomena, and this is referred to as genetic drift. Clearly the importance of drift progressively diminishes as the population size increases.

2.5 Gene action

So far we have considered genotypes as assessed directly by their phenotypes in families and populations and have briefly considered some of the principles needed to understand their patterns of inheritance from generation to generation. However, as will become apparent in certain parts of this book, we also need to consider the path between the gene and the behavioral phenotype that is observed.

Almost all cells of a given organism contain the same amount and type of a material known chemically as *deoxyribonucleic acid (DNA)*, which is organized in the chromosomes. Experiments in microorganisms show clearly that DNA contains the information necessary to make new cells essentially *identical* to the parent cells. The essentials of this principle have also been shown in higher organisms. The amount of DNA per cell is not large; in the human cell the figure is about 6×10^{-12} gm. In spite of this very small quantity, the amount of information contained is enormous and is sufficient to direct the synthesis of a human individual.

DNA is made up of chemical units consisting of:

- A base belonging to the purines or pyrimidines; the two possible purines are adenine (A) and guanine (G), and the two possible pyrimidines are cytosine (C) and thymine (T)
- A pentose (five-carbon sugar) — deoxyribose
- A phosphate group

The DNA molecule is made up of *nucleotides* comprising one base, one sugar, and one phosphate group. For a given DNA molecule the phosphate and sugar groups are all the same and only the bases vary. Because only the bases A, G, C, and T can vary, apparently the information determining heredity resides in these and in the degree to which they can vary.

The quantity of these bases is constant in a given species but varies among species. However, in every species, the quantity of A = T, and of G = C. Thus it seems that A and T are always paired together, and similarly for G and C.

The overall structure of DNA was demonstrated in 1953 by Watson and Crick. The bases are attached to a sugar-phosphate backbone forming a chain of nucleotides:

base − sugar
        <
           phosphate
base − sugar <
           phosphate
        <
base − sugar

Watson and Crick found DNA to be a double chain of nucleotides:

> sugar − base − base − sugar <
phosphate   phosphate
> sugar − base − base − sugar <
phosphate   phosphate
> sugar − base − base − sugar <

The two chains are joined by the bases, and on chemical grounds only A and T pair as do G and C, and are twisted around each other in a double helix (Fig. 7). The nucleotide pairs are separated by 3.4 Angstrom units (Å; an Angstrom unit equals 10^{-7}mm), and the whole structure repeats itself in 10 pairs or 34 Å. Thus, because of the pairing rules A = T and G = C, if we know the sequence of bases on one helix, we also know the sequence on the other.

According to the Watson-Crick theory, the linear sequence of nucleotides is constant for a given species (except for minor heritable variations within species). It is the actual sequence of nucleotides that leads to variations in proteins which have as their primary structure a chain of amino acids. The genetic component of the phenotype we observe depends on this linear sequence. Furthermore, variations in the observed phenotype are likely to be due to minor changes in the sequence of nucleotides (if environmental effects can be eliminated). Therefore, the sequence of nucleotides can be regarded as a *genetic code*.

There are 20 essential amino acids that must be specified by the genetic code. Since there are 4 possible bases (A, T, G, and C), one- and two-letter codes are inadequate, specifying 4 and $4^2 = 16$ combinations only. The triplet code specifying $4^3 = 64$ combinations is a necessary minimum. The nucleotide triplet is referred to as a *codon*, and because a triplet code gives 64 different combinations, or words, of which only 20 are needed for amino acid synthesis, the code is referred to as a *degenerate code*. In fact some amino acids are coded by more than one triplet (Fig. 8). The names and abbreviations of the 20 essential amino acids are:

• Alanine	Ala	• Leucine	Leu
• Arginine	Arg	• Lysine	Lys
• Asparagine	AspN	• Methionine	Met
• Aspartic acid	Asp	• Phenylalanine	Phe
• Cysteine	Cys	• Proline	Pro
• Glutamic acid	Glu	• Serine	Ser
• Glutamine	GluN	• Threonine	Thr
• Glycine	Gly	• Tryptophan	Tryp
• Histidine	His	• Tyrosine	Tyr
• Isoleucine	Ileu	• Valine	Val

From the DNA code a linear message is *transcribed* to a form of RNA (or ribonucleic acid) called *messenger RNA* (*mRNA*). RNA is chemically

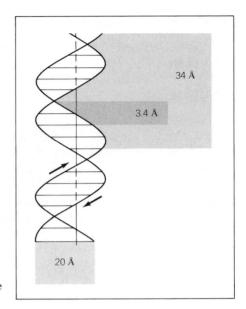

7. **The double helix of DNA.** See text for details.

very similar to DNA except that (1) it has ribose rather than the sugar deoxyribose, (2) it has the base uracil (U) substituting for thymine, and (3) it is single- rather than double-stranded. In *transcription* from DNA to mRNA, the following pairing rules therefore occur:

Base in DNA	Base in mRNA
A	U
T	A
C	G
G	C

There are three kinds of RNA upon which protein synthesis depends: messenger RNA, transfer RNA, and ribosomal RNA.

Messenger RNA (mRNA) is described above.

Before amino acids are assembled into protein chains they must be "activated" by the attachment of a special phosphoric acid group. They are then attached to another type of RNA called *transfer RNA (tRNA)*. In fact, there are as many tRNA molecules as there are triplets determining amino acids.

Alignment of the tRNA and mRNA so as to assemble protein chains in an orderly way is mediated by special particles in the cytoplasm of the cell called *ribosomes*, which are made from a third form of RNA called *ribosomal RNA (rRNA)*.

SECOND LETTER

		U	C	A	G	
FIRST LETTER	U	UUU⎫ UUC⎭ Phe UUA⎫ UUG⎭ Leu	UCU⎫ UCC⎬ Ser UCA⎪ UCG⎭	UAU⎫ UAC⎭ Tyr UAA CHAIN END UAG CHAIN END	UGU⎫ UGC⎭ Cys UGA CHAIN END UGG Tryp	U C A G
	C	CUU⎫ CUC⎪ Leu CUA⎪ CUG⎭	CCU⎫ CCC⎬ Pro CCA⎪ CCG⎭	CAU⎫ CAC⎭ His CAA⎫ CAG⎭ GluN	CGU⎫ CGC⎪ Arg CGA⎪ CGG⎭	U C A G
	A	AUU⎫ AUC⎬ Ileu AUA⎭ AUG Met	ACU⎫ ACC⎪ Thr ACA⎪ ACG⎭	AAU⎫ AAC⎭ AspN AAA⎫ AAG⎭ Lys	AGU⎫ AGC⎭ Ser AGA⎫ AGG⎭ Arg	U C A G
	G	GUU⎫ GUC⎪ Val GUA⎪ GUG⎭	GCU⎫ GCC⎬ Ala GCA⎪ GCG⎭	GAU⎫ GAC⎭ Asp GAA⎫ GAG⎭ Glu	GGU⎫ GGC⎪ Gly GGA⎪ GGG⎭	U C A G

THIRD LETTER

8.   **The genetic code.** See text for explanation.

The process of the formation of protein from the code carried by the mRNA is referred to as *translation*, so we can summarize what happens as:

$$\text{DNA} \xrightarrow{\text{transcription}} \text{mRNA} \xrightarrow{\text{translation}} \text{protein}$$

A great deal of additional information about this process can be found in many texts, but the details of the process, largely established in microorganisms, are of only peripheral interest to the behavior geneticist. On the other hand, with time there will certainly be a trend toward metabolic explanations of behavioral processes, so an understanding of the biochemical basis of gene action will assume progressively more importance. Even so, it should be clear that the hereditary unit discussed so far in this chapter has a definite structural and functional meaning.

So far as protein synthesis itself is concerned, most proteins are made only when they are useful. In other words, there are various means of *regulation* present. In fact, *regulator genes* have been discovered in microorganisms. These regulator genes determine whether the genes that specify types of proteins (*structural genes*) are active or otherwise. The regulator genes themselves are controlled by cytoplasmic events and thus are open to environmental influences; for example, if a certain amino acid is necessary for growth and is present in the environment, then the cells can cease making it and the enzymes necessary for its synthesis (enzyme repression). Undoubtedly this process of regulation of protein synthesis must be the basis of *differentiation*, which is the development of different cells and tissues. At different stages during development it is clear that various portions of DNA are active in different cells and tissues. Such temporal regulation of gene action must be carefully studied in order to understand behavior, and it is clear that genes do act in sequence during development so that a given gene may start one event and this may lead to a series of others. Gene-hormone interactions, for example, are probably involved in sex differentiation, the onset of puberty, and the development of learning in man. But the best known example of a gene-hormone interaction is for the molting patterns in Diptera and other insects, which depend on the production of the molting hormone ecdysone and its effect on certain loci.

The other important consequence of a knowledge of the physiological processes of a gene substitution is that their pattern can be modified by appropriate treatment. Thus in mice there is a condition, pallid, due to a single recessive gene on linkage group V. The mutant mouse has a lesion of the calcified otoliths in the inner ear. These otoliths normally move in response to an animal's change in position, and in this fashion neural responses are induced relevant to the organism's gravity response (Erway, Hurley, and Fraser, 1966). The *pallid* gene destroys otoliths in one or both ears, and so behavioral equilibrium is disrupted. It is interesting that otolith destruction can be produced by withholding manganese from the diet of normal mice; i.e., the phenotype can be induced environmentally — a phenomenon referred to as a *phenocopy*. Conversely, if gestating females bearing the *pallid* gene are given large supplements of manganese, the mutant offspring do not show the defect. Therefore, we have a gene-behavior relation capable of environmental control once the condition is understood.

A good example in man is phenylketonuria, which we have already considered in some other respects. Individuals homozygous for this recessive gene generally have IQs ≤30, occasionally higher. Their skin and hair pigmentation is generally lighter than that of the population from which they arise. Phenylketonuria is due to a deficiency or absence of the enzyme phenylalanine hydroxylase, which is essential in the metabolism of phenylalanine, an amino acid that is an essential dietary constituent. Normally, phenylalanine → tyrosine . . . → various metabolic breakdown products (Fig. 9). In phenylketonuria, this step is blocked; phenylalanine accumulates to a level 40 to 50 times that found in nonphenylketonurics, and this excess leads to mental deficiency. Hence, a likely treatment would be to feed a phenylalanine-deficient diet. This poses problems, as no known protein is deficient in

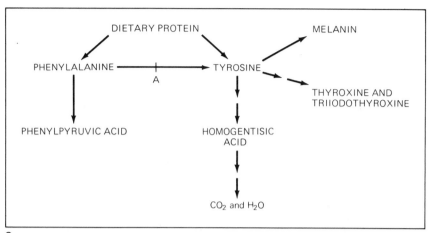

9. **Phenylalanine metabolism.** Normally phenylalanine is transformed to tyrosine and various other compounds derived from tyrosine. In phenylketonurics, when block occurs at A (i.e., phenylalanine hydroxylase is absent), the alternative pathway through phenylpyruvic acid assumes importance. The altered pigmentation of phenylketonurics is expected, because melanin formation depends partly on tyrosine. (After Harris, 1959.)

phenylalanine. However, such a diet can be synthesized by breaking down protein and reconstituting it without phenylalanine, but still containing other essential amino acids. The diet must begin early in life to have a major effect on IQ and is likely to be proportionately less effective if delayed beyond infancy. The treatment must strike a delicate balance between malnutrition (i.e., inadequacy of the essential amino acid phenylalanine) and intoxication.

Until recently, phenylketonuria was detected by a test that depends for its success on secretion of an abnormal product of phenylalanine, phenylpyruvic acid, in the urine of phenylketonurics (Fig. 9). However, phenylpyruvic acid secretion may be delayed for up to 5 to 6 weeks after birth. The test itself merely requires adding $FeCl_3$ to urine acidified with 1N HCl; if phenylpyruvic acid is present, the urine turns green. A more recent and more efficient procedure is the Guthrie test, based on the ability of certain strains of bacteria to grow only in a medium containing a high level of phenylalanine. To perform this test, blood is added to a bacterial culture lacking phenylalanine. If the culture grows, this indicates that the phenylalanine level of the blood is high, perhaps indicating phenylketonuria. Approximately 40 states of the United States now make the Guthrie test compulsory for all newborn babies, and it is commonly carried out in other countries.

Individuals heterozygous for the phenylketonuria gene can be detected by a *phenylalanine tolerance test*. This consists of feeding the fasting individual with phenylalanine and then testing serum levels of phenylalanine at various periods after fasting. Many persons heterozygous for the phenylketonuria gene ($p+$) metabolize phenylalanine more slowly than normals ($++$). Thus $p+$ individuals can often be distinguished from $++$, showing that at the biochemical level the p gene is not completely recessive. Heterozygote detection is of importance in *genetic counseling* (advice given to people about risks of having abnormal children on genetic grounds) and in other problems where genetic advice is given. Thus if two individuals are known to be $p+$, the chance of pp offspring from them is 25 percent.

In conclusion, understanding mechanisms of gene action behind a given behavioral phenotype is possible in a few instances. To find the molecular correlates of behavioral patterns is an exciting possibility, but one that will take many years of careful investigation for its full realization, especially in man. In a few cases in higher organisms it is possible to assess genetic code changes associated with different alleles at a locus. The future will undoubtedly reveal more examples.

General readings

Crow, J. F. 1951. *Genetics Notes*. Minneapolis: Burgess. A concise elementary text useful for beginners.

Dobzhansky, T. 1951. *Genetics and the Origin of Species*. New York: Columbia Univ. Press. A now classical book on genetics and evolutionary processes, with particular reference to species formation.

Lerner, I. M. 1968. *Heredity, Evolution, and Society*. San Francisco: Freeman. An elementary discussion of genetics and evolution stressing human society, and as such of importance to the behavior geneticist.

Srb, A. M., R. D. Owen, and R. S. Edgar. 1965. *General Genetics*, 2nd ed. San Francisco: Freeman. A good general textbook.

Stern, C. 1973. *Principles of Human Genetics*, 3rd ed. San Francisco: Freeman. A comprehensive text in human genetics assuming no prior knowledge of genetics.

Strickberger, M. W. 1968. *Genetics*. NewYork: Macmillan. Another good general textbook.

Single Genes and Behavior

Traits under the control of single genes are the best known simply because they are easiest to trace. This applies to all traits, whether morphological, physiological, or behavioral. Many such genes are rare and deleterious and so may not be of great importance in a population, but because their effects are easily traceable, the genotypes they control may provide information on behavioral variation in the species in question. First, there are genes that produce a visible alteration in appearance with a simultaneous change in behavior. For example, phenyl-ketonurics, discussed in Chapter 2, as well as having low IQs, tend to have lighter hair pigmentation than does the population from which they are derived. In other words, the gene has more than one effect (*pleiotropy*), and as shown in this and other chapters, pleiotropic effects involving morphological, physiological, and behavioral traits are common. Second, we may ask whether a single gene producing no known morphological effects can, primarily or even exclusively, produce behavioral changes. Outwardly this often appears to be the situation, but in fact detailed research often reveals associated physiological or biochemical variables.

This chapter discusses some behavioral traits for which single-gene effects are known, in some cases showing pleiotropic effects and in other cases not.

3.1   Nest cleansing by honeybees

Rothenbuhler (1964) carried out an elegant analysis of the nest cleansing of honeybee larvae killed by a disease, American foulbrood (pathogen, *Bacillus larvae*). The maintenance of the hygienic environment within a hive requires the opening of combs housing afflicted young and their immediate evacuation (Fig. 1). If this is not done, cadavers with their associated spores remain inside the hive as a continuous source of contamination. Genes at two independently segregating loci account for hygienic or nonhygienic behavior; one involves the uncapping of cells and the other the removal of their contents. Where *u* represents the recessive gene for uncapping behavior and *r* the recessive gene for removal, the genetic makeup of the bees of a hygienic colony is *uurr*.

No physical or physiological differences have been recorded between totally hygienic, partially hygienic, or totally nonhygienic honeybees, although detailed investigations might uncover some. From the genetic point of view, the example is of interest because the fractionation of hygienic behavior into two distinct components leads to the understanding of its genetic architecture. Surely both these acts enhance the survival and perpetuation of the reproductive unit built by these social insects — the hive and its inhabitants. We have here, therefore, a remarkable example of behavior controlled by two single-gene loci with obvious fitness effects.

3.2   Mating success in Drosophila

Cinnabar and *vermilion* are mutant genes in *Drosophila melanogaster* affecting eye colors, as their names imply. (*Cinnabar* is an autosomal recessive and *vermilion* a sex-linked recessive.) Flies with either mutant gene have bright red eyes, compared with the dull red eyes of the wild type. Bösiger (1957, 1967) compared the speed of mating of *D. melanogaster* with the *vermilion* and *cinnabar* mutants. After 12 days the following percentages of females, each of which was confined with a single male, were found to be gravid:

	Vermilion ♀ × vermilion ♂	Vermilion ♀ × cinnabar ♂	Cinnabar ♀ × vermilion ♂	Cinnabar ♀ × cinnabar ♂
Couples tested	200	302	200	325
Percent fertile	61.0	80.1	54.0	73.8

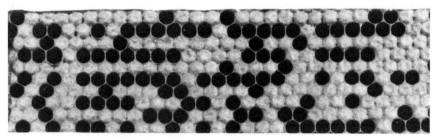

1. **Honey bee brood comb.** Cells house only live larvae and pupae, just before the emergence of young adults. Cells deliberately inoculated with bacterial spores have all been uncapped and eliminated. This is a sample of the work of a hygienic colony of genotype *uurr*. (From Rothenbuhler, 1964.)

In another experiment, groups of females were confined with males, and the percentages of the females that copulated after a lapse of different time intervals were recorded. The results are as follows:

Time (min)	Vermilion ♀ x vermilion ♂	Vermilion ♀ x cinnabar ♂	Cinnabar ♀ x vermilion ♂	Cinnabar ♀ x cinnabar ♂
0–5	12.9	48.3	0	13.0
5–10	32.3	65.5	21.1	39.1
10–15	35.5	79.3	36.8	43.5
15–20	35.5	82.8	42.1	52.2
20–25	38.7	86.2	47.4	56.5
25–30	38.7	89.7	47.4	56.5

In both experiments, when males had the *vermilion* gene the success rate was lower than when they had the *cinnabar* gene. We can therefore say that males with the *vermilion* mutant are disadvantaged compared with others of the same sex in respect to reproduction. Differential reproductive success rates of this nature are referred to as *sexual selection*. Many examples are discussed in this and in later chapters.

Bastock (1956) investigated the effects on mating success of the sex-linked recessive mutant *yellow* compared with the wild type in *D. melanogaster*. In her main experiments, a wild stock was crossed with a yellow stock for seven generations, so that the wild stock was genetically similar to the yellow stock except in the region of the *yellow* locus. Males with the *yellow* mutant are less successful in mating with females of the normal gray body color than are normal males. Bastock found that the courtship pattern of the males had been altered by the mutation from wild-type to yellow body color. Figure 2 illustrates this schematically by dividing courtship behavior into three component parts. "Orientation" normally occurs at the very start of courtship when the male follows the

2.   **Trisected courtship patterns of flies.** Each of the four rows (**A** to **D**) represents the behavior of a male *D. melanogaster* as time progresses from left to right: **1**, licking; **2**, vibration; **3**, orientation. (From Bastock, 1967.)

female, circles her, or stands in front of her. "Vibration" means wing movements and display following orientation, and this is followed by "licking" or contact between the male proboscis and the female genitalia. Both these are preludes to attempted mounts. Note that rows A and B contain the longer bouts of licking, and especially of vibrating, which are characteristic of wild-type males of this species. Rows C and D illustrate shorter intervals for everything but orientation; this is typical of the yellow-bodied *D. melanogaster* males.

Bastock's (1956) data illustrate another point: even in flies phenotypically yellow, background genotype may be relevant. Table 1 presents a comparison between an ordinary wild stock and the one that was crossed to the yellow stock for seven generations. In this latter wild stock the percentage mating success of the *yellow* × *yellow* flies is far lower than that of the wild × wild. In matings between *yellow* and wild flies the percentage success of the *yellow* male × wild female is lower than that of the wild male × *yellow* female. Thus in the crosses involving males with the *yellow* mutant, the percentage success is much lower than in crosses involving wild-type males; the genotypes of the females have little differential effect. However, before the wild stock was crossed with the *yellow* stock for seven generations, there was a significant difference between females as well as between males. The initial high female receptivity is therefore partly dependent on the genetic background. It can be argued that for matings to occur reasonably frequently between *yellow* flies, there would be selection for *yellow* females with high receptivity in view of the low level of stimulus offered by the *yellow* males; i.e., there is likely to be a balance between the level of female receptivity and the mating propensity of the males.

There have been many experiments where males of one or more genotypes are placed with females of one or more genotypes. Variations between genotypes in choice experiments are presumably due to differences in courtship behavior. Sturtevant (1915) carried out experiments, based on direct observation, in which males were offered two types of females (*male-choice* experiments), and others in which females were offered two types of males (*female-choice* experiments). Some data for a white-eyed (sex linked) strain and a wild strain are given in Table 2. Clearly, the wild-type males have an advantage in sexual selection over the white-eyed males leading to nonrandom mating. In order to quantify data of this nature, certain indices have been proposed in the literature. These indices give estimates of the strength of sexual selection and of *sexual isolation*, which comes from comparing the proportion of *homogamic* (like-to-like) matings and *heterogamic* (unlike) matings. Under random mating the proportion of homogamic and heterogamic matings is expected to be equal.

For the male-choice situation, let there be n_1 females of type 1 and n_2 of type 2, together with males of type 1. Further, let $x_{1,1}$ and $x_{1,2}$ be the numbers of type 1 and 2 females inseminated, respectively, and let $p_{1,1} = (x_{1,1})/n_1$ and $p_{1,2} = (x_{1,2})/n_2$, so that $p_{1,1}$ and $p_{1,2}$ represent the proportions of each type of female inseminated. An *isolation index* devised by Stalker (1942) is:

$$b_{1,2} = \frac{p_{1,1} - p_{1,2}}{p_{1,1} + p_{1,2}}$$

TABLE 1. Percentage Success in 1 Hour from Pair Matings Using Yellow-Bodied and Wild-Type *D. melanogaster*

Matings	Before crossing wild stock with *yellow* stock for seven generations	After crossing wild stock with *yellow* stock for seven generations
Wild male × wild female	62	75
Yellow male × wild female	34	47
Wild male × *yellow* female	87	81
Yellow male × *yellow* female	78	59

After Bastock (1956).

which ranges from +1 for 100 percent homogamic matings to −1 for 100 percent heterogamic matings and is zero if mating is at random. Simple χ^2 tests can be used on the raw data to see if deviations from $b_{1,2} = 0$ are significant. If the male is of type 2, a reciprocal index is:

$$b_{2,1} = \frac{p_{2,2} - p_{2,1}}{p_{2,2} + p_{2,1}}$$

Joint isolation indices based on combinations of the pairs of experiments with males of types 1 and 2 have been proposed. If there are equal numbers of females or couples of each of the two types, the *joint isolation index* is:

$$\frac{x_{1,1} + x_{2,2} - x_{1,2} - x_{2,1}}{N}$$

where $N = x_{1,1} + x_{2,2} + x_{1,2} + x_{2,1}$ the total number of matings (Malogolowkin-Cohen, Simmons, and Levene, 1965). If there are not equal numbers of females or couples, the arithmetic mean of the two indices $b_{1,2}$ and $b_{2,1}$ is used:

$$b_{1,2} = \frac{b_{1,2} + b_{2,1}}{2}$$

From female-choice data analogous indices can be computed. Sturtevant's data give a joint isolation index of 0.097 in the male-choice experiment and 0.026 in the female-choice experiment; there is, therefore, little evidence for sexual isolation, since both values are close to zero.

Bateman (1949) proposed an index that measures the relative mating propensity of females:

$$a_{1,2} = \frac{b_{1,2} - b_{2,1}}{2}$$

TABLE 2. Results from Male- and Female-Choice Experiments Between White-Eyed and Wild-Type D. melanogaster

	No. of females mated	
Male choice	Wild-type	White-eyed
Wild-type male	54	82
White-eyed male	40	93
	No. of males mated	
Female choice	Wild-type	White-eyed
Wild-type female	53	14
White-eyed female	62	19

After Sturtevant (1915).

This is positive if there is an excess of females of type 1 and negative if there is an excess of females of type 2 in a male-choice experiment in which the males are of type 1. A similar index can be derived from the female-choice experiment. These indices, therefore, measure sexual selection. Sturtevant's data show that the relative mating propensity of wild-type females compared with white-eyed females is -0.303 in the male-choice experiments, and the relative mating propensity of wild-type males compared with white-eyed males in the female-choice experiments is 0.558. There is, therefore, clear evidence of nonrandom mating due to differences in the vigor of sexual behavior: there is sexual selection.

In recent years, multiple-choice experiments have become common where males and females of types 1 and 2 are all placed together in an observation chamber. Several designs are available; one of the commonly used chambers, constructed by Elens and Wattiaux (1964), is shown in Figure 3. Direct observation is possible and quite a large number of flies can be introduced — say 60 or more virgin pairs — but this depends on the species. Copulating pairs do not generally move and so can be localized on the checkered canvas of the chamber. The Elens-Wattiaux technique permits the observation not only of the types of males and females in a mating, but also of the time at which a given mating takes place, its sequence among other matings, and the duration of copulation. Furthermore, from this design all the various indices described can be calculated. Remember, though, that the biological situation in a multiple-choice experiment is different from that in a male- or female-choice experiment.

A number of other genes have been shown to affect mating success in *D. melanogaster*, mainly due to variations in sexual selection. Thus, in males homozygosity for the autosomal recessive gene *scabrous* (*sca*), resulting in rough eyes, leads to low sexual vigor; in females, it produces enhanced receptivity compared with wild-type flies (McKenzie and Parsons, 1971). In other words, there is a threshold switch for mating in both sexes. This cancels out when *scabrous* flies are mated together, since the frequency of *scabrous* female × *scabrous* male matings is similar to matings between wild-type flies. A parallel example of such a balance has already been described for yellow-bodied flies. The low vigor of *scabrous* males with wild-type females is due to the fact that the males are blind (Crossley, unpublished).

Eye color is relevant to mating success in some instances. Normally *D. melanogaster* homozygous for the *v/bw* (*vermilion/brown*) gene have pale sherry-colored eyes and an associated marked attenuation of visual acuity in terms of optometer response. The mutation at the *vermilion* locus results in a block in brown pigment synthesis, causing the flies to have bright red eyes. However, if the chemical kynurenine is added to

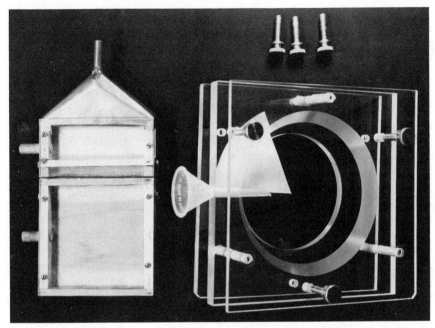

3. Alternate versions of Elens-Wattiaux chamber utilized for the direct scoring of Drosophila courtship and copulation. More than one chamber may be linked, either one atop the other or laterally. The older chamber (**left**) is plugged laterally with wooden pegs to prevent loss of moving air, a weak current being either sent from the copper apex or drawn into it. The flies producing the airborne odor are confined in the rectangular chamber nearest the apex, and a screened (on two sides) dead space separates them from the main, square chamber. It is in the square chamber that matings are scored. The dismantled plastic chamber (**right**) contains only a sector of the filter paper constituting the elevated circular center of its floor.

the diet, the *vermilion* block is bypassed, and brown pigment is formed in the eye. Table 3 shows the effect of kynurenine in enhancing mating success of male flies homozygous for the *v/bw* gene compared with those not treated with kynurenine (Connolly, Burnet, and Sewell, 1969). It seems likely that the mating disadvantage shown by flies lacking the eye pigment is due to a sensory defect accompanying the absence of screening pigment in the compound eye. This lack can be alleviated by biochemical supplements. Therefore, Connolly, Burnet, and Sewell (1969) suggest that the role of vision in the courtship of D. *melanogaster* has been underestimated — a result in agreement with work on the stock homozygous for the *scabrous* gene. Comparison of the courtship behavior of males with pigmented and nonpigmented eyes shows that the inferior courtship of *v/bw* males is attributable to difficulties in establishing and maintaining contacts with females. Thus *v/bw* male flies were found to have a significantly shorter bout length (the sum of licking plus vibration) than those to whose diet kynurenine was added. Generally, there was a close correlation between mating success and eye-pigment density. It is also of interest that only the presence of brown pigment is involved in mating success, since the absence of red pigment, as in *bw* flies, does not lead to an attenuation of optometer response and does not affect courtship latency or duration. Of interest is an earlier observation by Parsons and Green (1959), who showed a general correlation between brown eye-pigment density and fitness, such that the fitness of *v/bw* flies increased in competition experiments with increasing amounts of kynurenine. Various behavioral variables are therefore associated directly with biochemical changes and fitness changes. In other words, where investigated adequately, the likelihood of finding a purely behavioral trait associated with effects closer to the gene itself is always possible.

The final mutants of *D. melanogaster* to be considered are the sex-linked mutant genes that produce *Bar* and white eyes. *Bar* eyes are narrower than normal eyes and the mode of inheritance is dominant. In a mixture of *Bar* and wild-type flies, the *Bar* males are less successful in mating; their disadvantage is, however, reduced when only few of them are present, and increases as their frequency in relation to wild-type males becomes greater. With white-eyed males, mating success is greater when white-eyed males are rare or are predominant. Ehrman et al. (1965) found a similar situation in experiments with *Drosophila pseudoobscura*. It may indeed be that preferential mating correlated with frequency will prove to be of considerable importance in evolutionary processes, if it is at all widespread. More about this matter later (see Section 8.4).

Sexual discrimination is a trial-and-error affair among the drosophilids reported upon here. Males court females (sometimes, even

TABLE 3.   Results of Competition Between Kynurenine-Treated *v/bw* Males and *bw* Males and Between Kynurenine-Treated *v/bw* Males and *v/bw* Males

Total no. of mating competitions	Description of male	No. of males mating	$\chi^2{}_1$ for 1:1 ratio
126	*bw*	52	
	Kynurenine-treated *v/bw*	74	3.5
83	*v/bw*	15	30.12
	Kynurenine-treated *v/bw*	68	($P<0.001$)

After Connolly, Burnet, and Sewell (1969).

other males) of any species and try to repeat courting and mating. The acceptance is controlled mainly by the female, and so are other "turning points" in the courtship-mating-insemination sequence. Even so, as is shown in Sections 4.2 and 13.1, given receptive females, mating speed differences between males rather than females are common. Furthermore, duration of copulation is mainly male-determined, at least in *D. melanogaster* (MacBean and Parsons, 1967) and *D. pseudoobscura* (Kaul and Parsons, 1965). The female has three sperm-storing organs and mates a second time only if her supply of stored sperm is diminished (Manning, 1962). For a description of the courtship and mating behavior of *Drosophila*, see Spieth (1952); for a pictorial description, see Ehrman and Strickberger (1960) and Ehrman (1964).

3.3 Mating success in mice

Albinism in mice and other creatures (e.g., one type in man) is frequently controlled by an autosomal recessive gene. Levine (1958) compared the relative mating success of homozygous black agouti male mice with albino mice. Both strains were maintained separately by brother-sister matings and so were inbred, and all males were proved fertile both at the beginning and at the end of the experiment. The procedure was first to pair at random 10 male albinos with 10 male black agoutis. Each pair of males was then placed in a pen with a single albino female. Ten litters were obtained from each pen, giving a total of 100 litters. The results are shown in Figure 4.

Three types of litters were produced: those containing only albino offspring, only black agouti offspring, or mixed offspring (some albino and some black agouti). This last type of litter was the result of double inseminations. There was no statistical difference in litter size among the three types of litters. It was found that 76 percent of the litters were fathered solely by the albino males, 12 percent of the litters were fathered solely by the black agouti males, and 12 percent of the litters were the result of double inseminations. Within the mixed litters, albino offspring were more than twice as common as black agouti offspring. Of a total of 552 mice born in the 10 cages, 458 were fathered by the albino males while only 94 were fathered by the black agouti males. Interpretation is not easy because selective fertilization favoring the sperm of males belonging to the same strain as the females cannot be ruled out. Levine (1958) observed albino versus black agouti fights and noted a predominance of successfully aggressive albino males. Each fight was watched until one opponent assumed a "submission reaction," i.e., sitting on his hindlegs with forelegs extended in a defensive position. One wonders about the correlation between fighting ability and repro-

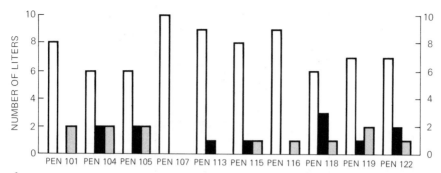

4. **Results of competitive matings in mice** of different inbred strains. **White columns** represent number of litters fathered solely by albino males. **Black columns** represent number of litters fathered solely by black agouti males. **Grey columns** represent number of litters fathered by both types of males (mixed litters). (From Levine, 1958.)

ductive success in these mice. This then may represent a true example of sexual selection in mice. Even so, the possibility that albino females may have some preference for albino males (homogamy) cannot be ruled out. A final point about the mice utilized in Levine's experiment and in many other experiments: unknown to him and ascertained some years later, *all* the mice he used were homozygous for the *rd* (retinal degeneration) allele and, therefore, blind (Sidman and Green, 1965). Even so, this does not alter the conclusion that sexual selection probably occurs. A further general discussion of behavior in mice is presented in Chapter 9.

3.4 Single-gene effects in man

HUNTINGTON'S CHOREA

What of the simple genetics of behavior in man, in whom breeding experiments cannot be carried out? A good example is Huntington's chorea, a heritable and invariably fatal disorder. (Chorea is a neurological disease marked by muscular twitching, from the Greek word for "dance.") The onset of Huntington's chorea is obscure, its primary metabolic lesion unknown, and its deterioration characteristically progressive from choreic movements to ataxia, dementia, and death. The progressive dementia is characterized by the degeneration of ganglion cells of the forebrain and corpus striatum. As Figure 5 and other pedigrees indicate, Huntington's (named after three generations of physicians practicing and maintaining longitudinal family records in Connecticut) chorea is the result of heterozygosity for an autosomal dominant gene with delayed age of onset. Symptoms usually do not appear until reproductive age has been reached or passed, so that although the condition

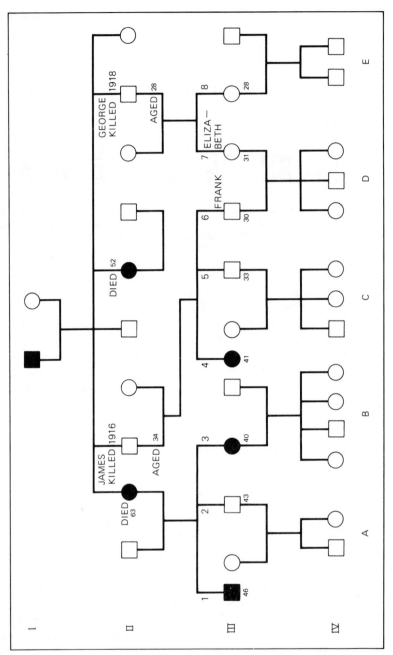

5. **Autosomal dominant transmission** evidenced in Huntington's chorea traced through four generations. Small numbers represent some ages at the time this pedigree was compiled (when all 14 children in generation IV were under 15 years of age). (From Roderick, 1968.)

is fatal, those who carry the gene are in most cases able to produce progeny before they are even aware of their affliction. The average age of onset is 35 years, with a range mainly from 15 to 65 years of age for the initial twitching, although some childhood cases are known. Potegal (1971) has demonstrated spatial-motor defects in patients with Huntington's disease. They are inaccurate in egocentric spatial localization wherein the position of a target in space must be defined in terms of its distance and location from the observer, e.g., "straight ahead" or "a yard to my left."

The gene for Huntington's chorea was brought to the United States by three young men who departed from Bures St. Mary, Suffolk, England, with the same ship convoy in 1830. They "left town" because of difficulties incurred by their unusual, even outrageous behavior (Vessie, 1932), and all three wed and fathered children in their new country. There are now more than 7000 people afflicted with Huntington's chorea in the United States; the incidence of this disorder is approximately 1 in 25,000 and occurrences have been reported throughout the world.

Consider the pedigree constituting Figure 5. The probabilities of inheriting the gene H for Huntington's chorea in generation IV are:

- Either member of family A: $1/2 \times 1/2 = 1/4$

- Any member of family B: $1/2$

- Any member of family C: $1/2 \times 1/2 = 1/4$

- Any member of family D (assuming two doses of the gene, HH, to be lethal, three possible types of mating exist):

$$Child$$

(a) $1/4\ Hh\ \times\ 1/2\ Hh\ \times\ 2/3\ Hh^* =\ 2/24\ =\ 4/48$
(b) $3/4\ hh\ \times\ 1/2\ Hh\ \times\ 1/2\ Hh\ =\ 3/16\ =\ 9/48$
(c) $1/4\ Hh\ \times\ 1/2\ hh\ \times\ 1/2\ Hh\ =\ 1/16\ =\ 3/48$

$$Total = 16/48 = 1/3$$

*$Hh \times Hh = 1HH$ (dies?): $2Hh$ (afflicted): $1hh$ (normal) $= 2/3\ Hh$: $1/3\ hh$.

- Either member of family E: $1/4 \times 1/2 = 1/8$

Consult Thompson and Thompson (1973), Fuhrmann and Vogel (1969), and especially Porter (1968) for more details and more evidence of the analysis of genetic patterns in nonexperimental subjects.

Before any symptoms develop in a person who had a parent with this disease, the probability that this person has the gene is $\frac{1}{2}$. (This gene is rare enough to allow the assumption that the affected parent was not homozygous or that the homozygosity is lethal.) Therefore, before diagnosis is possible, the probability that the person in question will have an afflicted child is $\frac{1}{2}$ (the probability the person in question has the gene) × $\frac{1}{2}$ (the probability the child will inherit the gene if the parent has it) $= \frac{1}{4}$.

Once diagnosis is certain, the probability of a child's manifesting the disease becomes $\frac{1}{2}$ (if a parent has it) or 0 (if the parent is not afflicted). See Falek and Britton (1974) on the psychology of this tense situation. Sometimes gaps occur in pedigrees such that affected persons are derived through unaffected individuals. The likely explanation is the late age of onset in a potentially affected parent who died of other causes before initial symptoms of Huntington's chorea appeared.

LACTASE DEFICIENCY

By stretching our definition of behavior genetics a bit, we are able to include in this chapter consideration of the role culture has played in the evolution of the three allelic genes controlling lactase deficiency and milk consumption in human populations. The emerging story (McCracken, 1971; Gottesman and Heston, 1972; Kretchmer, 1972) is fascinating. Lactose is the primary sugar found in milk; it is metabolized by lactase, an enzyme produced in the villi of the small intestine. Simply put, the reaction is:

$$\text{lactose (disaccharide)} \xrightarrow{\text{lactase}} \text{glucose + galactose (monosaccharides).}$$

The end products can subsequently be absorbed into the human circulatory system, but if lactase is absent, lactose passes through the intestines without providing any nutritional value. This passage sometimes results in bloating, cramps, and diarrhea.

Three alleles (L, l_1, and l_2) seem to occupy the autosomal locus that controls lactase production. Both l_1 and l_2 are recessive to L, the wild-type or normal allele, and l_2 is recessive to l_1. LL, Ll_1, or Ll_2 individuals produce lactase both as children and as adults; l_1l_1 and l_1l_2 individuals do not produce lactase as adults; and l_2l_2 is a very rare combination, usually lethal, because milk cannot be digested even in infancy. In Northern Europe, 80 to 100 percent of all adults are LL, Ll_1, or Ll_2, while just the reverse is true in, for example, Oriental, Amerindian, Southern European, Australian Aborigine, and African populations (l_1l_1 or l_1l_2). Note that these l_1l_1 and l_1l_2 adults can manage the digestion of sour milk products such as sour milk itself, yogurt, and cheese.

McCracken (1971) suggests:

> It is hypothesized that prior to the domestication of animals [400 generations ago for the start of the domestication of sheep and goats?] and the development of dairying, the normal condition for all men was adult lactase deficiency, but with the introduction of lactose into the adult diet in certain cultures, new selective pressures were created that favored the genotype for adult lactase production.

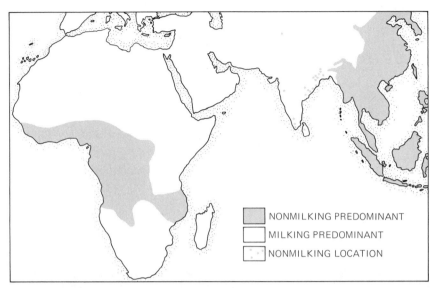

NONMILKING PREDOMINANT

MILKING PREDOMINANT

NONMILKING LOCATION

6. **Milking and nonmilking areas.** Map shows predominant traditional patterns in Africa and part of Asia. (From Simoons, 1970.)

Simoons (1970) cautions that one must not assume that where milkable animals were kept, people did in fact milk them and, furthermore, consume the milk as adults. Finally, lactase activity may be inducible, i.e., lactase may be produced at a rate in parallel with the demands of the diet — the more lactose consistently ingested, the more lactase synthesized. This is known as adaptive enzyme formation, a phenomenon not always observed in response to apparently significant dietary challenges. At any rate, culture is surely a major factor in the evolution of the one human species, and culture implies behavioral adaptation, at least some of which are likely to be under genetic control. Look at Figure 6 and judge for yourself the magnitude of the advantage of milk as a supplementary nutrient for adults and whether it seems reasonable that this particular advantage could eventually have altered the appropriate gene frequencies involved.

Chromosomes and Behavior

In Chapter 3 the effect of single genes on simple behavioral traits is considered. In this chapter we extend the genetic unit under consideration to the chromosome. Before discussing behavior as such, we give a short account of the various types of gross chromosomal changes commonly found. A good detailed account is in Herskowitz (1973) and a more extended but easily readable account of chromosomes generally is provided by White (1973).

4.1 Chromosomal changes

Changes involving unbroken chromosomes are common. Although most sexually reproducing organisms have a diploid chromosome complement, or genome, the occurrence of triploids and tetraploids (three and four complete chromosome sets instead of two) is common in plants. In Drosophila, triploid and tetraploid females occur and haploid/diploid somatic mosaics have been found. (A chromosomal *mosaic* is an individual with tissues of differing chromosomal constitutions due to an abnormal somatic cell division(s) early in fetal life.) In man complete triploidy is lethal, but individuals who are diploid/triploid mosaics may survive, though they are physically and mentally defective. A change in a complete set of chromosomes is referred to as *euploidy*. In contrast, *aneuploidy* is the addition or subtraction of single chromosomes from the genome. Aneuploids may arise as a result of aberrations during the process of cell division at the time of gamete formation (referred to as meiosis in Section 2.1). In gametogenesis chromosomes normally disjoin in a regular fashion to enable one representative from each chromosomal pair to be included in each daughter cell. For example, *nondisjunction* of the fourth chromosome of *Drosophila melanogaster* yields individuals with either one or three fourth chromosomes, the former individual being *monosomic* and the latter *trisomic*. In man, as we shall see, individuals trisomic for one of the smaller chromosomes may have Down's syndrome, characterized by gross morphological and mental impairment. (For a detailed account of chromosome behavior during meiosis, the reader is referred to a text such as White, 1973.)

In contrast to the modifications of chromosome number discussed above are changes involving *chromosome breakage* (Fig. 1), of which there are four possible types: deletion, duplication, inversion, and translocation.

Deletion, or removal, of a gene locus or a group of loci is most often lethal in the homozygous form. From an evolutionary point of view deletions are relatively insignificant.

Duplication of a gene locus, which may occur in various ways (White, 1973), may cause an imbalance of gene activity, reducing the viability of an organism. However, since some organisms can tolerate duplications of chromosomal material, duplications may play an evolutionary role. If a particular locus were duplicated, one of the twin loci could mutate to an allele having an entirely different function without reducing fitness. Presumably the unchanged alleles at the other locus could adequately perform the original function of the locus. In this way evolutionary change may occur — and indeed has been postulated to occur in the evolution of the four genes for the hemoglobin molecule in man (for details see Herskowitz, 1973). The genetic control of some other complex molecules has evolved in a similar way.

An *inversion* occurs when a chromosome breaks in two places, and the segment between the breaks then rotates 180° leading to a reversal of gene order with respect to that on the unbroken chromosomes. Inversions occur spontaneously during various chromosomal movements during cell division. The significant effects of inversions are due to the fact that during meiosis homologous chromosomes pair exactly, gene by gene, which leads to the formation of characteristic loops during meiosis in individuals heterozygous for inversions (Fig. 2). During meiosis, chromosome breakage and rejoining of homologous partners routinely occurs and is referred to as *crossing over*. In Figure 2 this process is illustrated for a heterozygous inversion and shows the consequence to be two normal chromosomes not showing the effect of crossing over,

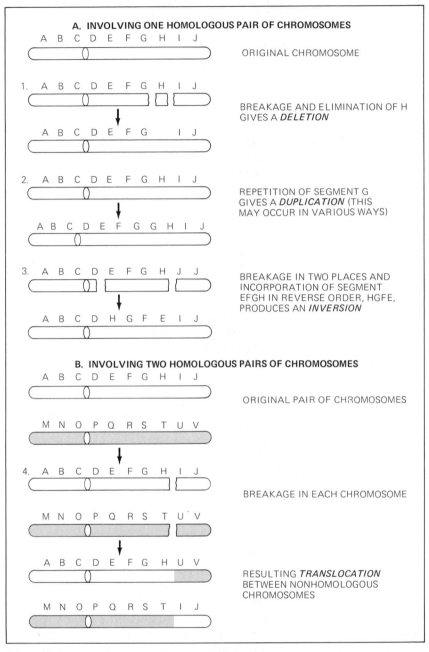

1. **Chromosome breakage.** Origin of the four principal types of structural change by means of chromosome breakage: deletion, duplication, inversion, and translocation.

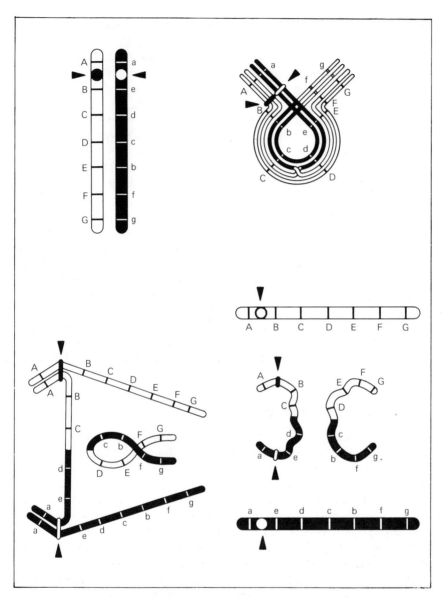

2.   **Crossingover in an inversion heterozygote. Upper left.** Two chromosomes differ-
ring by one paracentric inversion. **Upper right.** Pairing at the earliest stage of meiosis
depicted here. **Lower left.** After the first meiotic division a chromatid bridge and acentric
piece of chromosome are formed. **Lower right.** Two viable and two nonviable chromosomes, the result of crossingover at the completion of meiosis. **Black triangles** indicate
centromeres. (Reprinted from T. Dobzhansky: *Genetics and the Origin of Species,* 3rd ed.,
revised. New York, Columbia University Press, 1951, p. 125. by permission of the pub-
lisher.)

and two abnormal structures showing the effect of crossing over, one (without a centromere) which becomes lost during the meiotic process, and the other (with two centromeres) which breaks in cell division and so also is eventually lost. If an organism is heterozygous for an inversion, therefore, crossing over within the inverted segment is generally ineffective in contrast to crossing over exclusive of inversion heterozygosity, where crossing over leads to recombination of genes (Section 2.1). Therefore, the genes of an inverted segment are transmitted as a single unit in inversion heterozygotes, since only those chromosomes not showing the effect of crossing over remain. This is a point of some considerable evolutionary significance, especially in Drosophila, as will soon be shown (Dobzhansky, 1970).

A *translocation* occurs when two nonhomologous chromosomes break simultaneously and exchange segments. If an organism becomes homozygous for such a rearrangement, some of its genes have been transferred to a completely different chromosome, and linkage relationships of the genes are considerably changed, as shown in Figure 1.

With this brief background we can now turn to an assessment of chromosomal effects on behavior. It should be stressed, however, that not all the above chromosomal changes are important in this regard, according to our current knowledge. To date, inversions and chromosome number changes appear to form the main categories of importance.

4.2   Inversions in Drosophila

Many species of Diptera have two or more inversions in natural populations in such frequencies that these cannot be explained by recurrent mutation. When a population has two or more genetic alternatives in these frequencies, it is said to exhibit genetic *polymorphism*. The occurrence of polymorphism indicates a situation of particular genetic and evolutionary interest, since there must be a balance of selective forces maintaining the inversions in the population. Thus any behavioral differences associated with inversions could be of major evolutionary significance. Experimental work, in particular in *D. pseudoobscura*, has shown that inversion heterozygotes (often called *heterokaryotypes*) often have a fitness superior to that of inversion homozygotes (*homokaryotypes*). (A *karyotype* is the chromosomal constitution of an organism.) The introduction of two sequences, Standard (ST) and Chiricahua (CH), of chromosome III of *D. pseudoobscura* into population cages at 25°C leads to ultimate inversion frequencies at about 0.7 ST and 0.3 CH irrespective of the frequencies initially (Wright and Dobzhansky, 1946). In other words, the inversion frequencies arrive at an *equilibrium*

— a result in contrast with the Hardy-Weinberg situation discussed in Sections 2.3 and 2.4. We therefore should ask which of the various factors listed in Section 2.4 is important. It is known from an extensive series of experiments by Dobzhansky and his associates (references in Parsons, 1973) that various components of fitness, such as innate capacity for increase, population size, productivity, and egg-to-adult viability, show heterokaryotype superiority over the corresponding homokaryotypes. In other words, the fitnesses of the karyotypes differ.

Because ST and CH inversions essentially segregate as single genes, it is important to consider briefly the conditions under which two alleles, *A* and *a*, are polymorphic at a locus. In Section 2.3 the Hardy-Weinberg law under random mating is discussed. The additional complication to be considered now is that the fitnesses of the three genotypes *AA*, *Aa*, and *aa* are not necessarily equal, as assumed heretofore. We therefore let the fitnesses of genotypes *AA*, *Aa*, and *aa* be 1-s, 1, and 1-t respectively, so that the genotypic proportions before and after selection become:

	AA	Aa	aa	Total
Fitnesses	1-s	1	1-t	
Frequencies before selection	p^2	$2pq$	q^2	1
Frequencies after selection	$p^2(1-s)$	$2pq$	$q^2(1-t)$	$\overline{W}$

where $\overline{W}$ is the average fitness of the population after selection. Letting p' and q' represent the gene frequencies of A and a in the next generation, then:

$$p' = \frac{p^2 - p^2 s + pq}{\overline{W}} = \frac{p - sp^2}{\overline{W}}$$

and

$$q' = \frac{pq + q^2 - q^2 t}{\overline{W}} = \frac{q - tq^2}{\overline{W}}$$

The reason for dividing by $\overline{W}$ is to ensure that $p' + q' = p + q = 1$. At equilibrium, gene frequencies are fixed from generation to generation. If we write the change in gene frequency from generation to generation as Δp, then at equilibrium it is expected that:

$$\Delta p = p' - p = 0$$

or

$$\Delta p = \frac{p - sp^2}{\overline{W}} - p = \frac{pq(tq - sp)}{\overline{W}}$$

Noting that $\overline{W} = 1 - sp^2 - tq^2$, it is not difficult to show that $\Delta p = 0$ when $p = 0$, $q = 0$, or $tq = sp$. The first two solutions are trivial when the population is either all AA or aa, which is not a polymorphic situation. In other words, either gene A or a is lost while the other is fixed. It is ostensible and can also be demonstrated algebraically that if fitnesses are (1) $AA > Aa > aa$, then A will be fixed and a lost (i.e., $p = 1, q = 0$), and (2) $aa > Aa > AA$, then a will be fixed and A lost (i.e., $p = 0, q = 1$).

The solution $tq = sp$ gives equilibrium gene frequencies $p = t/(s + t)$ and $|q = s/(s + t)$, which therefore depend only on the selective values s and t. This means that irrespective of the initial values of p and q the same equilibrium is expected. Clearly, the only circumstances under which these equilibria can exist are for $s, t > 0$ or $s, t < 0$, since otherwise one or the other equilibrium gene frequencies would then be negative, which is impossible. For these two conditions the stability of the equilibria must be examined. A stable equilibrium occurs if after there is a small displacement from the equilibrium gene frequency, as may occur by chance in a finite population, the population tends to return to that gene frequency in subsequent generations. It can be shown (see Parsons, 1973, for a brief account) that if $s, t > 0$, a stable equilibrium is expected. This corresponds to the situation where $Aa > AA, aa$ in fitness; in other words, there is a heterozygote advantage over both homozygotes, called *overdominance*. Conversely, if $s, t < 0$, which means that $Aa < AA, aa$ in fitness, then a small displacement from the equilibrium gene frequency is accentuated, generation by generation, and eventually one allele or the other is fixed. This is an unstable equilibrium. All these conclusions can be derived algebraically. The important conclusion is, assuming random mating with unequal genotypic fitnesses, that if a heterozygote is more fit than its corresponding homozygotes, a stable equilibrium is to be expected with an associated polymorphism; then the equilibrium gene frequencies depend only on selective values.

These theoretical calculations show that one situation (but not the only one) where polymorphism is expected is where heterozygotes (heterokaryotypes) are more fit than the corresponding homozygotes (homokaryotypes). For the situation of inversion polymorphisms, we are dealing with the gene complex making up the block of genes locked into the inversion rather than a single locus. Because effective recombination is suppressed in inversion heterozygotes, complexes are built up in which the genes interact to produce *coadapted gene complexes* by natural selection. One may ask whether there is evidence of such coadaptation. The answer comes from comparisons of pairs of chromosomal heterozygotes derived from different ecogeographical regions often not very distant from each other (Dobzhansky, 1950). On the whole, heterokaryotype superiority breaks down completely even though it still exists within ecogeographical regions. In other words, the gene arrangements within the inversions are unique to specific geographical populations, owing to the mutual adjustment of gene complexes within these populations to give high heterozygote fitness, which Dobzhansky calls *coadaptation*. Clearly, gene complexes from different populations have not had the same opportunity for mutual adjustment of their genic contents — i.e., they have not been selected for coadaptedness, so that heterozygosity for inversion sequences from different localities is not expected to lead to high fitnesses.

Turning now to behavioral examples, Brncic and Koref-Santibañez (1963, 1964) studied the relation between sexual selection and chromosomal inversions in the South American species, *D. pavani*, which is found mainly in the southern part of the continent. In most natural populations the proportion of heterokaryotypes exists in fairly uniform frequencies. Mating activity was assessed using virgin females of *D. gaucha*, a sibling species of *D. pavani*. (*Sibling* — sister or brother — *species* are closely related and often morphologically indistinguishable.) Pair matings were observed for 30 minutes, and Brncic and Koref-Santibañez scored (1) pairs that mated during the period of observation, (2) those that courted but did not copulate, and (3) those that were sexually inactive during the observation period. The frequency of heterokaryotypes was significantly higher among males that courted and/or mated during the first few minutes of being placed with the females. These results indicate that the heterokaryotypes are superior at least up to the first mating. It can therefore be argued that superiority in mating activity of the heterokaryotypes is likely to be one important factor in the maintenance of this polymorphism in natural populations of *D. pavani*. It is an example of overdominance, as discussed in the theoretical considerations above.

We should also consider the extensive studies of Spiess and his associates (Spiess, 1962; Spiess and Langer, 1961, 1964a,b; Spiess, Langer, and Spiess, 1966; Spiess and Spiess, 1967) on homo- and heterokaryotypes in *D. pseudoobscura* and its sibling species, *D. persimilis*. In *D. pseudoobscura*, great differences in mating speed were found between homokaryotypes derived from stocks collected at Mather, California. Homokaryotypes for Standard (ST), Chiricahua (CH), Tree Line (TL), Pike's Peak (PP), and Arrowhead (AR) inversions were used (Fig. 3). The experimental procedure involved the direct observation of 10 pairs of 6-day-old flies in mating chambers over a 1-hour period at 25°C. Apart from the pairs AR, ST and CH, TL, all other pairings differed significantly at 60 minutes. For AR and ST rapid matings occurred; CH and TL were intermediate, and PP slow. The rapidity of acceptance, mounting, and insemination (other things being equal) in-

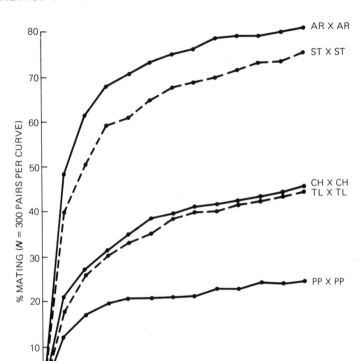

3.   **Homokaryotype homogamic matings.** Cumulative percentage curves for matings during 1 hour's observation. AR, ST, CH, TL, and PP represent different inversions in the third chromosome of *D pseudoobscura.* (From Spiess and Langer, 1964a.)

creases the fitness of the carriers of a given karyotype. With rapid mating, the deposition of fertile eggs by the females is likely to be accomplished without delay, and the females depleting their sperm supply are ready to mate again and receive a fresh supply. This is consistent with the widely accepted definition of natural selection: "Natural selection occurs when the carriers of some genotypes contribute more surviving progeny to the succeeding generations in relation to what the carriers of other genotypes contribute" (Dobzhansky, 1964). Spiess and Langer (1964a) point out that the observed frequencies of the inversions at Mather occur in approximately the same sequence as the mating frequencies, with AR and ST being the most frequent, and PP the least. From this it is tempting to suppose that mating speed is a major factor in maintaining the observed frequencies of the chromosomes in this population, and so is indeed an important component of fitness. Spiess and Langer (1964b) summarize the results of their studies as follows: "If mating speed is constant for each karyotype under 'competitive conditions,' net adaptive values (relative fitnesses) of karyotypes will be frequency dependent [see Sections 3.2 and 8.4]. In natural populations these karyotypes controlling mating behavior must contribute to a major portion of the population's total fitness."

Considering mating speed behavior between karyotypes, for both heterokaryotypes and homokaryotypes, our initial conclusion is that mating speed is almost entirely male-determined (Kaul and Parsons, 1965, 1966; Spiess, Langer, and Spiess, 1966). For example, Kaul and Parsons (1966) showed this to be so in two series of choice experiments consisting of one female with three males and the reverse, three females with one male. The mean period elapsing to the first mating was 0.53 minute in the experiments with the three females and 1.40 minutes in those with the three males. The likely interpretation is that in experiments with three males, competition which occurs among males delays mating, whereas in the reverse situation with three females, the one male tends to mate more rapidly, having no competition from other males. The second conclusion that emerges is that the male heterokaryotypes consistently have a higher mating speed than the homokaryotypes. Spiess, Langer, and Spiess (1966) studied mating speeds for a number of combinations employing 10 pairs of flies per mating chamber and found that male heterokaryotypes had a consistently higher mating speed than the corresponding homokaryotypes. Females displayed no such consistent superiority, which if it did exist would imply variations in receptivity. Clearly then, the overdominance (often called *heterosis*) displayed is due to the greater activity or persistence in courtship of males or to greater female acceptance of heterokaryotype males probably because of the males' increased sexual activity.

In *D. pseudoobscura*, therefore, male mating speed may well be an important component of fitness, and it must be considered in relation to all other components of fitness, some of which are cited earlier in this section — innate capacity for increase, population size, productivity, and egg-to-adult viability. The many associations among all of these components in a given population have been relatively unexplored (see Parsons, 1973), but must be of considerable importance in the study of the fitness of organisms. However, Parsons (1974) concludes, after a review of available evidence, that male mating behavior is a very important component of fitness, at least in *D. pseudoobscura*, *D. pavani*, and probably also *D. melanogaster*. However, in *D. persimilis*, a sibling species of *D. pseudoobscura* in which mating speeds have also been studied, Spiess and Langer (1964b) found a less one-sided situation. Certain females accept males readily, others tend to refuse them, and certain males court more actively than others. In other words, the differences found can be interpreted in terms of the relative intensities of the copulatory tendency of males and the acceptance (or, conversely, the avoidance) tendency of females.

4.3 Karyotype variations in man

DOWN'S SYNDROME

Trisomy for one of the smallest of the human autosomes is called trisomy-21, trisomy-G (Fig. 4), or Down's syndrome (or, anachronistically mongolism because of an Oriental appearance about the face and eyes). One in 600 to 700 newborns from all human populations is afflicted with this syndrome. It is characterized by congenitally retarded mental, motor, and sexual development, and a number of physical stigmata. Longevity is abbreviated. The mental retardation ranges from IQs of less than 20 to less than 65, and therefore represents idiocy or imbecility. Behaviorally, these individuals are often happy and friendly; they often imitate well, but are inevitably retarded. Dingman (1968) studied psychological test patterns in patients with Down's syndrome and noted no systematic behavioral differences between individuals with Down's syndrome and other mentally retarded patients; the differences he recorded are apparently due to retarded mental development.

Most of the time the presence of the extra chromosome number 21 (or G-group chromosome, as it is also categorized because chromosome pairs 21 and 22 were until recently morphologically indistinguishable) is due to an error or errors in meiosis. (Hungerford et al., 1970, 1971, presents compelling evidence that the supernumerary chromosome is actually number 22; this evidence, based on length, comes from testicular biopsy of chromosomes in one of the stages of meiosis called pachytene, when chromosomes are relatively short and thick.) The extra chromosome that characterizes Down's syndrome presumably arises by nondisjunction. This nondisjunction is probably restricted primarily to females because the frequency of affected individuals increases rapidly with maternal age. For a woman 45 years old at pregnancy the risk of producing a child with Down's syndrome is 1/50 compared with 1/3000 for a woman of 20. Presumably the increase in nondisjunction with age is due to a change in the environment of the oocytes induced by increasing age (see Penrose, 1963, for a detailed account of work on this syndrome). About 2 percent of those affected with Down's syndrome may have chromosomal breakages, such as translocations, involving this crucial G chromosome. One example is an individual with this syn-

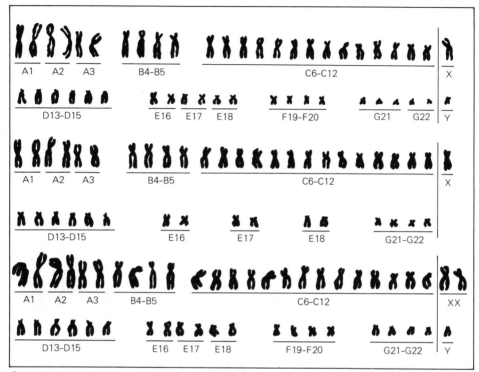

4. **Abnormal human karyotypes** prepared from chromosomes in human white blood cells. **Top:** Male with Down's syndrome. Note the additional chromosomes to set number 21. This chromosome abnormality is phenotypically expressed as a form of idiocy or imbecility. **Middle:** Female with Turner's syndrome. Note the single X chromosome. Phenotypic expression of this chromosome abnormality results in retarded sexual development and sterility. **Bottom:** Male with Klinefelter's syndrome. Note the *XXY* genotype expressed phenotypically as severe mental retardation and lack of male secondary sex characteristics, development of some female secondary sex characteristics, or both. (From *Biology Today*, 1972, pp. 259–260. CRM Inc., Del Mar, California.)

drome and only 46 chromosomes — no extra chromosome at all. Such an individual has two G chromosomes and an extra long D-group chromosome (chromosome 14 or 15). This suggests a translocation between a third G and a D chromosome leading to a new, large chromosome consisting of almost all the material of each (i.e., D and G). Thus, the affected individual has the material of three G chromosomes, as occurs through nondisjunction. Since the translocation can be inherited, for Down's syndrome due to translocation, a familial pattern of inheritance may be expected with carrier individuals having 45 chromosomes. On the other hand, some one half of such cases represent new gross mutations.

In passing it is of interest that Down's syndrome may not be restricted to man. A syndrome resembling Down's has been reported in *Pan troglodytes*, a chimpanzee (McClure, Belden, and Pieper, 1969). In Figure 5 is depicted the animal's karyotype, and Figure 6 shows the results of behavioral tests on this young female indicating retarded growth and neurological development compared with other nursery animals of her species. She also exhibited general inactivity and at 40 weeks of age was still unable to sit up or move about. Such investigations involving animal models of human conditions (see Leader, 1967) are recommended for behavior geneticists as well as for pathologists, who have been fruitfully employing them for some time.

SEX CHROMOSOME ABERRATIONS

A general conclusion that will become apparent is that autosomal aberrations appear to cause more severe effects (morphologically and behaviorally) than do X or Y chromosomal aberrations (but see Fig. 7). An

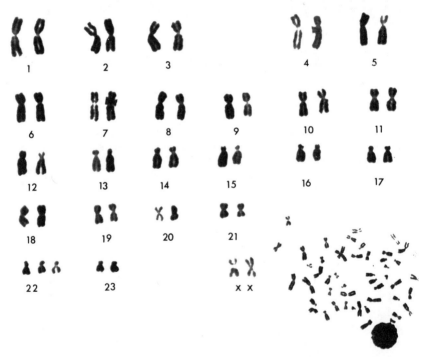

5.   **Blood cell karyotype of trisomic chimpanzee.** This animal, trisomic for chromosome number 22, exhibited mental retardation and other stigmata associated with Down's syndrome in man. (Courtesy of Dr. Harold McClure, Yerkes Primate Research Center, Emory University, Atlanta, Ga.)

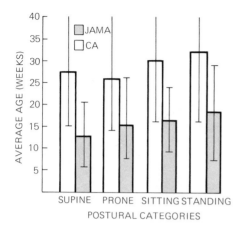

6. Behavioral development of trisomic chimpanzee (Jama) compared with average ages at which 50 percent of 14 tested control chimpanzees (CA) completed the 34 behavioral items here condensed into four categories. (After McClure et al., 1969.)

interpretation is presented later in this section. For instance, Turner's syndrome, or gonadal dysgenesis (Fig. 4 *top* and 8), is characterized by absence of an X or a Y chromosome and, therefore, a karyotype with 45 chromosomes (written X0). In appearance, those with Turner's syndrome are female, and behaviorally they are characterized by hypertension, normal verbal intelligence but visual infantilism in space perception, and a specific space-form defect — what can be called a degree of space-form blindness or more technically, partial congenital agnosia (Schaffer, 1962; Money, 1970). Klinefelter's syndrome (Fig. 4 *bottom*) is characterized by an extra X and, therefore, a karyotype with 47 chromosomes (written XXY). In appearance these individuals are male, but sterile and of low libido. The syndrome is not necessarily marked by mental retardation, but it may be. XXY males are socially inadequate, often quitting school and other activities requiring social contacts. Some XXY males are more than socially inept: they are antisocial and may require institutionalization. Many reports list passivity, dependency, withdrawal from reality, limited interests, and poor impulse control as Klinefelter personality characteristics. Klinefelter's and Turner's syndromes are among the commonest products of sex chromosome aberrations, and involve the addition or deletion of whole chromosomes with somatic effects rather minor compared with those of Down's syndrome. Turner's syndrome occurs with a frequency of about 3 per 10,000 among newborns, and the frequency of Klinefelter's syndrome is about 2 per 1000 newborns.

The twins depicted in Figure 8 have Turner's syndrome and were 17 years old when this picture was taken. These girls were short, but enjoyed good mental and physical health (having graduated from high school as average students), although neither had yet menstruated. No

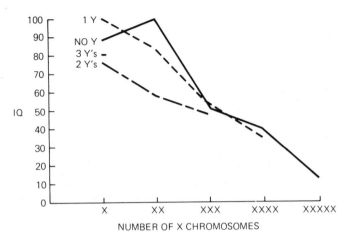

7.   Sex chromosome abnormalities and IQ. Effects on mean IQ of abnormal sex chromosome aneuploidies. (Compiled by Vandenberg, 1974, from Moor, 1967.)

uterus could be palpated in either girl, and thyroid therapy did not bring about menarche. According to Money and Mittenthal (1970):

> As with many of the other symptoms that may be associated with Turner's syndrome, space-form disability does not occur in 100 percent of cases (more likely in about 75 percent); and it occurs with varying degrees of severity. Its most likely explanation is that it is an effect of the genetic defect of the syndrome on the development and functioning of the cerebral cortex. On an intelligence test, this deficit shows up in the nonverbal and numerical items; verbal ability is not affected.
>
> There may be another direct effect of genetics on behavior in Turner's syndrome, namely with respect to personality. There is no commonly agreed upon name for the feature of personality shared by many Turner girls, which might be identified as inertia of emotional arousal. It is constituted of compliancy, phlegmatism, stolidity, equability, acceptance, resignedness, slowness in asserting initiative, and tolerance of personal adversity.
>
> The indirect effects of genetics on personality in Turner's syndrome are mediated through body morphology and function, by way of the body image and the person's interaction with her social environment. Shortness of stature is the number one problem shared by all Turner girls. It makes itself felt at an early age. The number two problem, also shared by all patients, is pubertal failure, and the issue of deciding the timing of induced puberty. This decision requires a weighing of the demand for extra height against the demand for more maturity of appearance. Cosmetic deformities constitute, at their worst, a very severe problem: but they are at their worst in relatively few cases. The effects of cosmetic handicap, however bad, differ from shortness and pubertal infantilism in at least one dimension: shortness and physical infantilism elicit "infantilizing" social

responses from other people of all ages. Thus, the major indirect effect of genetics on personality in Turner's syndrome is to lower the threshold for retardation of social development. The closer she approaches teenage, the more the Turner girl encounters situations and pressures that tend to arrest or impede her social maturation.

In psychosexual differentiation, absence or impairment of the X chromosome, in all (or mosaically in some only) of the body's cells in Turner's syndrome, does not interfere with feminine gender identity. Nor is feminine gender identity impeded or impaired by the absence of the gonads and their hormones in fetal life. In adolescence, for the full maturation of psychosexual femininity, it is necessary first to give hormonal substitution therapy with estrogen, in order to bring about sexual maturation of the body. It is also necessary to give at least minimal psychologic guidance with respect to the postponement of adolescent estrogenization in favor of a possible extra increment in adult height.

The genetic and hormonal (fetal and pubertal) impairments of Turner's syndrome do not have any adverse direct influence on the patient's interest or ability to marry, nor to be motherly with children. On the contrary,

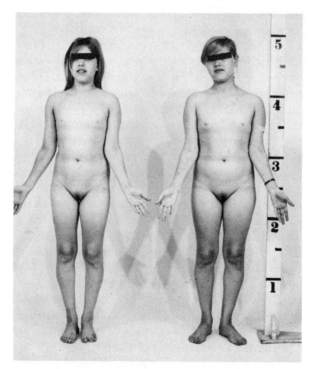

8.   **XO, Or Turner's syndrome.** Monozygotic twins with 45,XO karyotype. (From Riekhof, P. L., et al. Monozygotic twins with the Turner syndrome. *Am. J. Obstet. Gynecol.* *112:* 59–61, 1972.)

Turner patients have a heterosexual inclination (to the exclusion of homosexual inclination) and an interest in maternalism equal to that of matched normal controls.

Psychopathology is not a significant feature of Turner's syndrome, though it does occur, just as it does in a randomly selected population. Personality pathology in the parents, and/or their inability to cope with the implications of the diagnosis and prognosis of the syndrome, constitute in the long run more of a psychologic hazard to the girl herself than do the actual deficits and impairments of her body.

Campbell and coauthors (1972) observe: "The incidence of psychiatric abnormalities in Klinefelter's syndrome is far greater than in the general population." (Examples of this are seizure disorders, speech disorders, electroencephalographic abnormalities, schizophrenia, paranoid states, and deviant sexual behavior.) They tell the sad story of the youngest known Klinefelter patient, who exhibited psychopathology for which he was placed in a psychiatric hospital at the age of 3 years. His behavior involved temper tantrums; hyperactivity; withdrawal; drooling; brief attention span; grimacing; few, if any, vocalizations — mostly grunts; with aggressiveness directed against his disturbed (enough to be medicated) but intelligent parents or against his own body, e.g., hair pulling and swallowing, skin picking to the extent of causing open lesions, and severe head banging. It was in the psychiatric hospital that the boy's XXY chromosomal complement was identified, and when his parents were informed of the biological origin of their child's condition, they seemed relieved of some guilt.

Studies should surely be undertaken of the behavior of XXY tortoiseshell cats (mostly sterile) and of X0 mice (always fertile), as recommended by Morton (1972) (XX = female and XY = male in these mammals, as usual). One tortoiseshell XXY cat, Lucifer, was reported to have not the slightest sexual inclination. He was treated as a kitten by other tomcats; his presence did not disturb these toms even when females in season were present (Bamber and Herdman, 1932). However, not all such cats are sterile (Jude and Searle, 1957; also see Thuline and Norby, 1961). One tortoiseshell tom fathered at least 65 offspring, so behavioral variability exists. And note a report of testicular hypoplasia accompanied by the XXY sex chromosome complement in two rams — ovine counterpart of Klinefelter's syndrome (Bruere, Marshall, and Ward, 1969). These two unrelated rams both exhibited testicular hypoplasia, were small, and "showed strong male libido to ewes in oestrus and performed the physical actions of ejaculation." Meiotic aneuploidy has been reported in a strain of mice harboring a latent virus infection of the ectromelia (DNA virus causing mousepox) type in its gonads (Schroder, Halkka, and Brumer-Korvenkontia, 1970). Russell (1961), in an excellent review article on the genetics of mammalian sex chromosomes, reports on

X0 and XXY mice — on their existence, not on their behavior. Morton (1972) points out that recent technical advances, e.g., in autoradiography, allow the detection of relatively tiny deletions, duplications, and inversions; now perhaps, new attempts should be made to localize genes that alter behavior without altering morphology and anatomy.

And what of the notorious XYY human male? In 1967, Price and Whatmore filed the following report about a maximum security hospital in Scotland:

> All the patients admitted to this hospital have severely disordered personalities and they have been classified according to whether the cause is known or not. For example, some have brain damage which followed infections, others are epileptics, and others suffer from a psychosis. The largest group of patients have no known cause for their personality disorders. All the men with an XYY complement were classified in this category and eighteen other men have been randomly selected from this group for comparison with the nine XYY males. Seventeen of the eighteen control males were known to have an XY sex chromosome complement, the remaining being one of twenty-seven who had not been willing to be investigated when the chromosome survey was carried out.
>
> There are three ways in which the XYY males differ importantly from the controls. First, although the patients in the two groups have penal records of comparable length, those of the XYY males include considerably fewer crimes of violence against persons. Thus, the nine XYY males had been convicted on a total of ninety-two occasions, but only eight of these convictions (8.7 percent) had been for crimes against persons, while eighty-one (88.0 percent) had been for crimes against property. In contrast, the eighteen control males had been convicted on 210 occasions, and forty-six of these (21.9 percent) had been for crimes against persons while 132 (62.9 percent) had been for crimes against property. Second, the disturbed behavior of the XYY patients showed itself at an earlier age. This is reflected in a mean age at first conviction of 13.1 yr., compared with a mean age of 18 yr. for the control patients, a difference which is significant at the 5 percent level. Third, in the families of these patients the incidence of crime among the siblings of the XYY patients is significantly less than among those of the control patients. Thus, only one conviction is recorded among thirty-one sibs of the XYY patients while no less than 139 convictions are recorded for twelve of sixty-three sibs of the control patients.
>
> The distribution of intelligence quotient among the XYY males probably reflected the distribution among the patients of the hospital as a whole. Seven were considered to be mentally sub-normal, but it is worth noting that the pattern of behavior among the two whose intelligence quotients were not unusually low conformed with those of the other seven.
>
> The picture of the XYY males that emerges from examination of those detained at the State Hospital is of highly irresponsible and immature individuals whose waywardness causes concern at a very early age. It is generally evident that the family background is not responsible for their

behavior. They soon come into conflict with the law, their criminal activities being aimed mainly against property, although they are capable of violence against persons if frustrated or antagonized. Their failure to respond to corrective measures leads to a sentence of prolonged detention in safe custody at an earlier age than is usual for offences of this kind. All nine men with an XYY chromosome complement conform fairly closely to this broad description and it seems reasonable to suggest that their antisocial behavior is due to the extra Y chromosome.

But is this a conclusion reached by too many too fast, as Levitan and Montagu (1971) carefully caution? Is the only relatively consistent characteristic shared by XYY males their greater than average height? Rimoin and Schimke (1971) point out that surveys of newborns suggest that XYY may occur in as many as 1 in 300 live births and that this is much more frequent than the incidence of troublesome or even just subnormally intelligent tall men. Note that this high XYY frequency is occurring in the absence of transmission of the chromosomal abnormality from father to son (Melynk et al, 1969). Finally, we are obliged to document the very recent (1972) report by Gardner and Neu entitled "Evidence Linking an Extra Y Chromosome with Sociopathic Behavior," in which lawyers and their cohorts are advised to consider the evidence for this "linkage" in assessing legal responsibilities. (They ought simultaneously to consider the troublesome but very rare XXYY males too.)

Clearly, as Figure 7 indicates, there are other sex chromosome abnormalities in addition to those already considered that have behavioral effects. XXXY and XXXXY individuals show Klinefelter's syndrome — IQ decline is positively associated with the number of additional X chromosomes. The same type of IQ decline occurs for XXX, XXXX, and XXXXX females. Thus the triple-X female, who occurs with a frequency of about 1 in 1000, tends to show subnormal mental capacity. The few females reported with XXXX and XXXXX all showed severe mental deficiency. Triple-X females are fertile, and one would expect them to have children in proportions $\frac{1}{4}$XX:$\frac{1}{4}$XY:$\frac{1}{4}$XXX:$\frac{1}{4}$XXY. However, it seems that the abnormal XXX and XXY karyotypes are rarely found in offspring of triple-X females, perhaps owing to directed meiotic segregation whereby two of the mother's X chromosomes segregate preferentially into functionless polar bodies and the third goes to the egg nucleus. A similar unexpected lack of abnormal karyotypes is found in the offspring of XYY males. Thus the effects of nondisjunction do not last long, in terms of generations, in the progenies of those fertile individuals having abnormal karyotypes.

The final category of chromosomal variants to be considered is the genetic mosaics that occur due to nondisjunction in somatic cells leading to cell lines with different chromosomal constitutions. Some are female mosaics (X0/XX), some male mosaics (XY/XXY), and some intersex mosaics (X0/XY). Table 1 gives some idea of the types known. The phenotypes are clearly variable depending on the proportions of the karyotypically differing tissues in individuals, and this depends on the time in development at which the abnormal cell division or divisions occurred, the localization of the differing tissues in the body, subsequent cell migrations, and finally, pure chance. For mosaics with cells of different sex (*gynandromorphs*), intersexuality may occur, depending upon the developmental factors noted above. In some cases this most unsatisfactory situation can be partly rectified by removal of a gonad or by hormone treatment supportive of the expression of but one sex.

The point has been made that autosomal chromosome number changes appear to cause more severe effects on behavior than do X or Y chromosomal aberrations, as shown by the example of Down's syndrome. Furthermore, there is an apparent lack of individuals trisomic for the larger autosomes which are no doubt represented by aborted fetuses. In a proportion of normal female cells (XX) but not male cells (XY) there is a chromatin-positive DNA body situated at the nuclear membrane. This is referred to as a sex-chromatin body or a Barr body, named after its senior discoverer (Barr, 1959). In females, Barr bodies can be seen in many tissues, including the epidermis, oral mucosa, and the amniotic fluid surrounding female fetuses. It has been postulated by Lyon (1962) and others that the Barr body represents an inactivated X chromosome. Individuals with more than one X have sex-chromatin bodies and are called sex-chromatin positive, while individuals with only one X are sex-chromatin negative. In other words, irrespective of the number of X chromosomes only one is fully active, the remainder being largely inactive, although as shown in Figure 7, individuals with

TABLE 1. Human Sex Chromosome Mosaics

Female	Male	Gynandromorph (mixed sexes)
X0/XX	XY/XXY	X0/XY
X0/XXX	XY/XXXY	X0/XYY
XX/XXX	XXXY/XXXXY	X0/XXY
XXX/XXXX	XY/XXY/?XXYY	XX/XY
X0/XX/XXX	XXXY/XXXXY/XXXXXY	XX/XXY
XX/XXX/XXXX		XX/XXYY
		X0/XX/XY
		X0/XY/XXY
		XX/XXY/XXYYY

After Stern (1973).

three or more X chromosomes do tend to have a degree of mental deficiency. Generally, the rule is, therefore:

number of sex-chromatin (Barr) bodies =
$$\text{number of X chromosomes} - 1$$

The other rule about sex determination in man, although not yet formally stated in this text, is that irrespective of the number of X chromosomes, the presence of a Y leads to a male phenotype (even if abnormal, as in the case of Klinefelter's syndrome).

Because few genes are known on the Y chromosome, it is not surprising that XYY individuals occur without gross morphological abnormalities. Individuals with Turner's syndrome are female without a Barr body, while those with Klinefelter's syndrome are male *with* a Barr body. In the case of XXXY Klinefelter's syndrome, *two* Barr bodies are expected and found. Because of the ease of staining a few cells in material scraped from the oral mucosa, in which Barr bodies can be readily studied, Barr bodies provide important population information on the frequency of abnormal males and females — at least those abnormal as regards sex chromosomes.

Many Genes and Behavior

This chapter marks our entry into genetic and behavioral matters of greater complexity than those so far discussed, and involves the beginning of our consideration of traits under more complex control than that of genes or chromosomes that can be relatively easily followed from segregation data.

5.1 Biometrical genetics

In order to analyze the variations of such traits, the methods and techniques of biometrical genetics must be used. The aim is to separate the total variance of a trait into genotypic and environmental components. Some traits fall into an intermediate category, being partly controlled by genes whose segregation can be followed and partly the result of variation that makes tracing impossible. Essentially we are moving toward a consideration of traits for which the mode of inheritance is multifactorial or polygenic. In some cases, as we shall see, it is possible, using special breeding techniques, to localize genes controlling a quantitative trait to specific chromosomes and even to regions of chromosomes. One of the main methods comes from the *directional selection* experiment, in which individuals are selected at high or low extremes of a distribution in the hope of forming separate high or low lines in subsequent generations (Fig. 1). Provided a trait has some genetic basis, there should be a

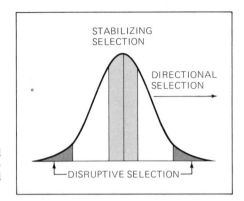

1. **Directional, disruptive, and stabilizing selection.** Sections of a normal population distribution favored under these three selection regimens.

response, since the selection of extreme phenotypes implies the selection of a proportion of extreme genotypes. For theoretical considerations, inappropriate here, see Falconer (1960) and Lee and Parsons (1968).

Little work on localizing effects to chromosomes for quantitative traits has been done except in certain species of *Drosophila*. Some species of this genus have the advantages (1) that their chromosomes are well marked with precisely located genes and can therefore be used in genetic analyses and (2) that the rapid generation time of 2 to 3 weeks permits rather complex breeding programs to be carried out in reasonably feasible amounts of time. When we realize that mice produce only four or five generations per year, it becomes immediately clear why few detailed genetic, as opposed to biometrical, studies of behavioral traits have been carried out in mammals, including man.

As an indication of the diversity of behavioral traits that are controlled polygenically, the following can be cited:

- Drosophila species. Locomotor activity, chemotaxis, duration of copulation, geotaxis (gravity-oriented locomotion), mating speed, optomotor response, phototaxis, preening, and level of sexual isolation within and between species. Evidence for actual localization of genetic activity to chromosomes exists for duration of copulation, geotaxis, and level of sexual isolation.

- Rodents (mainly mice, rats, and guinea pigs). Susceptibility to audiogenic seizures, running speed, activity, sexual drive, early or late onset of mating, emotional elimination (defecation and urination), fighting, alcohol preference, and various measures of learning such as maze brightness in running toward a food reward and the conditioned avoidance response wherein a mouse learns to avoid a shock by responding to a signal (light or buzzer). In only a few cases has any approach at the level of genetic loci been made; but some biochemical and physiological variables associated with behavior have been studied.

• Man. Specific major genes are known for some of the sensory perceptual traits, e.g., taste deficiency and color blindness. In the normal range of traits such as intelligence, temperament, emotional behavior, specific abilities, and neuroticism, polygenic control has been established. Outside the normal range, major genes are known, e.g., phenylketonuria in which IQ is very low. No localization of genetic activity to chromosomes has been carried out; man is a species where breeding experiments cannot be conducted.

This chapter discusses some examples of traits at least partly under polygenic control in Drosophila, lovebirds, rodents, and dogs, but further consideration of man is deferred to later chapters.

5.2   Geotaxis in Drosophila

Gravity-oriented locomotion (geotaxis) in *Drosophila melanogaster* represents the most complete genetic analysis of a behavior trait so far (Hirsch and Erlenmeyer-Kimling, 1962; Hirsch, 1963, 1967). A vertical 10- or 15-unit plastic maze is used (Fig. 2). Flies are introduced on the left-hand

2.   **A 15-unit geotactic maze** in a vertical position. Flies are introduced in the vial at the left and are collected from various vials at the right.

side and collected from one of the vials on the right, after being attracted through the maze by the odor of food and by lighting in the form of a fluorescent tube on the right-hand side. Conditions of maximum objectivity are provided, since there is no human handling of the large number of flies involved, once they are introduced into the maze. Rapid responses to selection for both positive and negative geotaxis are found (Fig. 3), although the actual total response to selection is greater for negative geotaxis. Using methods applied by Mather (1942) and Mather and Harrison (1949) in bristle number selection experiments, Hirsch and Erlenmeyer-Kimling (1962) assayed the role of the three major chromosomes possessed by *D. melanogaster* in the response to selection over a number of generations.

A brief description of the method is appropriate. It is possible to

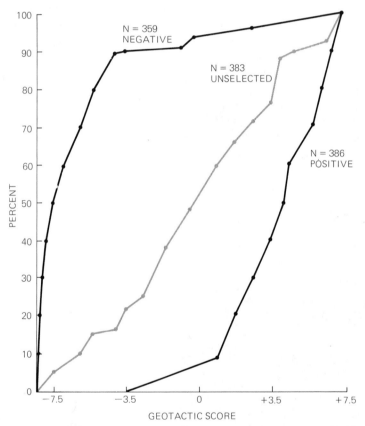

3. Geotactic scores of flies in a 15-unit maze. Cumulative percentages of flies achieving assorted scores for an unselected foundation population and for two selected lines. (After Hirsch, 1963.)

combine in all ways chromosomes extracted from a selection line with those of the control line, by using stocks carrying dominant genes with inversions inhibiting crossing over in the relevant chromosomes, one chromosome at a time. In this way the individual effects of chromosomes and their interactions can be studied with respect to a quantitative trait. Therefore, a multiple tester stock: $A/+$, B/C, D/E is utilized. A is a dominant gene on the X chromosome, B and C are dominants on chromosome II, and D and E on chromosome III, all genes being associated with inversions. These segregate essentially as whole units, since the presence of inversions in the heterozygotes prevents or diminishes crossing over. When females of these stocks are crossed to the selection line S to be assayed, daughters of the type A/S, B/S, D/S are obtained. These are then backcrossed to the selection line, and eight classes of progeny appear:

- A B D
- A B
- A D
-    B D
-    A
-    B
-       D
- Selected line itself (without A, B, or D)

These eight classes receive a representative S chromosome from their father and a variable number of S or tester (T) chromosomes (A, B, D). Each of the three major chromosomes is therefore heterozygous or homozygous for an S chromosome in females. From the eight classes above, the individual effects of the chromosomes and their interactions can be studied. One limitation is that the technique is fully efficient only in detecting recessive genes in the S chromosomes, because the comparison for each chromosome is between the heterozygote T/S and the homozygote S/S. This means that dominants are not detected and incomplete dominants are revealed to the extent of their level of dominance.

The above procedures have been applied to the lines selected for geotaxis (Fig. 3). The mean effects of chromosomes X, II, and III in the unselected population are given in Table 1 and are compared with those selected for positive or negative geotaxis. In the unselected population, the X and II chromosomes have genetic activity for positive geotaxis, and chromosome III is negative, by comparison with the standard tester chromosomes. Selection for positive geotaxis produces little effect for chromosome II, but a positive effect is increased for chromosome X, and

TABLE 1. Mean Chromosomal Effects (with Standard Errors) on Geotactic Scores after Selection Based on Maze Depicted in Figure 3

Population	Chromosome		
	X	II	III
Selected for positive geotaxis	1.39 ± 0.13	1.81 ± 0.14	0.12 ± 0.12
Unselected	1.03 ± 0.21	1.74 ± 0.12	−0.29 ± 0.17
Selected for negative geotaxis	0.47 ± 0.17	0.33 ± 0.20	−1.08 ± 0.16

Means are based on 10 replicates in each case; each unit represents one notch on the maze. After Hirsch (1967).

for chromosome III a negative effect is changed to slightly positive. The effect of these three chromosomes is greater for negative geotaxis considering the three chromosomes together, as expected, since the total response to selection obtained is greater for negative geotaxis. Selection for negative geotaxis reduces the positive effects of chromosomes X and II and increases the negative effect of chromosome III, compared with the unselected stock. Therefore, the analysis so far confirms that there are genes distributed on all the three major *D. melanogaster* chromosomes which affect gravity-oriented locomotion. While this analysis was not carried further, assuming extrapolation from the work of Thoday (1961) and his colleagues on bristle numbers, it is in theory possible to locate precise regions of genetic activity on chromosomes, of which there are likely to be several.

5.3   Sexual isolation between species of Drosophila

Species of sexually reproducing organisms are genetically closed systems. They are closed systems because they do not exchange genes or do so rarely enough so that the species differences are not swamped. Races are, on the contrary, genetically open systems. They exchange genes by peripheral gene flow unless they are isolated by extrinsic causes such as spatial separation. The biological meaning of the closure of a genetic system is simple but important: it is evolutionary independence. Consider these four species — man, chimpanzee, gorilla, orangutan. No mutation and no gene combination arising in any one of them, no matter how favorable, can benefit any of the others. It cannot do so for the simple reason that no gene can be transferred from the gene pool of one species to that of another. On the contrary, races composing a species are not independent in their evolution; a favorable genetic change arising in one race is, at least potentially, capable of becoming a genetic characteristic of the species as a whole.

There are therefore problems in defining that pivotal concept race. According to Parsons (1972a), about all we can say is that:

> A race is a population in which the gene frequencies at some loci differ from one another. It is a *quantitative* and not a *qualitative* definition, as there are no biological isolating mechanisms between different populations. Thus gene pools of different races have differing gene frequencies. It must be stressed that because the definition is quantitative rather than qualitative, the amount of difference needed to accept that we have two different races is completely arbitrary.

Species are genetically closed systems because gene exchange between them is impeded or prevented by reproductive isolating mechanisms. The term *isolating mechanism* was proposed by Dobzhansky in 1937 as a common name for all barriers to gene exchange between sexually reproducing populations. According to Mayr (1963), isolating mechanisms are "perhaps the most important set of attributes a species has." It is a remarkable fact that isolating mechanisms are physiologically and ecologically a most heterogeneous collection of phenomena. It is another remarkable fact that the isolating mechanisms that maintain the genetic separateness of species are quite different not only in different groups of organisms but even between different pairs of species in the same genus. They fall naturally into two primary divisions: (1) geographical or spatial isolation and (2) reproductive isolation. In the case of geographical isolation, the populations involved are *allopatric*, which means that they are found in different territories, so gene exchange between them is minimal. Indeed, they may or may not be genetically similar. For a complete discussion of allopatric and *sympatric* (populations living in the same territory) speciation and of ecological factors in speciation, see Mayr (1947).

The classification of reproductive isolating mechanisms outlined below is a composite one. It is based on the classifications published by Mayr (1942), Muller (1942), Patterson (1942), Allee et al. (1949), Stebbins (1950), and Dobzhansky (1951).

1. Barriers to gene flow preventing the meeting of potential mates
   - Habitat or ecological isolation: populations found in the same general territory, but occupying different ecological niches
   - Seasonal or temporal isolation: sexual maturity or activity may occur at different times
2. Barriers to gene flow preventing the formation of hybrid zygotes
   - Mechanical isolation: as occurs if the genitalia of the two sexes do not correspond
   - Gametic isolation or prevention of fertilization: as occurs if eggs and spermatozoa do not meet or do not fuse normally

> • Sexual, psychological, or ethological isolation: greater mutual attraction of conspecific males and females than between males and females of different species

3. Barriers to gene flow that eliminate or handicap the hybrid zygotes that have been produced
   - Hybrid nonviability or weakness: lower viability of hybrid zygote compared with either parental species
   - Hybrid sterility: hybrids unable to reproduce because of nonproduction of functional gametes
   - Selective hybrid elimination or hybrid breakdown: hybrid products eliminated in the F_2 or later generations because they are adaptively inferior

These eight items have a common function. They all have but one net effect — either singly or collectively to prevent the exchange of genes between populations (Patterson, 1942).

There are behavioral components to a number of these mechanisms, and many of these are illustrated in Section 13.2, where the mechanisms involved in habitat selection in closely related species of *Drosophila* and the deer mouse *Peromyscus* are discussed. In this section we discuss sexual, psychological, or ethological isolation, and we do so incorporating the genetic detail Drosophila allows. This particular isolating mechanism is an efficient barrier in that when effective, sexual isolation eliminates any wastage of gametes and, more seriously, eliminates the need for food and space for developing hybrids that may in some way be inferior to nonhybrids in fertility or viability. The sterile hybrids produced within the superspecies (species complex) *Drosophila paulistorum* are one example of this. The semispecies (or subspecies or races of incipient species) of *D. paulistorum* show a pronounced sexual isolation, and matings *between* the semispecies succeed much less frequently than do those *within* a semispecies (Ehrman, 1961, 1965).

The genetic basis of the sexual isolation has been studied in the hybrids between the Centroamerican and Amazonian semispecies, among others. Crosses have been made in which the distribution of a certain pair of chromosomes was followed with the aid of mutant genes that served as genetic markers. The sexual preferences were studied in the F_1 hybrids between the semispecies and in a series of backcrosses to each of the parental semispecies. The evidence obtained shows that the sexual isolation is fostered by factors distributed on all of the three chromosomes the species possesses. The many genes controlling the sexual preferences produce additive effects, the sum of which makes the bar to crossing nearly complete between the semispecies. The plan followed in ascertaining the role of each of the two autosomes and of the

X (the first) chromosome in the genetic architecture of the sexual isolation barrier via a mating preference technique is outlined in Figure 4. The object was to transfer one marked chromosome into the nuclear and cytoplasmic background of an alien semispecies. Backcross progenies were obtained by crossing F_1 hybrid females carrying suitable marker genes in certain chromosomes to males of one or the other of the parental semispecies. The backcrosses were repeated in each of three successive generations, always selecting as female parents the carriers of semispecies-foreign chromosomes. For each combination of parental semispecies, two series of recurrent backcrosses were made, to each of these semispecies. The F_1 hybrid females between semispecies A and B obviously contain one A and one B chromosome in each pair; the F_1 hybrid males have the X of their mother, the Y of their father, and an A and a B autosome of each parent. In the backcrosses of the A/B hybrid females to A males, the B chromosomes, except the one with the gene marker, tend gradually to be replaced because only the one with the gene marker is selected for; in the backcrosses to B males, the A chromosomes tend to be eliminated. In the progeny of the third backcross, most of the flies carry chromosomes of one semispecies only, except the foreign chromosome with the genetic marker (and sometimes also a foreign Y chromosome), as itemized in Table 2.

The control crosses involved the use of sisters of the same females used in the experimental series, but not containing the foreign chromosomes with the genetic markers. The progenies were, consequently, like

TABLE 2. Fractional and Percentage Dilutions of Alien Chromosomes in Repeated Backcrosses* Between Semispecies A and Semispecies B, where Semispecies B is the Recurrent Parent

	Marked chromosomes		Unmarked chromosomes		Percent A in entire genome	Percent B in entire genome
	A	B	A	B		
F_1	$\frac{1}{2}$	$\frac{1}{2}$	$\frac{1}{2}$	$\frac{1}{2}$	50.0	50.0
BC_1	$\frac{1}{2}$	$\frac{1}{2}$	$\frac{1}{4}$	$\frac{3}{4}$	33.3	66.7
BC_2	$\frac{1}{2}$	$\frac{1}{2}$	$\frac{1}{8}$	$\frac{7}{8}$	25.0	75.0
BC_3	$\frac{1}{2}$	$\frac{1}{2}$	$\frac{1}{16}$	$\frac{15}{16}$	20.8	79.2
BC_n	$\frac{1}{2}$	$\frac{1}{2}$	$\frac{1}{2}^{n+1}$	$1 - (\frac{1}{2}^{n+1})$	$(0.5 + \frac{1}{2}^n)/3$	$1 - [(0.5 + \frac{1}{2}^n)/3]$
			Controls			
BC_2	0	1	$\frac{1}{8}$	$\frac{7}{8}$	8.3	91.7
BC_3	0	1	$\frac{1}{16}$	$\frac{15}{16}$	4.2	95.8
BC_n	0	1	$\frac{1}{2}^{n+1}$	$1 - (\frac{1}{2}^{n+1})$	$(\frac{1}{2}^n)/3$	$1 - [(\frac{1}{2}^n)/3]$

* See Figure 4 for outline of these interpopulation crosses incorporating mutant marker genes.
From Ehrman (1960a).

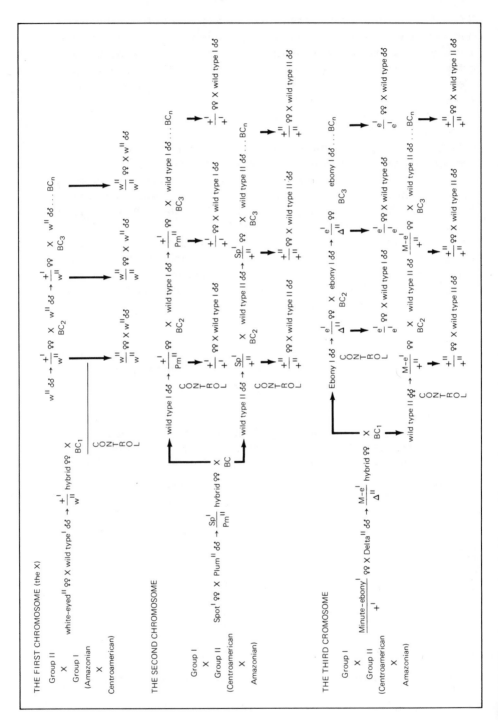

the experimental ones, except that they did not contain the foreign marked chromosome.

In most instances, the foreign chromosome contained only a single mutant gene, which served as a marker. This was, nevertheless, deemed a satisfactory experimental technique for two reasons: (1) whenever more than one marker was present, crossing over between homologous pairs was found to be suppressed in the intersemispecific crosses; (2) the semispecies involved in these experiments differed in at least one inversion in each of their five chromosome "arms" (Dobzhansky and Pavlovsky, 1962), so that they did not pair intimately and in this way allow for crossing over between homologous pairs. This does not, of course, exclude the possibility that some undetected crossing over occurs in the hybrids, but such crossing over is not frequent.

Females of the Centroamerican semispecies heterozygous for the second chromosome dominant gene *Spot* (a dark spot on the thorax) were crossed to the Amazonian semispecies males heterozygous for the second chromosome dominant marker *Plum* (an eye color). Among the F_1 hybrids, *Spot/Plum* females were selected to be the progenitors of the backcross progenies. *Spot*, *Plum*, and wild-type hybrid females and males were used for tests of their mating preferences. The females were placed together with equal numbers of wild-type males from the Honduras strain of the Centroamerican semispecies and from the Belem strain of the Amazonian semispecies. As shown in Table 3B, the 22 copulae observed were all with Centroamerican males. The males were given equal numbers of virgin Honduras females and Belem females. Of the 19 copulae observed, 17 were with Honduras and 2 with Belem females.

The *Spot/Plum* F_1 females were backcrossed to wild-type males of both parental semispecies. In two following generations, *Spot* or *Plum* females were backcrossed to wild-type males of the semispecies opposite from the marker gene. In every generation females and males carrying the mutant markers were tested for their mating preferences by placing them together with equal numbers of wild-type individuals of the opposite sex of the parental semispecies. All the results appear in Table 3B. In the backcrosses to the Centroamerican semispecies, the hybrids showed a decided preference for Centroamerican partners (except for the BC_1 females, which appeared to be neutral). In the backcrosses to Amazonian semispecies, the preference for the Amazonian partners rapidly became emphatic.

To test the effects on the sexual preferences of the foreign third chromosomes, Amazonian females heterozygous for the dominant

4. **Crosses and backcrosses** incorporating either a marked autosome or a marked sex chromosome in the species complex *D. paulistorum*. (From Ehrman, 1960a.)

TABLE 3. Direct Observation of Matings Testing Influence of Each of Three Chromosomes in Genetic Architecture of Sexual Isolation Between Two Strains of *D. paulistorum*

A. FIRST CHROMOSOME

	Number	Copulae with CA	Copulae with Am	χ^2	P
		Tests of Hybrid Females			
F_1	22	19	3	10.2	<0.01
		Backcrosses to Amazonian Parent			
BC_1	20	0	20	18.1	<0.01
BC_2	20	1	19	14.5	<0.01
BC_3	20	0	20	18.1	<0.01
		Tests of Hybrid Males			
F_1	20	8	12	0.5	0.70–0.50
		Backcrosses to Amazonian Parent			
BC_1	20	0	20	18.1	<0.01
BC_2	20	0	20	18.1	<0.01
BC_3	20	0	20	18.1	<0.01

B. SECOND CHROMOSOME

	Number	Copulae with CA	Copulae with Am	χ^2	P
		Tests of Hybrid Females			
F_1	22	22	0	20.0	<0.01
		Backcrosses to Centroamerican Parent			
BC_1	20	10	10	....	...
BC_2	20	19	1	14.5	<0.01
BC_3	20	19	1	14.5	<0.01
		Backcrosses to Amazonian Parent			
BC_1	20	4	16	6.0	0.02–0.01
BC_2	22	5	17	5.5	0.02–0.01
BC_3	22	2	20	13.1	<0.01
		Test of Hybrid Males			
F_1	19	17	2	10.3	<0.01
		Backcrosses to Centroamerican Parent			
BC_1	20	17	3	8.5	<0.01
BC_2	20	19	1	14.5	<0.01
BC_3	22	20	2	13.1	<0.01
		Backcrosses to Amazonian Parent			
BC_1	20	0	20	18.1	<0.01
BC_2	19	0	19	17.0	<0.01
BC_3	20	0	20	18.1	<0.01

TABLE 3. *(Continued)*

C. THIRD CHROMOSOME

	Number	Copulae with		χ^2	P
		CA	Am		
		Tests of Hybrid Females			
F_1	19	17	2	10.3	<0.01
		Backcrosses to Centroamerican Parent			
BC_1	20	18	2	11.3	<0.01
BC_2	20	19	1	14.5	<0.01
BC_3	20	19	1	14.5	<0.01
		Backcrosses to Amazonian Parent			
BC_1	20	2	18	11.3	<0.01
BC_2	20	1	19	14.5	<0.01
BC_3	20	1	19	14.5	<0.01
		Tests of Hybrid Males			
F_1	20	8	12	0.5	0.70–0.50
		Backcrosses to Centroamerican Parent			
BC_1	20	14	6	2.5	0.20–0.10
BC_2	20	19	1	14.5	<0.01
BC_3	20	19	1	14.5	<0.01
		Backcrosses to Amazonian Parent			
BC_1	20	2	18	11.3	<0.01
BC_2	20	2	18	11.3	<0.01
BC_3	20	2	18	11.3	<0.01

From Ehrman (1961).

marker *Delta* (wing venation) in the third chromosome were crossed to Centroamerican males which carried on one of their third chromosomes the dominant *Minute* (bristles) and the recessive *ebony* (body color). In the F_1 generation, the *Delta/Minute-ebony* females were used as progenitors of the backcross progenies, and their sibs were used for tests of the mating preferences of the F_1 hybrids (*ebony* is approximately 50 crossover units from *Minute* and is used here as a check on the suppression of crossing over in the hybrids). The results are reported in Table 3C. The F_1 hybrid females accepted Centroamerican males, while the F_1 hybrid males appeared to be neutral.

The backcross progenies showed that the sexual preference of the hybrids was for the semispecies of the recurrent parent. It appears that the third chromosome, like the second chromosome, does not by itself determine sexual preferences in these crosses.

To test the effects of the X chromosome, Amazonian females homozygous for the sex-linked gene for white eyes were crossed to wild-type Centroamerican males from the Honduras strain (no gene markers are available in the X chromosome of the Centroamerican semispecies). The gene *white* in *D. paulistorum* is incompletely recessive, heterozygous females being identifiable by their dilute red eye color. The tests of the sexual preferences of the F_1 hybrids are reported in Table 3*A*. The hybrid females preferred Centroamerican males despite their Amazonian cytoplasm (received from their female parent). The F_1 hybrid males (having white eyes) were not clearly preferred by females of either semispecies.

The F_1 hybrid females were backcrossed to white-eyed Amazonian males for three successive generations. The results are reported in Table 3*A*. No backcrosses to the Centroamerican parent could be made because of the lack of gene markers. Both females and males in the backcross progenies preferred and were preferred by Amazonian partners. Now, the backcross females carried one Amazonian and one Centroamerican X chromosome. The backcross males had, however, the Centroamerican X chromosome, or at least the part containing the wild-type allele of the gene *white*. They also had an Amazonian Y chromosome.

It appears that in the Amazonian × Centroamerican hybrids, sexual preference is decided by which semispecies contributes more than half the genome, no one chromosome being clearly more important than the others. A numerical presentation of this is offered in Table 2.

The sexual isolation analyzed here, in which matings between females and males of the different *D. paulistorum* populations are much less likely to succeed than matings within a population, is apparently controlled by polygenes scattered in every one of the three pairs of chromosomes. As such, the similarity with geotaxis (discussed in Chapter 8) is obvious.

Extensions of these experiments illuminate a fascinating variation of hybrid behavior. F_1 hybrid females from a cross between Andean-Brazilian and Amazonian semispecies were observed to accept no males courting them (Ehrman, 1960b). Most crosses between these two semispecies fail because of the powerful sexual isolation barrier. However, after repeated and lengthy attempts, viable male and female hybrids were obtained. It should be emphasized that these are normal males and normal females as far as external and internal anatomy are concerned. Yet the genic endowments contributed by the parents of these hybrids are so discordant that the hybrids are virtually unable to perform successfully the mating rituals normal in this species.

A study of the behavior of living flies under a microscope in special observation chambers shows that the hybrid females will not accept any males which court them regardless of how vigorous or persistent the courtship is. They have been observed to reject consistently the males of both parental semispecies, as well as their own hybrid brothers. They accomplish this by assuming the posture of rejection of the courtship which is characteristic of *D. paulistorum:* the female lowers her head and elevates the tip of her abdomen so that the vaginal orifice is inaccessible to an approaching male. Only twice have Andean-Brazilian males been seen to rush in so quickly that they succeeded in mounting the hybrid females; however, in one case it took the female 2 minutes to repel the male by shaking violently from side to side, and in the second instance, it took only 1 minute 47 seconds. Copulation normally takes an average of 17 minutes 12 seconds in this species. Furthermore, dissection of the female reproductive tracts involved showed that no sperm was transferred to the females in these two instances.

The hybrid males are of less interest in this respect because they are completely sterile. Even so, they are rarely successful in courting females, and they have been placed and observed with mature females of both parental semispecies, as well as with their own hybrid sisters. These males have been observed in a total of only seven copulae, whereas a normal *D. paulistorum* male begins courting again immediately after dismounting one female and may inseminate several females per day.

It is here suggested that the disharmonies in the sexual behavior of the hybrid females may serve as a very effificient isolating mechanism between the incipient species. These hybrid females, though potentially fertile, in the sense that their ovaries do produce normal and mature eggs, would probably never mate. This would make the appearance of backcross progenies impossible solely on behavioral grounds.

If populations have diverged genetically, developing different coadapted complexes as a result of having adapted to different environments, then gene exchange between these populations is likely to produce ill-adapted genotypes. Natural selection acts to build and reinforce the barriers to gene exchange between populations whose hybridization results in reproductive wastage. The appearance of hybrids with inferior fitness is, in this way, minimized or avoided altogether (Fisher, 1930; Dobzhansky, 1940, 1970). Alternately, Muller (1942) assumes that reproductive isolation arises as an accidental by-product of genetic divergence. As populations become adapted to different environments, they become different in progressively more and more genes. Reproductive isolation arises because the action of many genes is pleiotropic. Some gene differences selected for different reasons or resulting from random genetic drift (Wright, 1955; Dobzhansky and Spassky, 1962) may thus have isolating side effects.

That selection can indeed produce, or at least strengthen, reproductive isolation has been demonstrated experimentally by Koopman (1950) and by Knight, Robertson, and Waddington (1956). Koopman set up experimental populations in cages containing the two sibling species, *Drosophila pseudoobscura* and *D. persimilis*. Each species was homozygous for a different recessive mutant gene with easily visible external effects, so that each species and their hybrids were easily distinguishable. Adult hybrids were discarded each generation and new populations were initiated with nonhybrid progenies. In this way Koopman was selecting the progenies of intraspecific, and excluding those of interspecific, matings. In a surprisingly small number of generations (five to six) he obtained strains of *D. pseudoobscura* and *D. persimilis* which showed nearly complete sexual isolation between the species. In what were essentially three replicate population cage experiments, Koopman recorded the following reductions in percentages of progenies from heterogamic matings in his experimental cages:

Replicate populations	Generation	Percent hybrids
I	1	22.5
	5	5.1
II	1	49.5
	5	1.4
III	1	36.5
	6	5.2

The results of Knight, Robertson, and Waddington (1956) are, perhaps, even more dramatic. These authors obtained by selection a significant, though of course incomplete, sexual isolation of strains *D. melanogaster* that originally showed no such isolation. After seven generations of selection by destroying the progeny of interstrain matings, these authors were able to obtain an average reduction of from 66 to 38 percent in the proportion of such hybrid offspring produced.

Wallace (1954) obtained results similar to those of Knight and his colleagues (1956) using a technique essentially like Koopman's, this time with the *straw* and *sepia* mutants of *D. melanogaster*. Seventy-three generations of selection against hybrid progeny produced a significant nonrandomness of mating. One type of female, *sepia*, gave a 9:1 ratio of homogamic to heterogamic matings with *sepia* males and with *straw* males, respectively. The *straw* females, however, mated as often heterogamically as homogamically.

Drosophila behavior is considered again at length in Chapter 8. Suffice it here to refer additionally to an analysis of incipient sexual isolation between two *D. willistoni* morphs, one preferentially pupating on moist food sources, the other on dry surfaces. De Souza, Da Cunha, and Dos Santos (1970, 1972) discovered that this behavioral polymorphism was fostered by a pair of autosomal genes, and furthermore, that under competitive conditions, males that pupate in dry spots are more successful in inseminating females that pupate in similar sites than they are in acquiring as mates females that had previously pupated directly in their humid food. This then, is the spectrum: from natural or artificial directional selection forever strengthened until "complete" sexual isolation (many examples provided in this section); to sexual isolation constructed by disruptive (diversifying) selection (Gibson and Thoday, 1962, vs. Scharloo, 1971), of the sort noted in Figure 1 and exemplified by the *D. willistoni* morphs; to the evolution of preferences for matings within rather than between strains of *D. melanogaster* raised so that they differ in numbers of stout hairs, bristles, on their integuments (Parsons, 1965); and finally, to the appearance of rudiments of sexual isolation based upon no selection at all — as an accidental by-product of adaptation to development and life at different temperatures in different environments (Ehrman, 1964, 1969).

5.4 Isolation in lovebirds

Dilger (1962a,b) managed to obtain hybrids between two species of African parrots sometimes referred to as lovebirds, *Agapornis roseicollis* and *A. fischeri*, which breed and live well in captivity. These are the attractive birds employed in trained-bird acts to ride miniature trains, push small wagons, deliver mail, etc. because they learn new behaviors quickly. They can also learn to open cage doors and evade capture. So can hybrids between them, but these same hybrids consistently have difficulty in preparing nests. *A. roseicollis* females carry strips of nesting material (paper, bark, or leaves) tucked between feathers on the lower back or the rump. Several such strips are carried at a time, and in a given trip to the nest, pieces that slip out are retrieved. *A. fischeri* females, on the other hand, transfer pieces of bark, paper, or leaves, and more substantial materials such as twigs, one item at a time to the nest site in their bills. Hybrid females almost always attempt to tuck nesting material in the feathers, but are never successful — indeed when the hybrid first begins to build a nest it acts completely confused (Fig. 5). Among the reasons for this are that the hybrid bird seems unable to let go of the strip while tucking, those strips that are tucked fall out, the wrong spot for tucking is used, the strip is grasped awkwardly making proper tucking impossible, tucking movements gradually merge into preening ones, inappropriate objects are tucked, and sometimes the bird attempts to get its bill near its rump by running backward. In fact, the hybrids are successful only in carrying material in their bills, and it takes 3 years before the hybrids perfect even this behavior. Then they still are less efficient than *A. fischeri* (Fig. 5). This long learning period is in contrast

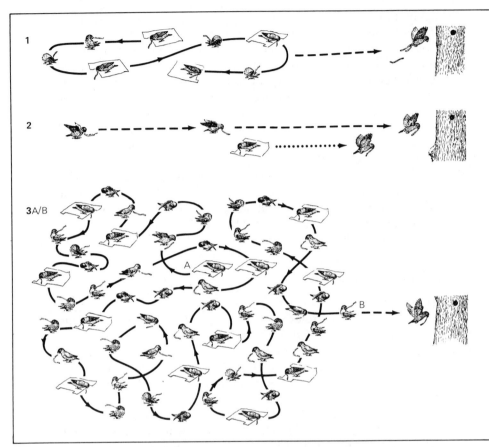

5.   **Nest building by hybrid lovebirds.** Offspring of *A. roseicollis* × *A. fischeri* inherit patterns for two different ways of carrying nest-building materials. **(1)** the hybrid inherits patterns for carrying strips several at a time, in its feathers. From the peach-faced lovebird, *A. roseicollis.* **(2)** From Fischer's lovebird it inherits patterns for carrying strips one at a time, in its bill. **(3)** When the hybrid first begins to build a nest it acts completely confused. Lines from **A** to **B** and from **A′** to **B′** indicate the number of activities necessary for the hybrid to

to the extreme rapidity these hybrids exhibit in learning the various other trick behaviors listed above.

No data are given for F₂ or backcross generations, but Dilger considers that the data so far suggest that this behavior is controlled polygenically. Clearly, a detailed analysis needs to take into account the various components of behavior involved in nest building. The hybrids are intermediate for other behaviors too. "Switch sidling" is a common precopulatory display of males involving a male's initial lateral approach to his potential mate during which he moves first toward, then away from, her while reversing his own direction often for each approach. This

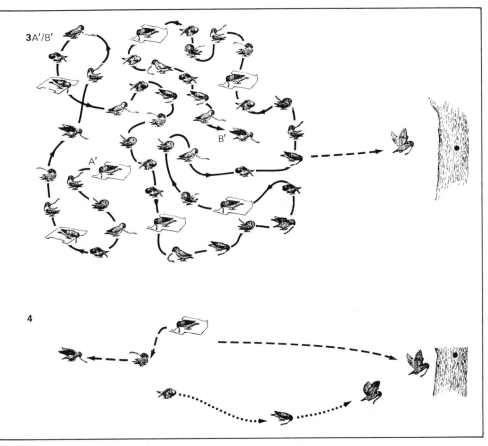

get two strips to the nest site, a feat achieved only when the strips are carried singly in the bill. (**4**) It takes 3 years for the hybrid bird to perfect bill-carrying behavior and even then it makes unsuccessful efforts to tuck its nest materials in its feathers. (From W. C. Dilger. The behavior of lovebirds. Copyright © 1962 by Scientific American, Inc. All rights reserved.)

composes about 32 percent of the precopulatory activities of *A. roseicollis* females mated with *A. roseicollis* males, and 51 percent in the case of *A. fischeri* females and *A. fischeri* males, while the figure for hybrids mated with each other is intermediate (40 percent). However, when the hybrid males are mated to parental-type females, the situation is quite different: 33 percent for *A. roseicollis* females and F_1 hybrid males, and 50 percent for *A. fischeri* females and F_1 hybrid males. It seems that this is a situation where the female's response is very important in determining what sort of behavior is elicited from attentive males. Furthermore, the rest of the precopulatory displays of hybrid males shows the same geneoral sort of pattern, but not as perfectly.

The behavior in lovebirds can therefore be interpreted at this point as being under polygenic control, although additional crosses are desirable for more detailed genetic analyses. Learning is important for nest building, since hybrid behavior slowly changes over 3 years. Then, for the precopulatory displays of males, it is not merely the male's behavior under study but also the reaction of males to differing females. The behavior of lovebirds therefore illustrates some of the unique complications that set behavior genetics aside from other subdivisions of genetics. This is surely an example worthy of more investigation.

5.5 Emotionality in rodents

When confronted with a novel and unexpected situation, rats and mice may freeze, defecate, urinate, or simply explore their new environment. These behaviors, singly or in combination, are often used as measures of emotionality. Hall (1951) selected for high and low rates of urination and defecation in rats and produced two lines, which he called "emotional" and "nonemotional," according to their urination and defecation rates. Broadhurst (1960) carried out a selective breeding program over a number of generations in rats for high (which he referred to as reactive) and low (nonreactive) defecation scores. Defecation score was defined as the number of fecal boluses deposited in an arena in exactly 2 minutes. High and low lines were established which diverged rapidly (Fig. 6A). The result is hardly surprising since various biometrical tests reveal heritabilities (see Section 2.2) between 0.5 and 1 for the trait. At the time of publication this work was recognized as being of considerable importance because it demonstrated more completely than earlier experiments the feasibility of applying biometrical methods to quantitative behavioral traits in experimental organisms. Although the analysis was not taken down to the chromosomal level, as has been done for some of the Drosophila examples, the basic assumption of biometrical genetics — a genetic architecture based upon a number of polygenes (multifactorial inheritance) — certainly is true here. The difficulty of reaching the chromosomal level has already been pointed out in Section 5.1.

In a selection experiment it may be desirable to study any correlated responses to selection, as these may provide information of both behavioral and genetic importance. Broadhurst obtained simultaneous information on ambulation score, defined as the number of marked areas in the arena entered by a rat in exactly 2 minutes (Fig. 6B). There is an increase in the score of both lines, but the increase in the nonreactive line is greater. Thus selection for defecation score has a marked effect on a trait not under direct selection. Two possibilities, those of pleiotropic effects and of linkage between polygenes affecting the two traits, are

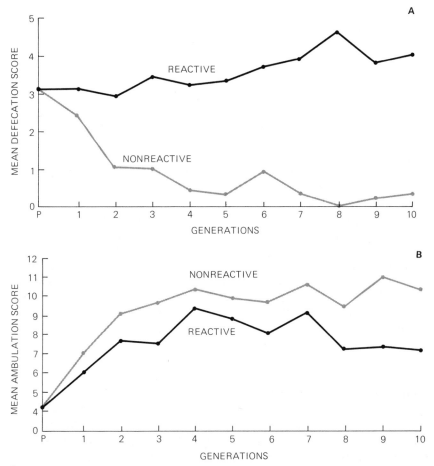

6. **Emotionality in rats. A.** Mean defecation scores in rats selectively bred over 10 generations. **B.** Mean ambulation scores. Correlated responses in the two lines selected for high and low defecation scores in **A.** (After Broadhurst, 1960.)

hypotheses worthy of consideration. In fact, one may conclude that a number of genes must affect both these behaviors: defecation and ambulation scores.

The same two behavioral traits have been studied extensively by others, particularly, in mice, by DeFries and Hegmann (1970), who applied sophisticated biometrical methods (see Chapter 9). The conclusions can be regarded as similar to Broadhurst's work, in that heritabilities were reasonably high, responses to selection were found, and a negative correlation between defecation and activity was found. A polygenic mode of inheritance explains the data best, just as is the case for the rat data above. It may in fact be concluded that many quantitative traits are under polygenic control, although some exceptions occur where behavioral traits can be associated with discrete loci. These mainly concern the pleiotropic effects of loci controlling coat color variations (see Chapter 9 for further details).

5.6 Some behavioral traits in dogs

Scott and Fuller (1965) have published the results of a number of lengthy experiments on behavioral differences among breeds of dogs. Intrigued by great differences among the breeds as well as individual differences among specimens of the same breed, they set out to investigate the importance of heredity, and in some instances they were able to come to some tentative conclusions about modes of inheritance. Their experimental design was to vary systematically the genetic constitution of the dogs while keeping all other factors as constant as possible. Five pure breeds were studied: wirehaired fox terrier, American cocker spaniel, African basenji, Shetland sheep dog, and beagle. Considerable breed differences were found for all behavioral traits studied. Perhaps the most interesting and extensive study was between the cocker spaniel and the basenji. The discussion in this section considers these two breeds and various derived crosses almost exclusively. It is based on the work of Scott and Fuller (1965).

Cocker spaniels have been selected in the past for nonaggressiveness and their ability to relate well to people. Basenjis on the other hand are highly aggressive, although not so much as wirehaired terriers. In contrast with cocker spaniels, young basenjis reared under standard conditions are very fearful of humans at 5 weeks of age, as shown by running away, yelping, snapping when cornered, and generally acting like wild wolf cubs. It is probable that in the African jungle villages, such wariness has considerable survival value. They were used in hunting by Pygmies and several other African tribes. The name *basenji* in the Lingala trade dialect of the Central Congo means "people of the bush"; the dogs were so called because they belonged to the bush people. They can be regarded as a general-purpose hunting dog not fitting into any of the conventional divisions of the European breeds. Even so, under laboratory conditions, where they are observed to be timid at an early age, basenji puppies tame rapidly with handling and human contact. Another feature is that basenjis are relatively barkless dogs compared with other breeds. They bark only when extremely excited and then soon terminate barking. At night in native villages they often produce a tremendous noise, referred to as yodeling or wailing.

This behavior was observed in Scott and Fuller's kennels. Figure 7 pictures cocker spaniel and basenji breeds and the hybrids between

7.   **Hybrid dogs.** Upper left: Basenji male ×
cocker spaniel female. Upper right: Their male and
female offspring. Lower left: Cocker spaniel male
× basenji female. Lower right: Their male and female offspring. Males are always to the
left of females. (From J. P. Scott and J. L. Fuller. *Dog Behavior: The Genetic Basis.* © 1965 by
the University of Chicago. All rights reserved.)

them. A summary of the characteristics of the two breeds appears in Table 4. The most likely modes of inheritance based on the crosses are given in Figure 8. The two breeds were crossed reciprocally to give two F_1s from which were derived two F_2s. From the F_1 based on a cocker spaniel female parent, a backcross was made to the cocker spaniels, and for the F_1 based on a basenji female parent a backcross was made to the basenjis.

The relative wildness of the basenjis is expressed in two behavioral characteristics. One is avoidance and vocalization in reaction to handling as young puppies, and the other is the tendency to struggle against restraint, which is especially marked in leash training. For avoidance and vocalization in reaction to human handling, the handling test can be regarded as a mild to strong stimulation that may cause a young puppy to be fearful. A majority of cocker spaniels shows no fearful behavior, whereas all basenjis show some fearful behavior. Since the F_1 behavior is similar to basenji behavior, one or more dominant genes are likely to be implicated, and the observed data, taking into account all crosses, fit a single dominant gene better than two dominant genes. Thus, in basenjis wildness is controlled by a dominant gene, and its contrasting gene, tameness, is recessive in cocker spaniels. According to Scott and Fuller, the struggle against restraint during leash training can be explained by a

TABLE 4. Characteristics of Basenjis and Cocker Spaniels

Characteristic	Basenji	Cocker spaniel	Most likely mode of inheritance
Wildness and tameness			
Avoidance and vocalization in reaction to handling	High	Low	One dominant gene for wildness
Struggle against restraint	High	Low	One gene with no dominance
Playful aggressiveness at 13 to 15 weeks of age	High	Low	Two genes with no dominance
Barking at 11 weeks			
Threshold of stimulation	High	Low	Two dominant genes for low threshold
Tendency to bark a small number of times	High	Low	One gene with no dominance
Sexual behavior (time of estrus)	Annual	Semiannual	Basenji type as a recessive gene
Tendency to be quiet while weighed	Low	High	Two recessive genes for high tendency

After Scott and Fuller (1965).

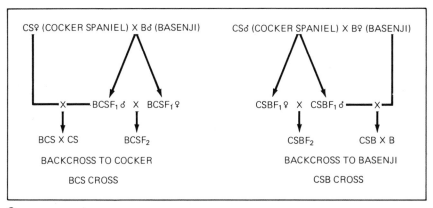

8. **Crosses between cocker spaniels and basenjis** and the derived crosses. (After Scott and Fuller, 1965.)

one-gene difference showing no dominance. However, the situation is somewhat more complex, since there are large differences between the two possible F_1 populations — basenji female × cocker spaniel male compared with cocker spaniel female × basenji male — such that the hybrids tend to behave like their mothers suggesting the possibility of maternal effects (see Section 6.6).

Playful aggressiveness was assessed by handling tests given at 13 to 15 weeks of age. The puppies characteristically rush toward the handler, leaping against him or up at his hands, and nipping playfully. When patted, the puppy usually turns around and paws or wrestles with the hand presented, chewing on it gently. Scott and Fuller concluded that the mode of inheritance of playful aggressiveness could not be explained simply and that while two genes without dominance can explain the data, the possibility of a more complex type of inheritance cannot be excluded.

Barking capacities were assessed by a dominance test where each pair of littermates was allowed to compete for a bone for 10 minutes. During this period, vocalizations were recorded including barking at 5, 11, and 15 weeks of age. Figure 9A shows that the maximum amount of barking occurs for all breeds tested at 11 weeks except for Shetland sheep dogs in which it is higher at 15 weeks. At 11 weeks cocker spaniels do most barking and basenjis least. The strain of basenji tested is not, however, completely barkless (Figure 9B). This offhandedly rather simple behavioral trait has two aspects as analyzed by Scott and Fuller: (1) the threshold of stimulation initiating barking, which is very high in the basenji and very low in the cocker spaniel; and (2) the tendency to bark only a few times (basenji) rather than to become excited and bark

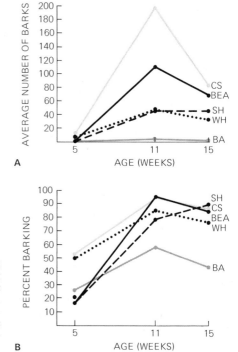

9.   Barking differences among canine breeds. Occurrence of barking during dominance tests at different ages. **A.** Average number of barks. **B.** Percentage of animals barking. BA, basenji; BEA, beagle; CS cocker spaniel; SH, Shetland sheep dog; WH, wolfhound. (After Scott and Fuller, 1965).

continuously (cocker spaniel). (The maximum number of barks recorded for a cocker spaniel in a 10-minute period was 907, or more than 90 per minute.)

For threshold of stimulation, Table 4 shows the F_1 to be similar to the cocker spaniel, indicating dominant inheritance for a low threshold. Assuming inheritance through a single dominant gene fits the data quite well, but the fit is better if two independent dominant genes are assumed as a theoretical expectation (Table 5). This does not rule out the possibility of a large number of genes. For this trait, since there is no great difference between puppies born of basenji mothers and those born of cockers, the possibility of sex-linked inheritance or learning by example from the mother is not important. In the case of the tendency to bark in excess, the F_1 animals are intermediate between the two parental strains, and the F_2 dogs are very much like the F_1 animals. The data can be explained by a single gene without dominance. That two genetic mechanisms are involved in barklessness is clear from the components of the trait. A dog will not bark to excess if it does not bark at all, so the presence of one trait is conditional upon the presence of the other. An important principle emerges — that a greater understanding of the genetic architecture of quantitative traits may come from subdividing them into components prior to genetic analysis — and we recommend it.

Basenji females come into estrus annually close to the time of the autumnal equinox, while most domestic breeds have estrous cycles at any season of the year and about 6 months apart. It was concluded that the basenji type of estrous cycle is inherited as a single-factor recessive, but the possibility of a somewhat more complex situation cannot be definitely excluded.

For the inheritance of the tendency to be quiet while being weighed at 14 to 16 weeks of age, results compatible with two genes were obtained, the cocker spaniel tendency being recessive to the basenji.

Table 4 shows that the inheritance of these behavioral traits can be explained in many cases on the basis of one or two genes, although there are indications of greater complexity. In other words, the situation is intermediate between the simple Mendelian inheritance patterns described in previous chapters and polygenic inheritance discussed earlier in this chapter. Scott and Fuller regard the result as rather surprising, arguing that a trait as complex as behavior might be expected to be affected by many genes, but our evidence so far indicates this to be not necessarily so. The two dog breeds concerned, having been isolated from each other for so long, no doubt were exposed to highly different selective pressures leading to the behaviors observed. If this led to genotypes homozygous for the traits discussed (and with little or no segregation occurring within breeds for the traits), then assuming only one or two major gene loci involved for each trait, the results obtained are reasonable. It should be noted, too, that other behavioral traits in dogs — generally of a more complex nature than those discussed — are difficult to interpret on a simple Mendelian basis because of the complex

TABLE 5. Percentage of Dogs Barking in 10-Minute Test Period Compared with Percentages Expected Under Two Theoretical Assumptions of One Dominant Gene or Two Dominant Genes

Breed or hybrid	Observed	Expected one dominant gene	Expected two dominant genes
Basenji	19.6		
Cocker Spaniel	68.2		
F_1	60.1	68.2	68.2
F_2	55.5	56.0	65.2
Backcross (F_1 × cocker)	65.1	68.2	68.2
Backcross (F_1 × basenji)	50.0	43.9	56.0

After Scott and Fuller (1965).

interactions with the environment which are frequent for behavioral traits.

In Chapters 6 and 7 and later in the book, we discuss more extensively the traits exclusively controlled by many genes. The material in this chapter therefore provides a descriptive bridge between traits clearly controlled by major genes and those controlled polygenically; the examples selected for this chapter cover this spectrum.

Genetic Analysis of Quantitative Traits: Experimental Animals

6.1 Quantitative genetics

Quantitative genetics is the study of variation of those traits where the genes responsible for the observed variation cannot normally be recognized individually. Some examples are presented in Chapter 5 and some basic principles are given in Section 2.2. The aim of quantitative genetics is to subdivide the phenotypic value we measure into component parts — the genotypic and environmental components. From this point of view, behavioral traits differ in no way from the more conventionally studied quantitative morphological traits in animals. Complete accounts appear in various texts on quantitative genetics, in particular Falconer (1960) and Mather and Jinks (1971). These authors employ different mathematical symbolism; the present discussion uses mainly the system credited to Falconer.

The level of variability for many behavioral traits within and between different environments is greater than for many morphological traits. For this reason, more attention must be paid to controlling the environment in which behavioral traits are studied than is necessary in the analysis of morphological traits, and in some cases the effect of the environment itself is of direct interest. In addition, learning and reasoning must be taken into account — one of the features that differentiates behavior genetics from other subdivisions of genetics (Chapter 1). The possibility of previous experience makes it essential to standardize experimental conditions extremely carefully; otherwise it may be difficult to obtain accurate genetic interpretations. In man, previous experience may be a factor of major importance that cannot be controlled as in experimental animals because of the impossibility of defining environments or carrying out breeding experiments, as can be done with experimental animals. Therefore, assessing the meaning of behavioral data in man is often very difficult, and this is a reason for the hotly debated issue of racial differences in intelligence (see Chapter 12). In many ways man can, therefore, be regarded as a special case. We consider experimental animals in this chapter and, armed with this knowledge, consider man in Chapter 7.

6.2 Interaction of genotype and environment

The simplest models of quantitative genetics assume that the joint effects of genotype and environment are additive. Under this assumption if one genotype has a value for a trait greater than another genotype in any environment, it has a greater value in all environments. This is an assumption of most of the simple theoretical quantitative inheritance models, and it is not necessarily valid in practice. Haldane (1946) discusses the relations that may exist between two genotypes, A and B, measured for a quantitative trait in two environments, X and Y. The four corresponding values of the trait are numbered from 1 to 4 in order of magnitude. There are 4! = 4 × 3 × 2 = 24 ways of arranging four items in a sequence. But if, arbitrarily, AX (genotype A, environment X) is designated as the largest measurement, there are then only six distinguishable logical arrangements, as illustrated in Table 1. From this we see:

- Arrangements 1a and 1b. $A > B$ in both environments. In 1a both values of $A >$ than the highest of B, i.e., A is always $> B$. In 1b, X > Y although within them A and B have the same relative ranking.

- Arrangement 2. $A > B$ in X, but $A < B$ in Y although X > Y. Haldane points out a possible example of this, where domesticated species (A) and wild (B) are in man-made (X) and natural (Y) habitats. Both species fare better in man-made habitats that offer protection than in the natural habitat, although wild types do relatively better in natural habitats than do domestic species.

- Arrangement 3. $A > B$ in X and Y, but $AX > AY$ and $BX < BY$. The environments X and Y have opposite effects on the two types of individuals. Haldane gives the example of normal (*A*) and mentally retarded (*B*) individuals in normal (X) and special (Y) schools.

• Arrangements 4a and 4b. The environments again have opposite effects on the two types of individuals, as in 3, but represent specialization such that *A* and *B* are each optimally adapted to their respective environments, X and Y. This represents the situation of evolutionary adaptation of individuals to their respective environments. The temperature preferences of races of *Peromyscus maniculatus* can be cited as a behavioral example. Races *P. maniculatus bairdii* (prairie mice) and *P. maniculatus gracilis* (deer mice) each display a preference for the artificial habitat most closely resembling its natural one (Harris, 1952). Furthermore, descendants of these mice raised in the laboratory also select similar environments, suggesting a genotypic role in the selection. Ogilvie and Stinson (1966) found the thermotactic optima of *P. maniculatus bairdii* and *P. maniculatus gracilis* to be 25.8°C and 29.1°C, respectively, showing that *P. maniculatus gracilis* has been selected for its warm, wooded environment and *P. maniculatus bairdii* for its cooler, field environment.

To finish with a laboratory example, we can consider some data of Henderson (1970) on the effect of early experience. The mean number of minutes required to reach food was assessed in six inbred strains of mice reared in both standard and enriched environments (Fig. 1). Some of Haldane's interactions emerge here if we consider the strains in pairs. Look at the data for the paired strains and try to see the types of interactions they represent.

Another type of genotype-environment interaction, little investigated

TABLE 1. Relations of Measurements of a Quantitative Trait in Two Genotypes (*A* and *B*) in Two Environments (X and Y)

Arrangement	Genotype	Environment		
		X	Y	
1a	*A*	1	2	*A*>*B* in X and Y
	B	3	4	
1b	*A*	1	3	
	B	2	4	
2	*A*	1	4	*A*>*B* in X, *B*>*A* in Y; X>Y
	B	2	3	
3	*A*	1	2	*A*>*B* in X and Y but BX<BY and AX>AY
	B	4	3	
4a	*A*	1	3	
	B	4	2	*A*>*B* in X; *B*>*A* in Y
4b	*A*	1	4	
	B	3	2	

Measurements are numbered 1 to 4 in order of magnitude. AX is always assumed to be the largest in each of the four types.
After Haldane (1946).

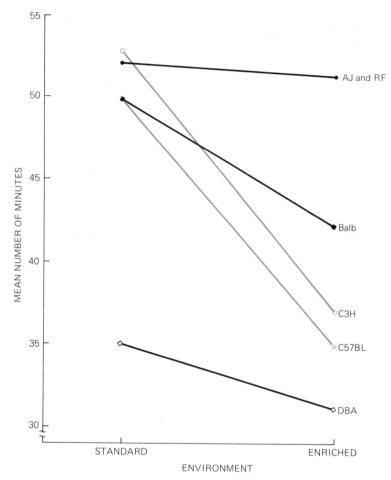

1. **Time required to reach food** for six inbred strains of mice reared in standard and enriched environments. One pair of strains, AJ and RF, achieved identical means. (From Erlenmeyer-Kimling, 1972.)

behaviorally, is the frequent occurrence of a high level of het-erokaryotype (and heterozygote) advantage in extreme environments (often temperature) compared with optimal environments. A good example comes from *Drosophila pseudoobscura.* Large fitness differentials are found between heterokaryotypes and homokaryotypes at the extreme temperature of 25°C leading to heterokaryotype advantage and a stable polymorphism, but at 16.5°C the fitness differentials disappear, and at 22°C an intermediate situation occurs (Wright and Dobzhansky, 1946; Van Valen, Levine, and Beardmore, 1962). Parsons and Kaul (1966) found substantial genotype-environment interactions between 20°C and 25°C for certain karyotypes of *D. pseudoobscura* for mating speed, such that there was more change between temperatures for homokaryotypes than for heterokaryotypes, leading to heterokaryotype advantage at the extreme temperature. We can say, because of this genotype-environment interaction, that the heterokaryotypes display more *behavioral homeostasis* than the homokaryotypes. Analogous results have been found for many fitness factors in many species (Parsons, 1971). For example, traits such as larval survival and viability show more homeostasis in heterozygotes than in homozygotes in *D. melanogaster* and *D. pseudoobscura* (for references, see Parsons, 1971, 1973).

The basic models of quantitative inheritance assume no genotype-environment interactions. Quantitative genetic theory becomes extremely complex if this assumption cannot be made. Indeed a major deficiency in quantitative genetic theory, so far as the behavior geneticist is concerned, lies in the frequent assumption of no interaction between genotype and environment.

As outlined in Section 2.2, the phenotypic value (*P*) of an individual consists of two parts, a genotypic value (*G*) determined by the individual's genetic constitution and an environmental deviation (*E*) which may be positive or negative, thus:

$$P = G + E \tag{6.1}$$

A fundamental feature of this formulation is that *G* and *E* are uncorrelated. The other parameter describing populations is variance: the phenotypic variance, assuming no correlation between genotype and environment, is:

$$V_P = V_G + V_E \tag{6.2}$$

where V_G and V_E represent the genotypic and environmental variances, respectively. If there is a positive correlation between genotype and environment, the genotypic variance is overestimated; if the correlation is negative, the variance is underestimated. In this chapter we look at some of the consequences of the simpler genetic models of quantitative inheritance, largely ignoring the possibility of such correlations. However, correlations between genotype and environment are shown to be important in Chapter 7, on human data, and the reader should bear in mind the possibility of such correlations as he reads this chapter.

6.3 Variation within and between inbred strains

Inbred strains maintained by brother-sister (sib mating) and other mating systems have been developed in species such as *D. melanogaster*, mice, and rats. Sib mating, being a form of inbreeding, leads to a pro-

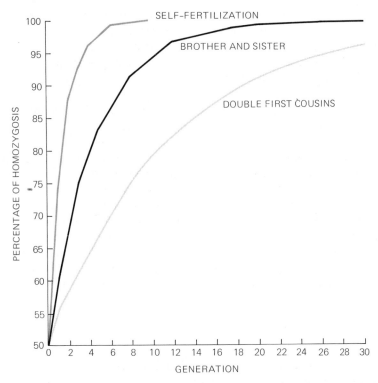

2.   **Homozygosis in successive generations** under three different systems of inbreeding. (From Fuller and Thompson, 1960.)

gressive increase in homozygosity in each generation. Figure 2 gives the percentage of homozygosis expected under three different systems of inbreeding: self-fertilization, sib mating, and double first-cousin mating. For a system such as sib mating, the expected rate of increase in homozygosity is high, and in fact the proportion of heterozygotes per generation is expected to fall by 19.1 percent of those in the previous generation, compared with 8 percent for double first-cousin mating. For self-fertilization the equivalent figure is 50 percent, i.e., the proportion of heterozygotes falls by one half every generation.

The rate of inbreeding, or the degree to which an individual is inbred, can be measured by the *inbreeding coefficient*, F (Falconer, 1960). Its detailed method of calculation need not concern us here; it expresses the probability that the two alleles at a locus are derived from a common ancestral allele. The more remote the common ancestor the smaller the value of F. F is calculated for any specific individual by tracing his lines of descent to the common ancestor of his parents. If we designate the number of steps from the inbred individual through each of his parents to the common ancestor as n_1 and n_2, respectively, his inbreeding coefficient is defined as:

$$F = (\tfrac{1}{2})^{n_1+n_2} \tag{6.3}$$

Thus, for first cousins $n_1 = n_2 = 2$ giving $F = 1/16$, for second cousins $n_1 = n_2 = 3$ giving $F = 1/64$, and for aunt-nephew or uncle-niece $n_1 = 1$, $n_2 = 2$ giving $F = 1/8$. Values of F may range from zero in a large random-breeding population to unity when complete homozygosity at all loci is obtained (some examples with pedigrees are given in Fig. 3).

Whether inbred strains ever become completely homozygous (*isogenic*) is debatable because the approach to homozygosity may be retarded if the heterozygotes are more fit than the corresponding homozygotes. Assuming that complete homozygosity is attained, all

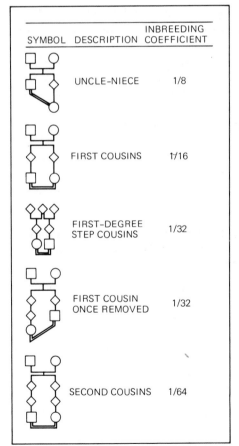

SYMBOL	DESCRIPTION	INBREEDING COEFFICIENT
	UNCLE-NIECE	1/8
	FIRST COUSINS	1/16
	FIRST-DEGREE STEP COUSINS	1/32
	FIRST COUSIN ONCE REMOVED	1/32
	SECOND COUSINS	1/64

3. **Inbreeding coefficients** under different breeding systems. (From Fuhrmann and Vogel, 1969.)

individuals within an inbred strain are genetically identical. This means that all variation within an inbred strain is environmental. Between strains, however, there are variations due to the different genetic constitutions of the strains as well as to the environment. Even if several inbred strains are set up from the same population, the genetic constitution of the strains is expected to differ, since by chance different loci are likely to be made homozygous in the different strains.

Table 2 gives some data for a behavioral trait in six inbred strains of *D. melanogaster*. The trait is the number of times out of 10 observations at 6-second intervals that each fly runs along (i.e., does not stand still) an observation tube. There are in total six sets of 10 observations per strain. To what extent is the variability in these data made up of variation within strains and variation between strains? Since the strains are inbred and are assumed homozygous, the variation *within* strains must be regarded as environmental. The variation *between* strains constitutes a genotypic component as well as the environmental component. To assess the variation within and between strains, a statistical procedure referred to as an *analysis of variance* must be carried out. Analysis of variance is a procedure that enables the total variability in a set of data to be attributed to specific causes. The detailed treatment of the data in Table 2 is given in Appendix 6.1. From this we see that the genotype variance $V_G = 3.53$ and the environmental variance $V_E = 2.05$.

It is reasonable to compute the proportion of the phenotypic variance that is genotypic thus:

$$\frac{V_G}{V_G + V_E} = \frac{V_G}{V_P} = 0.63 \tag{6.4}$$

TABLE 2.   Number of Times Out of 10 Observations at 6-Second Intervals that Individuals of Six Inbred Strains of *D. melanogaster* Ran Along Observation Tube

Oregon	Samarkand	Florida	6C/L	Edinburgh	Wellington
4	3	7	8	7	5
6	1	5	10	4	7
8	1	6	6	7	9
6	3	6	10	7	6
7	3	6	9	9	8
5	5	6	8	6	9
Total 36	16	36	51	40	44

Six flies were examined per strain. These 36 flies represent one testing session.
After Hay (1972).

This is referred to as the *heritability in the broad sense* or the *degree of genetic determination* of the trait. The degree of genetic determination must range between 0 and 1, since if it were 0, $V_G = 0$ and the trait would be determined entirely environmentally. If it were 1, the trait would be determined entirely genetically. The figure of 0.63 above indicates a relatively high degree of genetic determination.

It must be stressed that the degree of genetic determination so estimated is a characteristic of the actual inbred strains under the environmental conditions prevailing. If the experiment were done under different conditions, or with different strains, or both, different values might be obtained. If we wish to make inferences about the components of variance and the degree of genetic determination of a trait in a given population of an outbred species, we must set up inbred strains at random from the population. Then, in theory, the estimates obtained will relate to the parameters of the parent population rather than to the strains in the sample. This condition may not often be directly fulfilled, but to all intents and purposes, a set of inbred strains should permit more or less realistic estimates to be made. However, in all cases the environment must be as accurately defined as possible.

6.4 Components of genotypic variance

Let us now look at the genotype itself. If we consider two alleles at a locus A_1 and A_2, there are three possible alternative genotypes — A_1A_1, A_1A_2, and A_2A_2 — two of which are homozygotes and one a heterozygote. If the average measurement (*genotypic value*) of the heterozygote $A_1A_2 = (A_2A_2 + A_1A_1)/2$, we can say there is no dominance. We can regard the mean of the values of the homozygotes as zero, in which case there is no dominance (Fig. 4). This figure shows the three genotypes on a linear scale with the origin at the midpoint between the two homozygotes. The homozygotes are given values $-a$ and $+a$, and the heterozygote a value d on either side of the origin, depending on the size and magnitude of the heterozygote effect. It should be stressed that the averages are taken over all possible environments and with respect to other relevant loci.

We wish now to look at the expected contribution of this locus to the genotypic variance of the F_2 derived from a cross between two homozygous parental strains P_1 and P_2 carrying different alleles at this locus. The frequencies of the three genotypes in the F_2 are $\frac{1}{4}A_1A_1$, $\frac{1}{2}A_1A_2$, and $\frac{1}{4}A_2A_2$, and so the mean measurement of the F_2 is:

$$\Sigma p_i x_i = \tfrac{1}{4}a - \tfrac{1}{2}d - \tfrac{1}{4}a = -\tfrac{1}{2}d$$

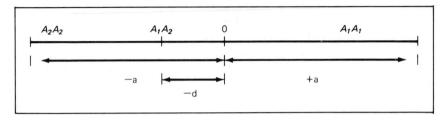

4.   **Genotypic variance.** Values for three genotypes A_1A_1, A_1A_2, and A_2A_2 arranged on a linear scale. Origin is at the midpoint between the two homozygotes. The heterozygote is on either side of the origin depending on the sign and magnitude of the heterozygote effect (d).

where p_i is the frequency of each class and x_i its phenotypic value. The contribution of the locus to the variance of the F_2 is then:

$$\Sigma p_i x_i^2 - (\Sigma p_i x_i)^2 = \tfrac{1}{4}a^2 + \tfrac{1}{2}d^2 + \tfrac{1}{4}a^2 - (-\tfrac{1}{2}d)^2 = \tfrac{1}{2}a^2 + \tfrac{1}{4}d^2$$

If, as in a cross between two inbred strains, there are many such loci contributing independently to the genotypic variance of the F_2, it can be written:

$$\tfrac{1}{2}\Sigma a^2 + \tfrac{1}{4}\Sigma d^2 = V_A + V_D \qquad (6.5)$$

where summation is over the various loci. The term $V_A = \tfrac{1}{2}\Sigma a^2$, being a function only of the difference between homozygotes, is the *additive genetic variance*. For $d = 0$, when the heterozygote is exactly between the two homozygotes, or

$$A_1A_2 = \tfrac{1}{2}(A_1A_1 + A_2A_2) = 0$$

the effect of the genes A_1 and A_2 is said to be additive. The quantity d is referred to as the *dominance deviation*, and $V_D = \tfrac{1}{4}\Sigma d^2$ is the *dominance variance*. Therefore, where there are deviations of the heterozygote from the mean of the two homozygotes, a dominance component occurs. The total phenotypic variance of the F_2 (V_{F_2}) can therefore be written:

$$V_{F_2} = V_A + V_D + V_E \qquad (6.6)$$

The same model permits the assessment of the contribution of such a set of loci to the variance components of other crosses, such as the various backcrosses to the parents and the F_3 generation. If enough data of this type are available, the quantities V_A, V_D, and V_E can be estimated, and this provides a basic description of the genotypic and environmental components of variation with respect to a quantitative trait.

Before we proceed further, mention must be made of one of the intractable problems of quantitative genetics: this concerns the *scale* on which a trait is measured. If a normal distribution cannot be assumed, it may be possible to find a suitable algebraic transformation that converts the data to an approximate normal distribution. The problem of scaling has not been satisfactorily resolved, although Mather (1949) has set up some scaling criteria based on relations between certain generations. The issue is too complex for a text of this nature, and the reader is referred to various published specialized texts (e.g., Falconer, 1960; Mather and Jinks, 1971). In theory, therefore, before any calculations on the relative importance of genotype and environment are carried out, the adequacy of the scale must be tested, and if it is found to be inadequate, a search should be made for a more suitable scale. The process of searching for a suitable scale is rather arbitrary, and in some cases a suitable scale cannot be found, so that genetic interpretations become difficult.

An example of a scaling criterion is that the variances should be independent of the mean in nonsegregating generations. For behavioral traits, which are often very sensitive to the environment, this criterion may be more difficult to satisfy than for less environment-sensitive characteristics such as morphological traits, but more experimentation is needed. Occasionally a suitable transformation is obvious. For example, if the variance is proportional to the mean in the P_1, P_2, and F_1 generations, taking logarithms provides a suitable transformation.

Broadhurst and Jinks (1961) summarize various behavioral experiments applying biometrical methods. As an example, we can take the data of Dawson (1932), who tested the inheritance of "wildness" in mice. He defined this in terms of the speed animals showed in running down a straight runway. A movable partition was used by the experimenter to prevent the mouse from moving backward. Unfortunately, this introduces a subjective element, but even so the experiment can be used to illustrate genetic points. Two strains were used, wild and tame, which although not highly inbred were very different, so it is reasonable to assess how they differ if we assume that they are likely to be homozygous for genes for wildness and tameness. The means and standard errors used in the analysis are given in Table 3. A transformation of the data was found to be necessary, and that fitting the data best was the logarithmic transformation. Variances estimated after taking logarithms are:

$$V_A = 0.026 \pm 0.012$$
$$V_D = 0.002 \pm 0.008$$
$$V_E = 0.020 \pm 0.005$$

These variance components show a common feature of many but not all traits: the additive genetic variance (V_A) is substantially greater than the dominance variance (V_D).

The ratio of the additive genetic variance to the phenotypic variance (V_A/V_P) can now be computed and is referred to as the *heritability in the narrow sense* (h^2) in contrast to the heritability in the broad sense or degree of genetic determination defined as V_G/V_P (Equation 6.4). Thus:

$$\frac{V_A}{V_P} = h^2 \tag{6.7}$$

The parameter h^2 is, therefore, a measure of the proportion of variation due to additive genes. It is a more useful concept than the degree of genetic determination because it is the gametes and the genes they carry rather than the diploid genotypes of the zygotes that are passed on from one generation to the next. From the predictive point of view, therefore, the ratio V_A/V_P is more useful. In animal and plant breeding programs, for example, the heritability in the narrow sense h^2 is a measure of the amount of genetic variability available as a basis for selective breeding. Comparisons between inbred strains (Section 6.3) do not provide an estimate of h^2 because it is not possible to obtain estimates of V_A and V_D by this method.

By taking the two homozygous inbred strains P_1 and P_2, and various crosses to give the F_1, F_2, BC_1, and BC_2 (backcrosses of F_1 to P_1 and P_2, respectively) generations, as well as all possible reciprocal crosses, a total of 14 crosses can be set up, which permits the separation of the genotypic variation into additive effects, dominance effects, epistasis, and reciprocal effects. *Epistasis* refers to interactions between genes at different loci, and a *reciprocal effect* occurs when a cross between two phenotypes differs according to whether a given phenotype is male or female. Thus a reciprocal effect is obtained if P_1 female $\times$ P_2 male differs from P_2 female $\times$ P_1 male. Such effects may be due to sex linkage or to maternal effects (Section 6.6). This more extensive analysis enables sex-linkage effects to be distinguished from maternal effects. Epistatic

TABLE 3. Running Speed (in Seconds) of Various Generations of Wild and Tame Mice

Generation	P_1 (wild)	P_2 (tame)	F_1	F_2	BC_1	BC_2
Males	6.7 ± 0.3	24.5 ± 1.0	7.6 ± 0.3	13.0 ± 0.6	6.6 ± 0.3	20.8 ± 1.6
Females	5.3 ± 0.3	25.3 ± 1.2	6.9 ± 0.3	11.8 ± 0.5	6.2 ± 0.5	18.7 ± 1.5
Both	5.9 ± 0.2	24.9 ± 0.8	7.2 ± 0.2	12.4 ± 0.4	6.4 ± 0.4	19.7 ± 1.4

After Broadhurst and Jinks (1961).

effects are normally small in quantitative data and are frequently ignored completely.

6.5 Diallel crosses

An extremely useful technique in behavior-genetics work in species where there is a ready supply of inbred strains is the diallel cross. This is the set of all possible matings between several strains or genotypes. For 4 strains there are 16 possible combinations:

Strain of female parent	Strain of male parent			
	A	B	C	D
A	AA	AB	AC	AD
B	BA	BB	BC	BD
C	CA	CB	CC	CD
D	DA	DB	DC	DD.

These are made up of six crosses, AB, AC, AD, BC, BD, and CD; their six reciprocals, BA, CA, DA, CB, DB, and DC, where the sex of the parents is transposed; and four kinds of offspring derived from the four parental strains, AA, BB, CC, and DD, which are arranged along the leading diagonal. In general, if there are n strains, the diallel table has n^2 entries made up of n parental strains, $n(n - 1)/2$ crosses, and $n(n - 1)/2$ reciprocals. Quite often not all these crosses are included; for example, the reciprocal F_1 and/or the parental strains may be omitted.

A variety of theoretical methods of analysis of diallel crosses is available depending somewhat on the information desired (see, for example, Griffing, 1956; Kempthorne, 1969; Mather and Jinks, 1971). The first detailed analysis of a behavioral trait was carried out on a 6×6 diallel cross between inbred strains of rats (Broadhurst, 1960) for defecation and ambulation scores (see Section 5.5 for definition). The analysis employed the methods of Mather and Jinks (1971) and Hayman (1958) and their colleagues. A heritability in the narrow sense (h^2) of 0.62 was obtained for the defecation score, and 0.59 for the ambulation score. Since the maximum value of h^2 is unity when the phenotypic variance is equal to the additive genetic variance, these values of h^2 show that the additive genetic variance component is relatively high for these traits.

A good example has come from Fulker (1966), who studied the mating speed of male *D. melanogaster* by taking single males from each of six inbred strains and testing them with six virgin females, one female from each of the inbred strains. The number of females fertilized in 12 hours was scored as assessed by the production of progeny. Since each male was given the same array of six females, the females can be regarded as a standard testing set so that we are dealing with the behavior of the male alone. (This contrasts, for example, with earlier data of Parsons, 1964, who studied single-pair matings within inbred strains and hybrids, a situation where interpretations can be confused by behavioral interactions between the two sexes.) Five males were tested for each of the six inbred strains and all possible hybrids between the inbred strains, thus making up a 6 × 6 diallel cross (Table 4). The entries in the table represent the number of females fertilized out of six possible. Two entries appear for each genotype because the diallel cross was replicated 2 weeks after the first cross. The diallel cross was analyzed by the rigorous methods of Hayman (1958) and Mather and Jinks (1971), which yield five parameters: one measuring additive genetic variation, two measuring dominance variation differing only in the presence of unequal gene frequencies, one indicating whether dominant or recessive alleles are more frequent, and one representing environmental variation. Conveniently, it turned out that the only parameters relevant were V_A, V_D, and V_E; furthermore no reciprocal effects were found. The variance components are: $V_A = 0.345$, $V_D = 0.328$, $V_E = 0.26$, giving $V_P = 0.933$. The degree of genetic determination = 0.71, and the heritability in the narrow sense $(h^2) = 0.36$.

The proportion of additive genetic variance is relatively low compared with many traits; the proportion of dominance variance is relatively high. Dominance in these data is in the direction of high mating speed.

TABLE 4. Replicated Diallel Cross of Mating Speed (Number of Females Fertilized Out of Six Possible) in Male *D. melanogaster*

Lines of fathers of males tested	Lines of mothers of males tested					
	6C/L	Edinburgh	Oregon	Wellington	Samarkand	Florida
6C/L	1.4*	3.6	2.2	3.2	2.6	3.0
	1.2	2.6	2.6	3.8	3.4	3.2
Edinburgh	4.0	3.0	3.7	3.4	3.2	3.2
	3.2	3.8	4.6	4.0	2.8	4.2
Oregon	2.3	3.4	1.8	3.4	2.4	2.8
	1.6	4.6	0.8	4.0	1.6	3.8
Wellington	3.2	4.4	3.8	3.0	2.4	3.6
	3.4	3.0	3.2	2.2	3.6	4.2
Samarkand	2.4	3.6	2.0	2.4	1.2	2.4
	3.2	4.0	2.2	4.6	1.2	3.8
Florida	3.3	4.0	3.2	4.6	2.0	2.8
	3.8	4.2	2.8	3.4	3.6	1.8

* Italic indicates inbred strain speed.
After Fulker (1966).

In fact, this is so marked that there is generally overdominance or heterosis in this direction. This is shown in Table 5. In all cases the means for the hybrids exceed the corresponding means for the inbred strain, showing the general occurrence of heterosis for fast mating speed. This result indicates that in natural populations, there is likely strong natural selection in the direction of fast mating. The importance of mating speed as a component of fitness is discussed further in Section 13.1.

Another method, the simplified triple test cross (TTC), has been developed for the analysis of quantitative traits. This method, related to the diallel cross, is more economical in that usually fewer crosses are needed. In its simplest form the design involves crossing n inbred strains to the two extremes of the n strains, producing a $2 \times n$ table. From the analysis of variance of this table, tests of significance for additive and dominance variation are obtainable. If scores for the n strains themselves are also available, a test of significance for epistasis is possible. The diallel cross described above lacks a rigorous test of significance for epistasis. This simplified TTC procedure has been applied to behavioral traits by Fulker (1972). It is of interest to compare the sizes (Table 6) of the simplified TTC and the full diallel cross and the half diallel cross (which omits reciprocal crosses). For $n = 3$ the simplified TTC requires as many crosses as the full diallel, and in fact the crosses are the same, but thereafter the TTC is more economical than the full diallel, the relative advantage increasing as n increases. The simplified TTC and half diallel require an equivalent number of crosses at $n = 5$, and thereafter the TTC is more economical.

TABLE 5. Mean Mating Speed Scores (Numbers of Females Fertilized Out of Six Possible) in Male *D. melanogaster*

	Hybrids of females of given strain with males of remaining five strains	Hybrids of males of given strain with females of remaining five strains	Inbred strains
6C/L	3.04	3.02	1.3
Edinburgh	3.74	3.63	3.4
Oregon	3.03	2.99	1.3
Wellington	3.68	3.48	2.6
Samarkand	2.76	3.06	1.2
Florida	3.42	3.49	2.3
Overall mean	3.93	3.93	2.4

After Fulker (1966). Based on data in Table 4.

The simplified TTC scheme is particularly useful when the aim is to survey genetic control widely; many inbred strains can be used, facilitating inferences about the base population. A limitation is that the n strains are being assayed only in relation to the genes carried by the two tester strains; but, provided the tester strains are extreme, there should be no loss of information about important loci. An obvious further advantage for behavior-genetics work comes from the possibility of replicating over a variety of environments, so that ambitious genotype-environment studies become feasible.

The disadvantage is the need to choose two strains that are extreme so far as phenotypic effects are concerned. This means that if a given genotype is extreme for only one of a number of phenotypic traits, the study is restricted to that one trait. In a diallel cross there is no such restriction, and many traits can be studied simultaneously, irrespective of which strains are extreme for them. On the other hand, it frequently turns out that certain strains are extreme for a number of traits, thus giving a complex behavioral phenotype. This issue is discussed further with reference to mice in Section 9.3.

Fulker considers empirically that for $n < 8$ the half diallel represents the best all-round design for the investigation of gene action controlling a number of behaviors. However, in comparison with the complete diallel cross, the simplified TTC and the half diallel do not permit the estimation of reciprocal differences. Even though reciprocal differences are not common in the behavior-genetics literature, it seems desirable to test for them routinely. In particular it may well be that younger animals show reciprocal effects that disappear as the animal becomes older, as has been found for temperature preference in mice (Silcock and Parsons, 1973; Section 9.3).

As an example of the simplified TTC, we can look again at Fulker's (1966) male mating speed data as presented in Table 4. The analysis of variance shows significant additive genetic and dominance components of variance for the complete diallel arrangement, and the test for epistasis is not significant. The variance components are: $V_A = 0.415$, $V_D = 0.3$, $V_E = 0.32$, giving $V_P = 1.035$. The degree of genetic determination =

TABLE 6. Comparison of sizes (Number of Crosses Required To Test n Strains) of Full Diallel, Half Diallel, and Simplified TTC Designs

		No. of strains (n)					
No. of crosses		3	4	6	8	12	20
Full diallel	n^2	9	16	36	64	144	400
Half diallel	$n(n+1)/2$	6	10	21	36	78	210
TTC	$2n + n$ parents $= 3n$	9	12	18	24	36	60

0.69, and the heritability in the narrow sense (h^2) = 0.40. These figures are in good agreement with those estimated previously from the 6 × 6 diallel cross.

Table 7 presents some data on rats from Broadhurst (1960) and Fulker, Wilcock, and Broadhurst (1972) analyzed by means of the TTC and the diallel cross. Of the four traits analyzed, defecation and ambulation are defined in Sections 5.1 and 5.5 and the diallel referred to is a complete 6 × 6 cross. The other two examples (avoidance and intertrial crossing) are taken from an 8 × 8 diallel cross. The subjects were given 30 1-minute trials of escape-avoidance training in a shuttle box divided into two equal compartments, one compartment of which was shocked after 8 seconds during which a buzzer was sounded. Crossing from one side of the box to the other terminated the buzzer or both buzzer and shock. If animals failed to cross, shock terminated automatically after 10 seconds. Intertrial intervals averaged 1 minute and were varied unsystematically between 40 and 80 seconds. The number of avoidances out of 30 was taken as the measure of avoidance. In the same experiment, a crossing from one compartment to the other was recorded as an intertrial crossing.

Considering the variance components, the agreement between the simplified TTC and the diallel cross is striking for ambulation, but

TABLE 7. Components of Variation for Behavioral Traits in Rats Analyzed by Means of TTC and Diallel Cross

Component of variation	Ambulation (Broadhurst, 1960)		Defecation (Broadhurst, 1960)		Avoidance (Fulker et al., 1972)		Intertrial crossing (Fulker et al., 1972)	
	TTC	Diallel	TCC	Diallel	TTC	Diallel	TTC	Diallel
V_A	15.2	19.7	0.083	0.131	24.12	19.44	0.28	0.22
V_D	2.5	1.5	−0.006*	−0.034*	−0.028*	−1.97*	0.03	−0.03*
V_E	5.8	4.9	0.166	0.160	5.42	3.94	0.24	0.17
Directional dominance	None	None	None	None	None	None	None	For low expression
Epistasis	None	None	None	None	None	None	None	None
Degree of genetic determination	0.75	0.81	0.32	0.38	0.82	0.82	0.56	0.52
Heritability in narrow sense	0.65	0.75	0.34	0.51	0.82	0.91	0.50	0.62

* Not significant.
After Fulker (1972).

somewhat less so for defecation, suggesting that the simplified TTC may be somewhat limited in value when the heritability is low. Both methods are in reasonable agreement for avoidance and intertrial crossing, although the diallel indicates significant directional dominance for low intertrial crossing. For all traits the additive genetic variance V_A accounts for by far the highest proportion of the genotypic variance, the dominance variance V_D being much smaller and in some cases negative, although not differing significantly from zero in these cases. No significant epistasis is detected. It is not, then, surprising that the degree of genetic determination and the heritability in the narrow sense are rather similar in most cases.

The diallel cross also provides a useful method for ascertaining the sex important in determining mating speed or duration of copulation. This can be illustrated with a 3×3 diallel table constructed from data on duration of copulation in three *D. pseudoobscura* strains — ST/ST, ST/CH, and CH/CH (Table 8; also see page 172). Here we are studying the durations of copulation for males of each strain with females of each strain. The experimental procedure was to shake unetherized flies together into vials as single pairs and record the duration of copulation (Kaul and Parsons, 1965). Inspection of the marginal frequencies shows that males of the CH/CH karyotype have the shortest duration of copulation, followed by ST/CH and ST/ST with the longest. The differences between strains are much smaller for females. Analysis of variance of the data in Table 8 shows significant effects for both sexes, but that for males is far greater than that for females. Therefore, the diallel method enables us to say that duration of copulation is primarily male-determined. MacBean and Parsons (1967) arrived at similar conclusions in *D. melanogaster*.

6.6   Maternal effects

The complete diallel cross enables the assessment of reciprocal effects which, as pointed out in Section 6.5, have not often been found in

TABLE 8.   Mean Duration of Copulation (in Minutes) in *D. pseudoobscura*

Female	Male			
	ST/ST	ST/CH	CH/CH	All strains
ST/ST	5.08	4.22	3.17	4.16
ST/CH	5.49	4.47	3.82	4.59
CH/CH	5.95	4.38	3.55	4.63
All strains	5.51	4.36	3.51	4.46

Each based on 78 observations.
After Kaul and Parsons (1965).

behavioral data. However, especially early in the life of offspring of animals such as rodents, the influence of the mother may well be apparent. Some comments on the designs needed to detect maternal effects are therefore necessary. Indeed, a major biometrical study that does not incorporate means of testing for maternal effects is rather limiting. We are more concerned here with mammals than with insects, in which there is no appreciable parental care of offspring.

There are two possible times at which a maternal effect can be exerted. The first is the prenatal period when the animal is in the mother's uterus and is physiologically dependent upon her; the second is the postnatal period, the time before weaning when the animal is in intimate contact with the mother and is still to some extent dependent on her. At this time too, learning may take place, both from the mother and from littermates.

The prenatal effect is assessed from reciprocal crosses and comparisons of the offspring. For example, using two parents, P_1 and P_2, the F_1s from crosses P_1 female $\times$ P_2 male and P_2 female $\times$ P_1 male are compared. A difference between these reciprocal crosses may be due to prenatal intrauterine environment. The complete diallel cross is suited to assess this possibility, since it involves breeding reciprocal crosses throughout. Even so, there is the possibility that sex linkage may lead to discrepancies between reciprocal crosses.

Fulker (1970) reanalyzed a study based on two parental strains and their reciprocal F_1 generations. The study compared open-field exploration in mice subjected to prenatal stresses and in unstressed mice. Stressful experiences such as mechanical vibration, swimming, and loud noise were applied to half the pregnant mice (DeFries, 1964). Maternal effects were found that acted in opposition to the additive genetic effects. This suggested to Fulker the possibility of a buffering mechanism moderating the expression of the offspring phenotype. It also shows the likely complexity of maternal effects. Fulker goes on to discuss models for assessing such maternal effects. This is surely an area where more experimentation is desirable.

Another possibility, not much discussed with relation to behavioral traits, is extranuclear or cytoplasmic inheritance. The amounts of cytoplasm contributed by sperm and ovum are very different, the contribution of the ovum being much greater, so that the cytoplasmic contribution of the male parent is negligible. If a maternal effect is established phenotypically, therefore, it is necessary to investigate the possibility that the effect results from cytoplasmic involvement rather than an intrauterine factor. To elucidate this, the technique of transplanting fertilized ova between strains is useful; the relevant techniques for mice are described by McLaren and Michie (1956, 1959). DeFries et al. (1967) recorded a small maternal effect for open-field behavior by means of ovarian transplantation, but a much larger effect was found for body weight.

Postnatal effects in rodents can be detected by transferring part or whole litters to a foster mother of an appropriate genotype, who then rears them until weaning. Three main types of postnatal environment can be envisaged, assuming two strains, A and B: (1) rearing by natural mothers, (2) rearing by foster mothers of the same strain as the natural mother, and (3) rearing of offspring of strain A by mothers of strain B and vice versa. Comparisons of traits from these various groups of offspring reveal the presence or absence of a postnatal maternal effect (see Broadhurst, 1967a, for further discussion).

The study of postnatal maternal effects by fostering could well be incorporated into routine behavior-genetics analyses. Indeed, it may be possible to examine such matters by means of a diallel crossing system. Any procedure would entail an extensive breeding program to ensure that the appropriate litters were available for fostering at birth or within a few days of each other, otherwise age differences would lead to complications in their subsequent handling. Clearly, by careful breeding and experimental designs, maternal effects can be investigated.

6.7 Components of genotypic variance in random mating populations

It is possible to partition the genotypic variation of random mating populations into additive genetic and dominance components, as is done in Section 6.4 for crosses between inbred strains. The model indicated in Figure 4 is assumed. However, instead of the genotypes A_1A_1, A_1A_2, and A_2A_2, having frequencies $\frac{1}{4}:\frac{1}{2}:\frac{1}{4}$ (as in the derivation of Equation 6.5), the frequencies $p^2:2pq:q^2$ apply, following the Hardy-Weinberg law where $p + q = 1$ (Section 2.3). If we give the homozygotes A_2A_2 and A_1A_1 genotypic values of $-a$ and $+a$, and the heterozygote A_1A_2 a value of d, as before, it turns out by simple algebra that the genotypic variance (V_G) is equal to the sum of the additive genetic variance (V_A) and dominance variance (V_D) as found for the models discussed so far. The expressions for V_A and V_D depend on the gene frequencies, as shown in Appendix 6.2.

6.8 Relations between relatives: correlation approach

Much research is concerned with the relations between relatives. For example, consider a trait in a sample of brothers and sisters. Equation 2.2 gives the formula for the variance $V(x)$ of a trait x_i. If the trait is to be measured on brothers and sisters and we let the values of the trait for the brothers be x_i and for the sisters y_i, a formula analogous to Equation 2.2 applies to the sisters:

$$V(y) = \frac{1}{n-1} \Sigma(y_i - \bar{y})^2 \tag{6.8}$$

This tells us nothing about the possibility of relations *between* brothers and sisters for the trait. To obtain this information, we need the sum of products between the two variables, and from this a quantity analogous to the variance, which is referred to as the *covariance*, can be calculated:

$$\text{Cov}_{x,y} = \frac{1}{n-1} \Sigma(x_i - \bar{x})(y_i - \bar{y}) \tag{6.9}$$

From this we obtain the *correlation coefficient* between the two variables:

$$r = \frac{\text{Cov}_{x,y}}{\sqrt{V(x)V(y)}} \tag{6.10}$$

Table 9 presents some data for heights of 11 pairs of brothers and sisters, from which $r = +0.57$ was obtained. Therefore, while brothers tend to be taller than their sisters, there is a positive correlation between their heights; in other words, in families where the boy is tall relative to other members of his sex, so is his sister relative to other females. The

TABLE 9. Heights (centimeters) of 11 Pairs of Brothers and Sisters and Computation of Correlation Coefficient (*r*)

	Family										
	1	2	3	4	5	6	7	8	9	10	11
Brother (x)	180	173	168	170	178	180	178	186	183	165	168
Sister (y)	175	162	165	160	165	157	165	163	168	160	157

Thus

$$\bar{x} = 175.36, \bar{y} = 162.45$$

$$\Sigma x_i^2 - \frac{(\Sigma x_i)^2}{n} = 478.54$$

$$\Sigma y_i^2 - \frac{(\Sigma y_i)^2}{n} = 428.73$$

$$\Sigma x_i y_i - \frac{\Sigma x_i \Sigma y_i}{n} = 259.18$$

hence

$$r = \frac{259.18}{\sqrt{478.54 \times 428.73}} = +0.57$$

For mode of computation, see Appendix 6.3.

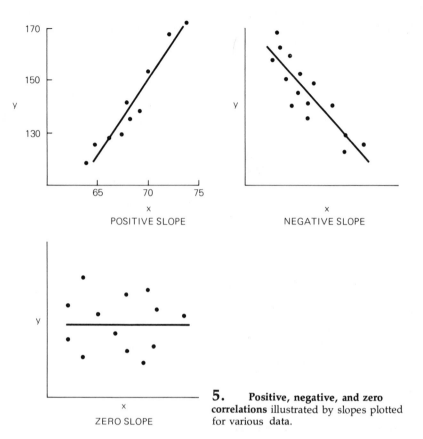

5. **Positive, negative, and zero correlations** illustrated by slopes plotted for various data.

range of r is from -1 when the correlation is completely negative to $+1$ when the correlation is completely positive. Graphically, a positive correlation gives a positive slope between x and y, and a negative correlation a negative slope (Fig. 5). In the absence of correlation between x and y ($r = 0$), the plotting of the data shows no obvious slope. Therefore, the value of $+0.57$ indicates a positive correlation between sibs. This value differs significantly from zero at the 5 percent level (see Section 2.4 for discussion of the concept of significance).

We are now ready to consider correlations between relatives in more detail — a valuable analytical tool in organisms that lack a ready supply of inbred strains. The method applies to man, but correlations between relatives are frequently calculated in experimental organisms also. Considering the correlation between one parent and offspring, summing over all loci, it can be shown (Falconer, 1960) that the covariance between parent and offspring is:

$$W_{OP} = \tfrac{1}{2}V_A \tag{6.11}$$

This is reasonable, since half the genes of any offspring are the same as those of a given parent and half are different. Thus, of the additive genetic variance (V_A) in the parent, half goes to the offspring. Note that there is no dominance component in this expression. This is logical since in transmission from parent to offspring, the gametes carrying genes, not the genotypes, are passed on from generation to generation (Section 6.4). The only situation where a dominance variance component (V_D) is expected is for full sibs. Full sibs, unlike other relatives, have two parents in common and thus have some *genotypes* in common. The covariance between full sibs is:

$$W_{SS} = \tfrac{1}{2}V_A + \tfrac{1}{4}V_D \tag{6.12}$$

Numerically, therefore, W_{SS} is expected to be a little greater than W_{OP} but not much, since V_A is usually much greater than V_D.

From the above two covariances the correlations between relatives can be obtained by dividing the covariances by the total phenotypic variance (V_P). This comes directly from the formula for the correlation coefficient (Equation 6.10). In that formula we can generally regard $V(x) = V(y) = V_P$. Hence the covariance is merely divided by the total variance. For the correlation between one parent and offspring, we therefore have:

$$r_{OP} = \frac{\tfrac{1}{2}V_A}{V_A + V_D + V_E} \tag{6.13}$$

which is equal to $\tfrac{1}{2}h^2$, where h^2 is the heritability in the narrow sense. The correlation between sibs is:

$$r_{SS} = \frac{\tfrac{1}{2}V_A + \tfrac{1}{4}V_D}{V_A + V_D + V_E} \tag{6.14}$$

which slightly overestimates $\tfrac{1}{2}h^2$.

For intelligence test scores Burt (1966) gives correlations (Table 10) as $r_{OP} = 0.49$ and $r_{SS} = 0.53$, from which one can conclude that the heritabil-

TABLE 10. Examples of Correlations Between Relatives for Intelligence Test Scores

Relation	Correlation coefficient	Coefficient of relationship
Parent-offspring	0.49	0.5
Siblings reared together	0.53	0.5
Uncle-nephew; aunt-niece	0.34	0.25
First cousins	0.28	0.125
Second cousins	0.16	0.03125

After Burt (1966).

ity is high, in fact of the order of unity. As will be seen later, this is a rather high estimate because of complexities such as genotype-environment interactions and deviations from random mating such as assortative mating.

Finally it should be noted that full-sib analysis alone is difficult to use, especially from the behavioral point of view, because of the possibility of bias due to the common environment of sibs reared together. Thus, for completeness:

$$W_{SS} = \tfrac{1}{2}V_A + \tfrac{1}{4}V_D + V_{Ec} \qquad (6.15)$$

where V_{Ec} is the variance component due to the common environment of sibs reared together. A full-sib analysis alone can do little more than set an upper limit to h^2. For behavioral traits where V_{Ec} could be high due to early experience, this mode of analysis should be viewed with caution and interpretations made accordingly.

In theory, correlations are obtainable between any sets of relatives, and as the relation between the relatives becomes more remote, the coefficient of V_A in the covariance becomes smaller:

- Half sibs, aunt-nephew, uncle-niece   1/4
- First cousin   1/8
- First cousin once removed   1/16
- Second cousin   1/32

The coefficient of V_A is referred to as the *coefficient of relationship* and reflects the share of genes due to common ancestry. It is related to the inbreeding coefficient F (Equation 6.3) and is twice the value of F for the relatives considered so far. Table 10 gives some correlation coefficients for intelligence test scores which show declining coefficients with decline in the coefficient of relationship. (In many cases the correlation coefficients exceed the coefficients of relationship; further discussions appear in Chapters 7 and 11.)

Before leaving the subject of correlations between relatives, for completeness we should consider the correlation between midparent and offspring. The midparent ($\bar{P}$) is defined as $\tfrac{1}{2}(P_1 + P_2)$ where P_1 and P_2 are the values of the two parents. Since it can be normally assumed that the variances of each parent are the same, namely V_P, then $V_{\bar{P}} = \tfrac{1}{4} \times 2V_P = \tfrac{1}{2}V_P$. Thus the correlation between midparent and offspring is:

$$r_{O\bar{P}} = \frac{\tfrac{1}{2}V_A}{\sqrt{\tfrac{1}{2}V_P V_P}} = \frac{\tfrac{1}{2}V_A}{\dfrac{V_P}{\sqrt{2}}} = \sqrt{2} \times r_{OP} \qquad (6.16)$$

This method is used less than the one parent-offspring relations because for both mother-offspring and midparent-offspring relations there is the possibility of maternal effects. Furthermore, the method assumes that the variances are equal in both sexes. For sex behavior traits, many of which are sex-limited, the midparent approach is clearly invalid. Also the presence of assortative mating, which seems common for behavioral traits, may lead to bias (see Section 7.6).

6.9   Relations between relatives: regression approach

Relations between relatives can be looked at in another way. The work of Galton and Pearson has shown that the sons of tall men tend to be tall — but not as tall as their fathers and yet not as short as the average of the population; in fact the height of sons tends to be about halfway between that of their fathers and the population average. Similarly, the sons of short men tend to be short, but not as short as their fathers, and on average have heights about halfway between those of their fathers and the population average. This trend back toward the population mean is exactly what is expected on the basis of additive genes. To look at this situation, some additional statistics must be mentioned. The correlation coefficient so far discussed does not imply any causal relation between the variables x and y even if such a relation exists. However, in some cases we can look upon y as a variable dependent on x. Frequently both approaches can be used, e.g., in looking at parent-offspring data. The correlation coefficient allows us to test only whether two variables are related. However, a procedure called *linear regression* allows us also (1) to predict the value the dependent variable y should have for any value of the independent variable x, and (2) to test how much of the variation in y actually depends on x.

Essentially we aim to find the values of a and b in the *regression line*, which has the formula:

$$y_i = a + b\,(x_i - \bar{x}) \tag{6.17}$$

This line is so constructed that the squared distance between it and all points on the graph is minimized. In Figure 6 Connolly's (1966) data on locomotor activity in *D. melanogaster* are plotted. The activity was assessed by measurements in an open-field apparatus. The procedure was to take 25 pairs of parents selected from a wild-type (Pacific) strain and mate in single pairs. From among the offspring of each of these matings two were selected and measured. Figure 6 gives the regression of offspring on midparent. The equation of the line is:

$$y = 15.56 + 0.51\,x$$

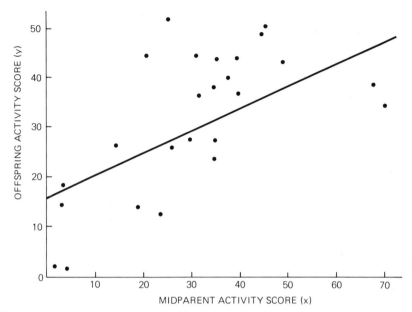

6. **Activity scores in D. melanogaster.** Regression of offspring scores on midparent scores. (From Connolly, 1966.)

showing a positive association between the midparent and offspring activities as indicated by $b = 0.51 \pm 0.10$, which is significantly greater than zero ($P<0.01$). Therefore the activities of the offspring depend in some way on the parental values.

The quantity b is the *slope* of the line, or the *coefficient of regression* of y on x, and is defined as:

$$b = \frac{\text{Cov}_{x,y}}{V(x)} \tag{6.18}$$

This may be compared with Equation 6.10, where the correlation coefficient between the two variables is given. The regression coefficient of y on x as the independent variable has $V(x)$ as the denominator; if x is regressed on y, then $V(y)$ is the denominator. For the correlation between x and y, where neither variable is considered dependent on the other, it is not unreasonable that the denominator should be $\sqrt{V(x)\,V(y)}$.

From Equations 6.10 and 6.11 the covariance between one parent and offspring is $\text{Cov}_{x,y} = \frac{1}{2}V_A$, so that from Equation 6.18, the regression of one offspring on parent is:

$$b_{OP} = \frac{\frac{1}{2}V_A}{V_P} = \frac{1}{2}h^2 = r_{OP} \tag{6.19}$$

See also Equation 6.13.

For the regression of offspring on midparent ($\bar{P}$) defined as $\frac{1}{2}(P_1 + P_2)$ it can be shown that:

$$b_{O\bar{P}} = h^2 \qquad (6.20)$$

In other words the regression coefficient is equal to the heritability. Thus it can be concluded that the heritability in the narrow sense of locomotor activity is 0.51 ± 0.10. Additional examples of the regression approach are given in Section 12.1, where intelligence in man is discussed.

6.10   Repeatability

We have reviewed a number of ways of estimating heritability. There is a considerable literature discussing experimental designs, precision of estimates, and how many individuals should be measured or how many families. On this last point, the issue is to find the best compromise between large families and many families. If the labor of measurement is the limiting factor, as in many experiments with Drosophila, then the limitation is in the total number of individuals measured. However, if breeding and rearing space is the limiting factor, as is the case with larger animals, there may be difficulties in obtaining a required total number of families or in obtaining adequate offspring per family. See Falconer (1960) for a detailed discussion of these problems.

A feature not greatly considered for behavioral traits is *repeatability*. There are in theory two ways by which the repetition of a trait can provide multiple measurements: temporal repetition and spatial repetition. Traits repeated in space are chiefly structural or anatomical and are found more often in plants than in animals, e.g., the measurement of a series of fruits. Spatial repetition in animals is chiefly found for traits that can be measured on the two sides of the body or on serially repeated parts, such as the number of bristles on the abdominal segments of Drosophila. Behavioral traits involve temporal repetition, e.g., of a behavior measured at certain specified ages in mice. There are difficulties, since for traits with a learning component, such as avoiding a shock in a conditioned avoidance apparatus, learning may occur between trials, some individuals learning more rapidly than others (Rose and Parsons, 1970). Because of this, repeated measurements are probably most reliable for traits without a learning component. Most behavioral traits in a species such as *D. melanogaster* have no detectable learning components. However, as it is easy to breed large numbers of this species, repeated measurements are not usually carried out. They could, however, tell us something about temporal changes.

To look at the measurement of repeatability, we use the term *special environmental variance* (V_{Es}) to refer to the within-individual variance resulting from temporary or localized circumstances, and the term *gen-*

eral environmental variance (V_{Eg}) to refer to the environmental variance contributing to the between-individual component arising from permanent or nonlocalized circumstances. The ratio of the between-individual component to total variance measures the correlation (r) between repeated measurements of the same individual and is known as the *repeatability* of the trait:

$$r = \frac{V_G + V_{Eg}}{V_P} \tag{6.21}$$

where V_P is the phenotypic variance. Falconer (1960) gives an example for litter size in mice, where variances between and within mice were 3.58 and 4.44, respectively, giving $r = 3.58/(3.58 + 4.44) = 0.45$. Knowledge of the repeatability sets upper limits to estimates of heritability.

The repeatability indicates the gain in accuracy obtainable from multiple measurements. If each individual is measured n times, and the mean of these n measurements is taken to be the phenotypic value of the individual $P(n)$, then the phenotypic variance can be shown to be made up of the genotypic variance, the general environmental variance, and $1/n$ of the special environmental variance:

$$V_{P(n)} = V_G + V_E + \frac{1}{n}V_{Es} \tag{6.22}$$

Increasing the number of measurements reduces the amount of variance due to special environment that appears in the phenotypic variance. This reduction of the phenotypic variance represents a gain in accuracy. The variance of the mean of n measurements as a proportion of the variance of one measurement can be expressed in terms of the repeatability, as follows (see Falconer, 1960):

$$\frac{V_{P(n)}}{V_P} = \frac{1 + r(n-1)}{n} \tag{6.23}$$

where r is the repeatability. In Figure 7, $V_{P(n)}/V_P$ is plotted against n for various values of r. This shows that multiple measurements are of greatest value for traits with low repeatability, but the gain falls off rapidly with number of measurements. Therefore, it is unlikely to be of much value to take more than three measurements. Since the repeatability of many behavioral traits is likely to be low, multiple measurements may be worthwhile in animals such as rodents where obtaining adequate numbers presents difficulties. However, the complications due to learning must be borne in mind. Furthermore, if measurements are taken at different developmental times, the trait itself may vary because

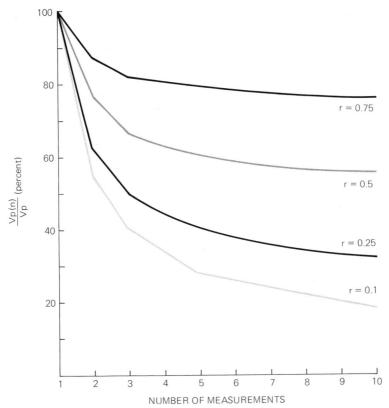

7. **Accuracy gained from multiple measurements** of each individual. Horizontal scale gives number of measurements (*n*) up to 10. Vertical scale gives variance of the mean of *n* measurements as a percentage of the variance of one measurement. Graphs for four repeatabilities (*r*) are given. (From Falconer, 1960.)

different genes are involved at different developmental times, so that genotype-development time interactions may be important. For example, when Silcock and Parsons (1973) studied thermotactic optima in mice at two ages (15 to 18 days and 55 to 58 days), developmental changes were found in that the older mice preferred lower temperatures. Relative changes between age groups differed between genotypes, indicating genotype-development time interactions. Hence the genetic identity of the trait at the two ages cannot be assumed (see Section 9.3 for further discussion).

6.11 Directional selection experiments for quantitative traits

The selection experiment arises out of the biometrical approach just discussed and consists of selecting and manipulating various chosen genotypes with respect to a trait from a population. We are mainly concerned with *directional* selection (Chapter 5, Fig. 1), where extreme individuals of a population are selected in the hope of forming separate high or low lines in subsequent generations. Examples discussed in Chapter 5 include geotaxis in *D. melanogaster* and emotionality as measured by defecation scores in rats.

If a quantitative trait has some genetic basis there should be a response to directional selection, since the selection of extreme phenotypes means that extreme genotypes are selected. Initially the *response to selection* (R) can be assessed by:

$$R = b_{O\bar{P}}S \qquad (6.24)$$

where $b_{O\bar{P}}$ is the regression of offspring on midparent as discussed in Section 6.9, and S is the *selection differential*. The selection differential is defined as the mean phenotypic value of the individuals selected as parents, expressed as a deviation from the mean phenotypic value of all individuals in the parental generation before selection was begun. It can be shown that the magnitude of S depends on both the proportion of the population included among the selected group and the standard deviation of the trait.

From Equation 6.20 we can write:

$$R = h^2 S \qquad (6.25)$$

This is not a surprising result, since the response to selection must be based on a component representing the selection differential combined with the heritability of the trait under selection. From this equation, if $h^2 = 0$, no response is possible since the trait is entirely environmentally determined. Conversely, the larger h^2 is, the greater is the expected response. In theory, the prediction of response is valid for one generation only, because a major effect of selection is to change gene frequencies and hence the genetic properties of offspring. Even so, in many experiments the predicted response is maintained for five or more generations. Much effort and statistical sophistication have been devoted to the estimation of the predicted response, which depends on an accurate estimate of the heritability (h^2). If the heritability had not been estimated before selection, the above equation provides an estimate of it as R/S. Given a response to selection over a number of generations, the response obtained gives an estimate of R/S that is referred to as the *realized heritability*.

The standard deviation of the trait is a property of the trait in the particular population. It is expressed in the units of the response, i.e., minutes to mating or to run a maze. On the other hand, responses to selection may be generalized if both the response to selection and the selection differential are expressed in terms of the phenotypic standard deviation, which we can write as σ_P. Thus R/σ_P can be regarded as a generalized measure of the response by which different traits and populations can be compared, and similarly S/σ_P can be regarded as a generalized measure of the selection differential by means of which we can compare different measures or procedures for carrying out selection. The standardized selection differential (S/σ_P) is called the *intensity of selection* (*i*). The equation for the response can be written as (compare with Equation 6.25):

$$\frac{R}{\sigma_P} = \frac{S}{\sigma_P} h^2$$

or

$$R = i\sigma_P h^2 \qquad (6.26)$$

Furthermore, noting that $h^2 = \sigma_A/\sigma_P$ where σ_A is the standard deviation of the additive genetic variance, then:

$$R = ih\sigma_A \qquad (6.27)$$

which can be used for comparing different methods of selection.

It is, therefore, clear that there are two main methods for improving the response to selection. The first is by increasing the heritability, which may be possible by reducing the environmental variation by selecting a trait that can be measured objectively and easily and generally trying to minimize random effects. Repeated measurements on an individual, as discussed in Section 6.10, can be useful under some circumstances. The second method of improving the response is by reducing the proportion selected so that those that are selected are extremes. There are some limitations to this. One important consideration is that the population size sets a lower limit to the number of individuals to be used as parents, since extremely high numbers may need to be measured to reduce the proportion selected to a very low level. Furthermore, if population sizes are very small this automatically leads to inbreeding and hence homozygosity, which reduces the variability on which selection can act. Another factor limiting the response is that many direct fitness traits such as fertility and viability are often adversely affected during selection. This can be explained by the appearance of extreme genotypes not previously exposed to the action of natural selection which often tend to be of rather low fitness.

During the directional selection process, extreme phenotypes are continuously favored and this leads to an increasing proportion of extreme genotypes likely to be homozygous. Ultimately, therefore, the rate of response to selection is expected to diminish. Frequently a plateau is attained and for a variable number of generations there is no response; occasionally, after some generations at the plateau, there is a rapid response for a few generations which is referred to as an accelerated response to selection. The rapid response is likely the result of recombination between linked genes controlling a trait, since when produced some of the recombinants are favored by selection and rapidly increase in frequency (Thoday, 1961).

The research value of the selection experiment lies in the evidence it may provide on the genetic architecture of a trait, including behavioral traits. Furthermore, it may enable learning about behavior itself, especially if the behavioral trait under analysis can be divided into components, some of which may be differentially affected by selection. Early studies showing that responses to selection can be obtained for behavioral traits in rodents are reviewed by Broadhurst (1960). Traits examined include audiogenic seizures in rats and mice; running speed in mice; and sex drive, maze-learning ability, cage activity, early or late onset of mating, and emotional elimination in rats (see Section 5.5). The occurrence of responses shows that there are heritable components for these traits, but further conclusions are difficult without additional genetic analyses on selectively bred strains. In Drosophila traits such as activity, duration of copulation, geotaxis, mating speed, and phototaxis show response to selection, and in some cases genetic analyses have been carried out (see Section 5.2).

As pointed out in Section 5.5, it is desirable to study correlated responses to selection. These should provide information on behavior itself as well as on its genetic control. Broadhurst obtained simultaneous information on ambulation score in his selection experiment for defecation score (Chapter 5, Fig. 6A). Eysenck and Broadhurst (1964) have taken this procedure further and list the results of over 50 tests, some behavioral and some physiological, many of which show correlated responses. Many of the correlated responses agree with what would be predicted from the dichotomy of emotionality occurring in the reactive and nonreactive lines as shown in Chapter 5, Figure 6B. Thus the selection experiment supplies not only information of genetic importance but also some hints about the biochemical and physiological bases of behavior.

6.12 Correlated traits

Correlated responses to selection may be positive or negative. An obvious morphological example in man is that genes increasing growth rate tend to increase both stature and weight, so that a positive correlation is expected between these two traits. Conversely, we have already noted the frequent negative correlation between defecation and ambulation scores in rodents. It is therefore important to know how a change of one trait by selection is associated with simultaneous changes in other traits. In genetic studies it is necessary to distinguish two causes of correlation between traits — genetic and environmental. The directly observable correlation between two phenotypic values for traits x and y is the *phenotypic correlation* (r_P). Similarly we can take r_A as the *genetic correlation* (the correlation between the additive genetic values of x and y) and r_E as the *environmental correlation* between the traits.

The phenotypic correlation can be written:

$$r_P = \frac{\text{Cov}_P}{\sigma_{Px}\,\sigma_{Py}} \tag{6.28}$$

where σ_{Px}, σ_{Py} are the phenotypic standard deviations of traits x and y. Since the phenotypic covariance is the sum of the additive genetic and environmental covariances, the phenotypic correlation can be written:

$$r_P = \frac{\text{Cov}_A + \text{Cov}_E}{\sigma_{Px}\,\sigma_{Py}} \tag{6.29}$$

Letting $e^2 = 1 - h^2$, we use the identities $\sigma^2_A/\sigma^2_P = h^2$ and $\sigma^2_E/\sigma^2_P = e^2$ to rewrite the denominator in Equation 6.29. First, however, these two identities give $\sigma_P = \sigma_A/h = \sigma_E/e$. Therefore,

$$r_P = h_x\, h_y\, \frac{\text{Cov}_A}{\sigma_{Ax}\,\sigma_{Ay}} + e_x\, e_y\, \frac{\text{Cov}_E}{\sigma_{Ex}\,\sigma_{Ey}}$$

$$= h_x\, h_y\, r_A + e_x\, e_y\, r_E \tag{6.30}$$

which shows how the genetic and environmental causes of correlation combine to give the phenotypic correlation. If both traits have low heritabilities, the phenotypic correlation is determined mainly by the environmental correlation; if they have high heritabilities, the genetic correlation is more important. Occasionally the two correlations have differing signs, so the sign of r_P is not necessarily a guide to that of r_A and r_E. To compute genetic correlations, the method of correlations between relatives or data from a directional selection experiment can be used. For the latter approach, we can cite correlations from five generations of selection for high and low activity in mice, with defecation score and body weight (DeFries and Hegmann, 1970). For activity and defecation,

$r_A = -0.86 \pm 0.14$, which is not surprising in view of data already cited, indicating a negative genetic correlation. For activity and weight, $r_A = 0.34 \pm 0.22$, which is positive but not significant. Further discussions appear in Section 9.2.

Appendix 6.1: Analysis of variance within and between inbred strains

The analysis of variance is a procedure whereby the total variability in a set of data can be attributed to specific causes. Considering the data in Table 2, a measure of variability or the variance (see Equation 2.2) comes from:

$$V(x) = \frac{1}{n-1} \Sigma(x_i - \bar{x})^2$$

For greater ease in computation this can be shown to be equivalent to:

$$\frac{1}{n-1} \left[\Sigma x_i^2 - \frac{(\Sigma x_i)^2}{n} \right]$$

For the data under discussion we therefore have:

$$\Sigma x_i^2 - \frac{(\Sigma x_i)^2}{n} = 4^2 + 6^2 + 8^2 + 6^2 + 7^2 + 5^2 + 3^2 + 1^2 + 1^2 + \ldots - \frac{223^2}{36}$$

$$= 177.6488$$

which is referred to as the *corrected sum of squares*. Dividing by $n-1 = 35$ gives the overall phenotypic variance $= 5.0757$.

Since there are six totals for the strains, we can examine the variability between strains by computing:

$$\frac{1}{6}(36^2 + 16^2 + 36^2 + 51^2 + 40^2 + 44^2) - \frac{223^2}{36} = 116.1389$$

The sum of the squared quantities in this corrected sum of squares must be divided by 6; otherwise it is too large, since in obtaining it the summed scores for each inbred strain are squared, making them on average six times as large as single observations (simply because the summed scores are made up of six observations).

The variation within strains comes from subtracting the total corrected sum of squares from that for strains, thus giving a value of 61.5099. The analysis of variance is merely a table setting out this subdivision of causes of variation (Table 11). There are 5 degrees of freedom for strains (see Section 2.4 for definition) since there is a total of 6 strains; similarly the total number of degrees of freedom is 35, based on 36 observations. The number of within-strains degrees of freedom (30) is obtained by subtraction.

As already indicated for the corrected total sum of squares, the variances are obtained by dividing the corrected sum of squares by the number of degrees of freedom. These are frequently referred to as *mean squares* in analyses of variance. Clearly, the variance between strains is greater than that within strains, and the ratio of the variances is $23.2277/2.0503 = 11.33$. This is the basis of the *variance ratio* or *F test*. Values of $F =$ larger variance/smaller variance have been tabulated based on n_1 degrees of freedom for the larger, and n_2 for the smaller variance. In our example, $n_1 = 5$ and $n_2 = 30$. Consultation of standard statistical tables reveals that the above value of F is significant ($P<0.001$), confirming the high variability between strains. Hence the strains differ behaviorally.

The variance within strains is expected to be purely environmental; thus $V_E = M_2 = 2.0503$ (Table 11). However, the variance between strains is expected to contain a genotypic component as well and can be shown to equal:

$$V_E + rV_G = M_1$$

where $r =$ the number of replicates within each strain, namely 6. From this we obtain:

$$V_G = \frac{1}{r}(M_1 - M_2) = \frac{1}{6}(23.2277 - 2.0503) = 3.5296$$

Appendix 6.2: Components of genotypic variance in random mating populations

The genotypes A_1A_1, A_1A_2, and A_2A_2 have frequencies $p^2:2pq:q^2$ following the Hardy-Weinberg law. Giving the homozygotes A_2A_2 and A_1A_1 genotypic values of $-a$ and $+a$, and the heterozygote A_1A_2 a value d, which may be positive or negative as in Figure 4, the mean of the population (m) is:

$$m = ap^2 - 2pqd - aq^2 = a\,(p-q) - 2dpq$$

since $p^2 - q^2 = (p-q)(p+q) = p - q$

The variance due to segregation at this locus is given by:

$$
\begin{aligned}
&p^2 \times a^2 + 2pq \times d^2 + q^2 \times a^2 - m^2 \\
&= a^2(p^2 + q^2) + 2pqd^2 - [a(p - q) - 2pqd]^2 \\
&= 2pq[a^2 + 2ad\,(p-q) + d^2(1-2pq)] \\
&= 2pq[a + d(p - q)]^2 + 4p^2q^2d^2
\end{aligned}
$$

TABLE 11. Analysis of Variance of Data in Table 2

Source of variation	Degrees of freedom	Corrected sum of squares	Mean square (variance)	Expected mean square
Between strains	5	116.1389	23.2277	$M_1 = V_E + 6V_G$
Within strains	30	61.5099	2.0503	$M_2 = V_E$
Total	35	177.6488		

As in the cross between inbred strains, if there are many such loci each acting independently, the total contribution to the genotypic variance can be written:

$$V_G = \Sigma 2pq \,[a + d\,(p - q)]^2 + \Sigma 4p^2q^2d^2 = V_A + V_D$$

where $V_A = \Sigma 2pq\,[a + d\,(p - q)]^2$, and $V_D = \Sigma(2pqd)^2$, the summation being over all the polymorphic loci in question. V_A and V_D are, as before, the additive genetic and dominance variances. If $d = 0$ at every locus, then $V_D = 0$ as expected. Thus when there is no dominance, $V_A = 2pqa^2$, where a is half the difference between homozygotes.

It is not surprising that both V_A and V_D depend on the gene frequencies. Thus V_D has a maximum value when $p = q = \frac{1}{2}$, which can easily be checked arithmetically. However, the terms of V_A are at a maximum when $p = q = \frac{1}{2}$ only if as well $d = 0$. When $p = q = \frac{1}{2}$, V_A and V_D are the same as those previously derived for the variance of the F_2 between two inbred strains (see derivation of Equation 6.5). This is to be expected, since an F_2 between two inbred strains is equivalent to a Hardy-Weinberg population with the gene frequencies of p and q equal to $\frac{1}{2}$. This is because at all segregating loci in an F_2 $\frac{1}{4}A_1A_1:\frac{1}{2}A_1A_2:\frac{1}{4}A_2A_2$ is expected.

Appendix 6.3: Computation of correlation coefficient

Equation 6.9 gives the covariance between two sets of measurements x and y as:

$$\text{Cov}_{x,y} = \frac{1}{n-1}\,\Sigma(x_i - \bar{x})(y_i - \bar{y})$$

This can be shown to be equivalent to:

$$\text{Cov}_{x,y} = \frac{1}{n-1}\left[\Sigma x_i y_i - \frac{\Sigma x_i \Sigma y_i}{n}\right]$$

a form that is easier for computation.

The correlation coefficient (Equation 6.10)

$$r = \frac{\text{Cov}_{x,y}}{\sqrt{V(x)V(y)}}$$

can be expressed as:

$$r = \frac{\Sigma(x_i - \bar{x})(y_i - \bar{y})}{\sqrt{\Sigma(x_i - \bar{x})^2\Sigma(y_i - \bar{y})^2}}$$

Note the formula for $V(x)$ is given in Appendix 6.1, and as pointed out in the text there is an analogous formula for $V(y)$ (Equation 6.8). For computation, the forms in Appendices 6.1 and 6.3 are normally used. See example in Table 9.

General readings

Falconer, D. S. 1960. *Introduction to Quantitative Genetics*. Edinburgh: Oliver & Boyd. A well-presented account of principles mainly using the notation of this chapter.

Hirsch, J. (ed.). 1967. *Behavior-Genetic Analysis*. New York: McGraw-Hill. Most of the topics in this chapter are discussed.

Mather, K., and J. L. Jinks. 1971. *Biometrical Genetics*. London: Chapman & Hall. An advanced text useful for those with statistical training.

Parsons, P. A. 1967. *The Genetic Analysis of Behaviour*. London: Methuen. Some aspects of quantitative inheritance are discussed using behavioral traits as examples.

Genetic Analysis of Quantitative Traits: Man

7.1 Threshold traits

This chapter extends the discussion of Chapter 6 to man. We begin with a consideration of *threshold traits* — traits for which organisms can be phenotypically classified into those having a given trait and those not having it. Table 1 gives the approximate population incidence of some common congenital malformations, excluding the chromosomal abnormalities discussed in Chapter 4. These add to about 1.2 percent of total births and thus constitute a very important source of illness in present-day Western industrialized societies, where relatively few die in infancy. Anencephaly and spina bifida, abnormalities of the central nervous system, are disorders with behavioral consequences. Harelip with or without cleft palate, and clubfoot, especially if not corrected surgically, must necessarily have behavioral consequences. For harelip the frequency among sibs is 35 times the population incidence, and for anencephaly and spina bifida about 8 times — quite low compared with many congenital abnormalities. From these figures and information on other relatives, it can be argued that genetic factors play a part in the cause of these conditions. Even so, in no case does the incidence among sibs rise above 5 percent, which is low compared with the expected incidence of 25 percent for a simple recessive defect among the sibs of a person having the defect.

There is no clear-cut evidence for specific environmental effects as causative agents, but there are some associations between the incidences of the defects and certain socioeconomic and demographical parameters. In Scotland, Edwards (1958) showed that the frequency of anencephaly differs from 0.9 per 1000 among professional people to 3.6 per 1000 among skilled workers. There are significant variations among localities, and the variation among seasons in which birth occurs is 30 to 50 percent. Many congenital malformations vary in incidence for different birth orders and between sexes. Factors such as prenatal exposure to radiation, chemical agents, and infection, and injury at birth, may all be relevant, especially if we can argue from a number of experiments in animals (see Penrose, 1961, for review). For example, fairly heavy therapeutic irradiation during pregnancy has been shown to lead to severe microcephaly in the child, and of 205 children exposed to the atomic blast at Hiroshima during the first half of intrauterine life, 7 were microcephalic and mentally retarded. All these factors make a simple basis for the inheritance of these disorders unlikely.

Wright (1934) introduced a model for dealing with threshold traits in a study of the number of toes of guinea pigs. The assumption made is that threshold traits are inherited polygenically, in the same way as the quantitative traits discussed in Chapter 6. We can look upon quantitative traits such as those discussed in Chapter 6, for which we can all be assessed, as threshold traits. Regarding intelligence, for example, individuals can be classified as normal, borderline, or defective, according to predetermined cutoff scores on a test. Therefore, in the analysis of threshold traits it is reasonable to assume an underlying continuous variable, inherited in the same way as the continuously varying traits

TABLE 1. Population Incidence of Some Common Congenital Malformations, Based on British Surveys

Malformation	Incidence per 1000 births
Anencephaly	3
Spina bifida	3
Heart malformations	1
Harelip with or without cleft palate	1
Clubfoot	1
Pyloric stenosis	3
Congenital dislocation of the hip	1

After Carter (1965).

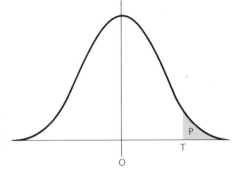

1. **Basic model for threshold trait.**
All individuals with a value of x exceeding T are affected. The proportion of affected individuals (P) is the area under the curve beyond T.

discussed in Chapter 6. In Figure 1 all individuals with a value of x greater than some threshold value (T) are assumed to have the trait. The added complication is that we are obliged to specify the relation between the distribution of x and the underlying continuous variable, and p, the proportion of people having the trait.

Threshold traits range all the way from those that show hardly any evidence of familial concentration to those that can be explained in terms of a single gene with reduced penetrance. A method of dealing with human threshold traits was developed by Falconer (1965). He used some of the concepts developed by animal and plant breeders to predict the outcome of directional selection experiments (Section 6.10) for behavioral traits. In such experiments, a proportion of a population is selected to provide subsequent generations. In the analysis of threshold traits, the analogy is the proportion of relatives of affected individuals who are themselves affected. If we take the *liability* to a certain disease in a population at large as represented by individuals above the threshold value (T), we then compare this liability with that among the relatives of those affected. In Figure 2 (distribution A) the vertical line indicates the threshold (T) in the population at large; distribution B gives the liability distribution of the relatives of affected individuals. The mean in distribution B is shifted toward the threshold, indicating the liability to the disease has a genetic component. We are therefore interested in the liability among the relatives derived from all those individuals with liability above T in Figure 2 (distribution A). The directional selection experiment analogy is to find the gain in liability due to this selection process of breeding only from such affected individuals. The difference between the means of the two distributions ($R-G$) gives the realized gain in liability based on selecting the relatives. In Figure 2 (distribution A), the difference between the mean of the general population and that of the selected individuals (A), or $A-G$, is equivalent to the selection differential of the directional selection experiment (Section 6.10).

The ratio of these two differences

$$\frac{R - G}{A - G} \qquad (7.1)$$

is the regression of relatives on affected propositi with respect to liability. This regression coefficient comes from the slope of the line obtained by plotting a number of $A - G$ values against $R - G$ values. We are therefore dealing with a simple parent-offspring regression situation for which we have already seen (Equation 6.19) that:

$$b_{OP} = \frac{\frac{1}{2}V_A}{V_P} = \frac{1}{2}h^2$$

which is also equal to the ratio given in Equation 7.1. Therefore, the regression of relatives on affected propositi gives an estimate of the heritability of the liability to a disease. An alternative way of writing Equation 7.1 is:

$$\frac{x_g - x_r}{a} \qquad (7.2)$$

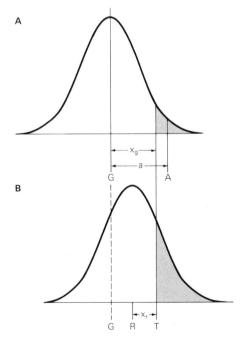

2. Inheritance of liability to diseases. Distribution **A** represents the general population, and distribution **B** represents the relatives of affected individuals compared with a fixed threshold (T). G is the mean liability of the general population. A is the mean liability of affected individuals in the general population. R is the mean liability of the relatives. $x_g = T - G$, or the difference between the threshold (T) and the mean liability of the general population; x_r = deviation of the mean liability of the relatives from T; and $a = A - G$.

where $x_g = T - G$, or the difference between the threshold (T) and the mean liability of the general population; $x_r = T - R$, or the deviation of the mean liability of the relatives from T; and $a = A - G$.

Falconer (1965) prepared a graphic representation (Fig. 3) of the incidence in the general population and in relatives who are sibs, parents, or

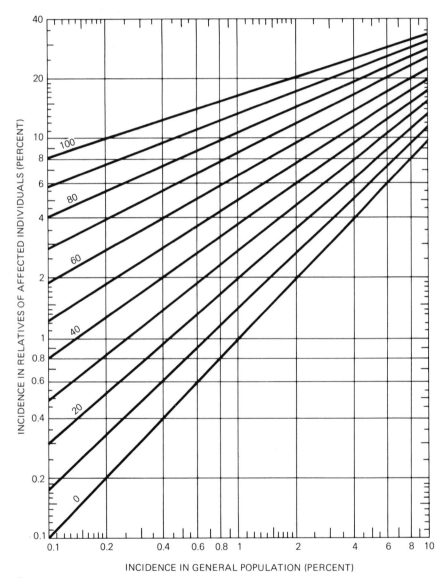

3. **Heritability of liability** for a threshold trait from two observed incidences, when the relatives are sibs, parents, or children. Numbers on the lines are the heritabilities (h^2) in percent. (After Falconer, 1965.)

children (first-degree relatives). The incidence among them is represented along the vertical axis. The population incidence is plotted along the horizontal axis. Both the horizontal and vertical incidence scales are logarithmic. To estimate heritability, first read along the horizontal axis to determine the point representing the population incidence; then read up to the point representing the familial incidence. The point where the two incidences intersect is the heritability.

Generally, we can write, for relatives of various degrees of relation:

$$b = rh^2 \qquad\qquad (7.3)$$

where r is the coefficient of relationship. This equation is strictly true only when the dominance variance $(V_D) = 0$, i.e., when dominance plays no role. For human populations this applies to many but not all those categories readily available for analysis. Full sibs (Section 6.8) are an exception, but in any case V_D is usually much lower than the additive genetic variance. Some values for r for more remote relatives are given in Section 6.8.

A detailed analysis using this method has been carried out on the disease diabetes mellitus (Falconer, 1967). Diabetics show a normal IQ distribution, but diabetic children seem to be relatively superior in verbal expression and retarded in performance (Rosenthal, 1970). Some studies purport to show an emotional basis for the onset of the disease, but others do not. Some have claimed a distinctive personality in diabetics, and the question of whether the disease produces abnormal psychological behavior has been raised. As Rosenthal comments, it is difficult to reach conclusions. A major problem in dealing with the disease is that its incidence is age-dependent, as shown in Table 2. The incidence is about 0.1 percent for persons up to 24 years of age and 3 to 5 percent for persons over 60 years of age. The incidence is to some extent a matter of the criteria used to define the disease, which may vary from an almost complete lack of insulin activity to a mild but consistently elevated blood sugar level of little clinical significance. The incidence increases if diag-

TABLE 2. Prevalence of Diagnosed Diabetes Mellitus, Based on United States National Health Survey, 1960

Age (years)	Males (%)	Females (%)
0–24	0.11	0.07
25–44	0.49	0.38
45–54	1.12	1.37
55–64	2.52	3.15
65–74	3.44	5.03
75 and over	3.15	3.88

After Rosenthal (1970).

nostic criteria are extended to include more than just the clinical manifestations of the disorder. Thus the incidence of clinical diabetes is about 3.5 percent in the United States, but this figure increases to over 6 percent if a glucose tolerance test is employed as the diagnostic criterion.

Neel et al. (1965) have written: "Diabetes mellitus is in many respects a geneticist's nightmare. As a disease, it presents almost every impediment to a proper genetic study which can be recognized." Some authors have suggested that single-gene inheritance, particularly that due to a single recessive gene with incomplete penetrance, may be the genetic basis — a view not held by Neel et al. (1965), who argue for multifactorial inheritance. However, Edwards (1960) points out that there is little difference between a model of a single gene with incomplete penetrance and multifactorial inheritance, unless the variability in penetrance is entirely environmental. Even though variations in diagnostic precision occur which would increase the environmental variance, it seems reasonable to consider diabetes a threshold trait.

Since in carrying out analyses of diabetes incidence, it is desirable to have both propositi and relatives close to the same age, a sib analysis is convenient. This has disadvantages, as pointed out in Section 6.8, because (1) environmental causes of resemblance (V_{Ec}) may be important for a particular dietary regimen, and (2) the dominance variance (V_D) may be relevant even though it is likely to be much smaller than the additive genetic variance (V_A). The heritability estimates, therefore, have these limitations. Figure 4 shows the variation in heritability estimates for three samples. The heritability declines from between 60 and 80 percent for persons under 40 years of age to between 20 and 40 percent for persons between 40 and 70 years old. Therefore, as has been suggested in the literature, the genetic etiology of the disease for younger people may differ from that for older people. Probably all that can be said at present is that the hereditary basis of the disease is multifactorial, with a threshold for its manifestation associated with the possibility that different genes may influence early and late onset. This example is discussed here in some detail because some of the diseases described in Chapter 11 reveal similar difficulties in analysis, e.g., variations in diagnosis and in age of onset.

7.2 Twin analysis: general considerations

Francis Galton was one of the first to emphasize the significance of twins for studies on human inheritance. Since then, twins have been studied extensively with a view to defining the relative importance of genetic and environmental effects on a wide variety of traits: morphological, behavioral, and pathological. The basic comparison is between

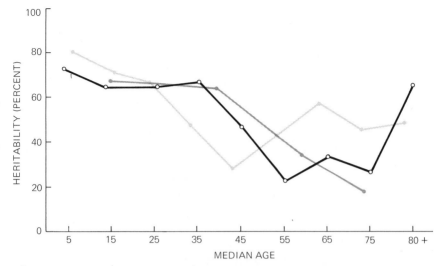

4. **Changes in heritability of liability** for diabetes mellitus with increasing age as estimated from sib correlations. **Dark gray line**, data on Canadian males; **solid line**, data on Canadian females; **light gray line**, data on both sexes in Birmingham, England. (After Falconer, 1967.)

monozygotic (MZ) or identical twins, who arise from one fertilization and therefore are identical genetically, and dizygotic (DZ) or nonidentical (fraternal) twins, arising from two fertilizations and thus genetically comparable to sibs. MZ twins are always of like sex, but DZ twins can be of like or unlike sex. MZ twins are the only humans who have identical genotypes; this explains the numerous twin studies that have been carried out. In many experimental animals, as we have seen, inbred strains are commonly developed, each strain composed of individuals with identical or nearly identical genotypes. There are problems in dealing with MZ twins, since there are likely to be special factors — especially relating to personality development and other behavioral traits of a learning and reasoning type — which yield a greater joint influence over MZ than DZ twins. This problem can be further examined by comparing differences between the members of a pair of MZ twins reared apart in different homes with the differences between MZ twins reared together in the same home. Such comparisons provide an assessment of the effect of the environment on twins reared in the same home. Although twin data have been extensively used in human genetic research, it is important to realize that twin studies can provide only limited information on the degree of genetic determination of a trait and can provide no information about modes of inheritance.

MZ twins are derived from a single fertilization, but even so, four different types of pregnancies are possible so far as uterine configurations are concerned, and two of these occur in DZ twins also:

- MZ or DZ with separate amnions, chorions, and placentas
- MZ or DZ with separate amnions and chorions, and fused placentas
- MZ with separate amnions and a single chorion and placenta
- MZ sharing a single amnion, chorion, and placenta

Overall the total frequency of twin births is between 1.0 and 1.5 percent, with some racial variation, Japan having an unusually low rate of 0.65 percent and United States blacks having a relatively high rate, with Africans higher still (Morton, Chung, and Mi, 1967). Almost all the racial variation is due to variations in DZ twinning rates. Variables such as socioeconomic factors and, particularly, maternal age affect twinning rates. Maximum twinning rates occur in the 35- to 40-year maternal age group, owing mainly to variations in the DZ rate, there being little variation in the MZ rate. There may be a small genetic component for twinning rates, particularly DZ twins, but Cavalli-Sforza and Bodmer (1971) regard much of the data as inconclusive.

For genetic studies it is critical that the determination of whether a twin is MZ or DZ be absolutely objective. Often the outward similarities of MZ twins makes this obvious compared with DZ twins, who are genetically no more alike than are other sibs. Even so, such a diagnosis could imply some subjectivity, and the only true criterion is genotypic identity, which cannot be checked directly. But there are so many known polymorphisms (e.g., for blood groups, enzymes, serum proteins, red-green color blindness, ability to taste phenylthiocarbamide) that when twins are classified with respect to an adequate number of them, the probability that a set of DZ twins will be identical for all is so low it can safely be discounted (see Mittler, 1971, for further details). Therefore, if identity is obtained for a large number of traits, the twins are likely to be MZ. Furthermore, various dermatoglyphic traits can be used as an aid in diagnosis. A direct test for MZ twins is the survival of skin grafts from one to the other, since there are known to be a large number of polymorphic loci for histocompatibility antigens responsible for graft rejection.

The principle of zygosity diagnosis using various polymorphic markers is as follows. The probability of DZ twins being identical is calculated separately for each locus. The nature of the calculation depends on the information available on the parental types. If the genotypes of the parents and twins are known exactly (taking into account other relatives, where available), exact probabilities can be computed. If, however, the parental genotypes are not known, probabilities can be worked out based on the gene frequencies in the population to which the twins belong. Detailed examples of the procedure appear in several texts (see especially Mittler, 1971, and Stern, 1973).

Even though critical studies differentiate with almost complete certainty between MZ and DZ twins, the simple use of visual criteria is almost as efficient as the use of blood groups and other polymorphisms. In large-scale twin studies, such as one carried out in Denmark, a simple questionnaire about the degree of similarity between twin pairs has been found to be 90 to 95 percent accurate in diagnosing zygosity. The typical questions relate to eye color, hair color and texture, height, weight, body build, the tendency to be mistaken by parents and by close and casual friends, and the twins' own opinion (Harvald and Hauge, 1965).

7.3 Twins in genetic studies: threshold traits

We consider in this section threshold traits and how they can be handled in twin data. In Section 7.4 we consider continuously varying traits. For threshold traits, we can say that a pair of twins is *concordant* if both members have the trait or both do not, i.e., if they are alike. The concordance frequency is the proportion of concordant twin pairs among all those which include at least one having the trait. Therefore, a significantly higher concordance frequency in MZ than DZ twins is considered evidence for a significant genetic component in determination of the trait. The significance of the data can be tested by a 2 × 2 contingency χ^2_1 (see Section 2.4), as shown for some data of Harvald and Hauge (1965) collected from the extensive Danish twin study for such human behavioral disorders as mental deficiency, epilepsy, and manic-depressive psychosis (Table 3). Writing the percentage concor-

TABLE 3. Twin Concordance for Various Mental Disorders

Disorder	Concordant pairs		Discordant pairs	Total pairs	χ^2_1	H
	No.	%				
Mental deficiency						
MZ	12	66.67	6	18	35.39*	0.67
DZ	0	0	49	49		
Epilepsy						
MZ	10	37.04	17	27	9.76†	0.30
DZ	10	10.00	90	100		
Manic-depressive psychosis						
MZ	10	66.67	5	15	20.84*	0.65
DZ	2	5.00	38	40		

* P<0.001.
† P<0.01.
After Harvald and Hauge (1965).

dances of MZ and DZ twins as CMZ and CDZ, respectively, in all cases CMZ is greater than CDZ. The χ^2_1 values are all highly significant, indicating the likelihood of a heritable component. Conveniently, in these data, like- and unlike-sexed DZ twins could be combined because they showed no significant differences in concordance; however, many analyses must consider data separately for like and unlike sex.

An index for estimating the degree of genetic determination (Holzinger, 1929) has been widely used. It is a statistic

$$H = \frac{CMZ - CDZ}{100 - CDZ} \tag{7.4}$$

that has been referred to in the literature as the heritability. However, to avoid confusion we refer to it as the H statistic, since it is a completely arbitrary quantity and difficult to relate to the estimates of heritability or degree of genetic determination based on quantitative traits discussed in Chapter 6. Referring again to Figure 2, if x_M is the deviation of the threshold of MZ twins from the threshold (T) of the general population (x_r in Fig. 2), and assuming the dominance variance (V_D) = 0, from Equation 7.2, the regression of MZ twins on the general population with respect to liability is:

$$\frac{x_g - x_M}{a} \tag{7.5}$$

which is equal to the heritability (h^2), since for MZ twins, the coefficient of relationship (r) is equal to unity (Equation 7.3). For DZ twins it can be shown that:

$$\frac{x_g - x_D}{a} = \tfrac{1}{2}h^2 \tag{7.6}$$

where x_D = deviation of the threshold of DZ twins from the mean threshold (T) of the general population (again it is assumed that $V_D = 0$). In this case the coefficient of relationship is $\tfrac{1}{2}$. The values of x_M and x_D are derived from the concordance frequencies.

Unfortunately, many surveys do not provide an estimate of population incidences, so that the available data are for MZ and DZ twins only as in Table 4. Because of this, heritabilities as in Equations 7.5 and 7.6 cannot be estimated. Cavalli-Sforza and Bodmer (1971) have provided a method for obtaining estimates of the degree of genetic determination as upper and lower limits based on two extreme assumptions. One is where the dominance variance is absent and the other is where the additive genetic variance is absent. Such limits are given for data (Table

4) from Harvald and Hauge (1965) based on the Danish Twin Registry. At these limits, the degrees of genetic determination differ by about 10 percent at most. The only nonsignificant comparisons between MZ and DZ twins are for cancer at any site and death from acute infection, which are very nonspecific categories. There is such a high level of heterogeneity in the category "cancer at any site" that little difference in concordance rates is expected. When cancer is defined as being at the *same* site, a higher degree of genetic determination is found, which is expected since this is a more homogeneous category.

The high degrees of genetic determination for mental deficiency and manic-depressive psychosis are in contrast to the other diseases listed in Table 4, even though they must be overestimates (Cavalli-Sforza and Bodmer, 1971, advance possible mathematical reasons for this). These extremely high values contrast with a series of rather specific diseases — arterial hypertension, tuberculosis, rheumatic fever, rheumatoid arthritis, and bronchial asthma — which cluster in the relatively high 0.5 to 0.7 region. The latter estimates are compatible with other comparable studies in suggesting important genetic components for these diseases. Environmental factors are known that may affect their incidence; for example, the incidence of arterial hypertension and of bronchial asthma is undoubtedly affected by stress. The high degrees of genetic determination in the behavioral traits mental deficiency and manic-depressive psychosis are considered further in Chapters 11 and 12.

TABLE 4.   Twin Concordance and Upper and Lower Limits for Degrees of Genetic Determination

Disease	Percent concordance		Limits of genetic determination	
	MZ	DZ	Upper $(V_D = 0)$	Lower $(V_A = 0)$
Cancer at same site	6.8	2.6	0.33	0.23
Cancer at any site	15.9	12.9	0.15	0.1
Arterial hypertension	25.0	6.6	0.62	0.53
Mental deficiency	67.0	0.0	1.0	1.0
Manic-depressive psychosis	67.0	5.0	1.05	1.04
Death from acute infection	7.9	8.8	−0.06	−0.06
Tuberculosis	37.2	15.3	0.65	0.53
Rheumatic fever	20.2	6.1	0.55	0.47
Rheumatoid arthritis	34.0	7.1	0.74	0.63
Bronchial asthma	47.0	24.0	0.71	0.58

All comparisons except those for cancer and acute infection are highly significant.
Data of Harvald and Hauge (1965), analyzed by Cavalli-Sforza and Bodmer (1971).
From *The Genetics of Human Populations* by L. L. Cavalli-Sforza and W. F. Bodmer. W. H. Freeman Company. Copyright © 1971.

7.4 Twins in genetic studies: continuously varying traits

Let us look first at the mean intrapair differences between MZ and DZ twins for an anthropometric trait, standing height. From studies by Newman, Freeman, and Holzinger (1937) we obtain:

	Number	Intrapair difference (cm)
MZ	50	1.7
DZ	52	4.4
Sibs	52	4.5

MZ twins are much more alike than DZ twins and sibs. Similar results have been obtained for many other continuously varying anthropometric traits, such as weight, and for behavioral traits, such as assorted measures of intelligence.

Intelligence quotient (IQ) is determined by a standard testing procedure, the Stanford-Binet Test. The IQ consists of the quotient, multiplied by 100, of the mental age of an individual as defined by the test, divided by his chronological age. A score of about 100 is the mean of the population, so that higher and lower scores represent higher or lower intelligence — assuming of course that the IQ test is a true measure of that essentially indefinable trait intelligence. Figure 5 depicts curves of intrapair differences in the Stanford-Binet IQ for MZ twins, DZ twins, and pairs of sibs. The smaller difference between MZ twins compared with DZ twins and sibs is clear, as is the similarity of the DZ twins and sibs. The probability that the Stanford-Binet IQ is at least partly under genetic control must be rated high at this stage.

Twin data can also be assessed from correlations between members of a pair. A statistical problem arises in that it is not possible to decide which twin measurement is x and which is y (see Equation 6.10, where the formula for the correlation coefficient is given). In this case, we use a different type of correlation coefficient, the *intraclass correlation coefficient*, which treats pairs symmetrically. It is defined as:

$$r = \frac{2\Sigma(x_i - \bar{x})(x_i' - \bar{x})}{\Sigma(x_i - \bar{x})^2 + \Sigma(x_i' - \bar{x})^2} \tag{7.7}$$

The measurements x_i and x_i' are the pairs of measurements taken in a completely arbitrary order (see Appendix 7.1).

Equation 7.4 gives a formula for an H statistic that has been widely used. This is expressed in terms of concordances of MZ and DZ twins. An exactly analogous H statistic can be derived from intraclass correla-

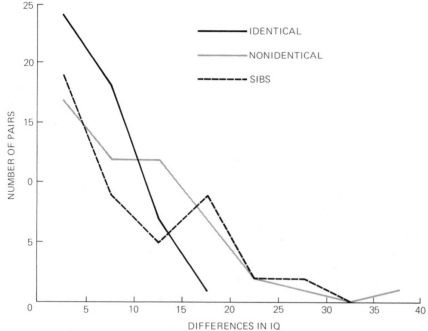

5. **Genetic basis of intelligence.** Curves are based on Stanford-Binet IQs of 50 pairs of identical (MZ) twins, 47 pairs of nonidentical (DZ) twins, and 52 pairs of sibs. (After Newman, Freeman, and Holzinger, 1937.)

tion coefficients. Thus, if r_{MZ} and r_{DZ} are the intraclass correlation coefficients of MZ and DZ twins respectively, then an H statistic is:

$$H = \frac{r_{MZ} - r_{DZ}}{1 - r_{DZ}} \qquad (7.8)$$

Obviously, if r_{MZ} is very much greater than r_{DZ} and is close to unity, H is close to unity; conversely, as in the case of an infectious disease, when r_{MZ} and r_{DZ} are expected to be about equal, H is close to zero.

The existence of MZ twins who have been reared apart permits an extension of this analysis so that the effect of two different environments on the same genotype can be investigated by comparing MZ twins reared apart (MZA) with those reared together (MZT). If r_{MZA} and r_{MZT} are the appropriate intraclass correlation coefficients, it is possible to estimate the effect of different environments on the same genotype. This is done in a relation analogous to Equation 7.8:

$$E = \frac{r_{MZT} - r_{MZA}}{1 - r_{MZA}} \qquad (7.9)$$

where E represents the environmental effect.

From Equation 7.16 (Appendix 7.1) it is clear that both the H and E statistics can be expressed in terms of variances of the differences between members of twin pairs, thus:

$$H = \frac{V_{DZ} - V_{MZ}}{V_{DZ}} \qquad (7.10)$$

and

$$E = \frac{V_{MZA} - V_{MZT}}{V_{MZA}} \qquad (7.11)$$

providing an alternative method of calculation.

The H statistic is closely related to a quantity based on the mean of the absolute differences rather than the variance of the differences as above. If M_{DZ} and M_{MZ} are the mean observed absolute differences for DZ and MZ twins, it can be shown that an estimate of H is:

$$H = \frac{M^2_{DZ} - M^2_{MZ}}{M^2_{DZ}} \qquad (7.12)$$

In other words, this H statistic is expressed in terms of the squares of the mean absolute differences between the two types of twins.

There is a question about the genetic meaning of the H statistic, even though it is the simplest and most convenient way of summarizing twin data on continuously varying traits. It does *not* provide a direct estimate of the heritability or the degree of genetic determination, as is shown by Cavalli-Sforza and Bodmer (1971). By various manipulations, they show that H can be written:

$$H = \frac{V_A + \frac{3}{2} V_D}{V_A + \frac{3}{2} V_D + 2V_E} \qquad (7.13)$$

If V_E were a representative environmental variance, the H statistic would *under*estimate the degree of genetic determination, which is:

$$\frac{V_A + V_D}{V_A + V_D + V_E}$$

Because there is no adequate way of estimating the effect on values of V_E obtained from twin data of limiting environmental variation to differences between twins, *there is no ready way of relating H values from twin studies to estimates of the degree of genetic determination obtained by other means.*

In conclusion, a significant value for the H statistic is likely to indicate that there is genetic variability in a population for a trait, but it is not possible to estimate the more exactly definable quantities such as the degree of genetic determination and the heritability. Furthermore, little can be said about the genetic architecture of the trait under study; for example, nothing can be said about the relative dominance of the genes controlling the trait.

7.5 Heredity and environment in man

Previous sections have presented indications of a high degree of genetic determination of some behavioral abnormalities in man. There is, especially for twins, a problem of a lack of control over the environment. This is a drawback in all studies on quantitative traits in man. In twin studies this problem is aggravated because of the difficulties inherent in comparing the environmental variation within twin pairs with that of unrelated individuals chosen at random and that of sibs who are not twins. The environment of relatives, especially sibs, is, of course, likely to be similar.

A method of assessing the effect of the similar environments of MZ twins, described in Section 7.4, has come from the small proportion of MZ twins who are separated at birth or soon after and then brought up apart (Table 5). This provides a unique natural experimental situation for comparing the performance of two identical genotypes in different families, i.e., in different environments. Relatively few such pairs have been found, and from three surveys the total is 116. From these, values of H (Equation 7.8) and E (Equation 7.9) are obtainable, providing estimates of hereditary and environmental determination (Table 5). Generally H for height $> H$ for weight $> H$ for the various behavioral measures (IQ and personality). In particular, the values for personality assessments are low, probably because of the low precision and arbitrariness of the tests. The E values are rather erratic, some being negative and some positive, in contrast with the H values, which are all positive. This suggests the greater importance of genotype than environment for all traits, including IQ and personality, but does not mean that the environment is irrelevant. Variations in E may be partly explained by differences between the samples. Another behavioral example showing a greater genetic component than an environmental one is smoking habit (see Section 2.4), since MZ twins brought up together and apart are in close agreement, but collectively differ from DZ twins. Twin studies may, therefore, give valuable information on the genetic and environmental components of quantitative traits, especially in the rare cases where MZ twins separated at birth are included.

We now extend the discussion beyond considerations of twins, as such, and consider family groups more widely. Taking the genotypic variance, in man we can include an additional component, V_{am}, the variance due to positive assortative mating (Section 2.3), which has the consequence that homozygous genotypes tend to be somewhat more common than under truly random mating. As a consequence, V_A is inflated by increasing the frequency of individuals who have extreme expressions for a trait (normally homozygotes). The genotypic variance (ignoring epistasis, as in Section 6.4) can then be written:

$$V_G = V_A + V_{am} + V_D \qquad (7.14)$$

Using appropriate breeding techniques in experimental animals, V_{am} can be made equal to zero. One effect of V_{am} in man is to increase the heritability, which becomes $(V_A + V_{am})/V_P$ because V_{am} acts by inflating the observed additive genetic variance.

TABLE 5. Estimates of Hereditary (H) and Environmental (E) Determination of Traits of MZ Twins Reared Apart and Together, and of DZ Twins

	H			E		
Trait	Shields (1962)	Newman, Freeman, Holzinger (1937)	Burt (1966)	Shields (1962)	Newman, Freeman, Holzinger (1937)	Burt (1966)
Height						
Females	+0.89			+0.67		
Males				+0.89	−0.54	+0.33
Both sexes		+0.81	+0.93		−0.64	+0.33
Weight						
Females	+0.57					
Both sexes		+0.78	+0.83	−0.62; +0.68	+0.27	+0.39
IQ						
Dominoes and vocabulary	+0.53					
Binet		+0.68			+0.64	
Other			+0.86	−0.04		+0.40
Personality						
Extroversion	+0.50			−0.33		
Neuroticism	+0.30			−0.36		
Woodworth Mathews Neuroticism Questionnaire		+0.30			−0.06	
Other						+0.95

Number of pairs studied: Shields, 44; Newman, Freeman, and Holzinger, 19; Burt, 53.
After Cavalli-Sforza and Bodmer (1971).

We turn now to estimating the environmental variance. It is possible in experimental animals, but not in man, to control the environment. One way of looking at the environmental variance in man is by subdividing it (after Cavalli-Sforza and Bodmer, 1971):

$$V_E = V_{ind} + V_{fam} + V_{soc} + V_{rac} + V_{GE} \tag{7.15}$$

where the variance components are defined as follows:

- V_{ind} is the variance among individuals within families. It is included in all families, but may vary from family to family. For example, the environmental variance for MZ twins may be less than for DZ twins, as the MZ twins, because of their genotypic identity, may *choose* similar environments. The environmental variance between DZ twins may differ from that between nontwin sibs, the latter including a birth-order component. Also there may be variations for families of different sizes.

- V_{fam} is the variance among families within socioeconomic strata. This inflates the covariance between parent and offspring. Some idea of its importance can be obtained from the correlation between foster parents and adopted children, but the selective placement often carried out by adoption agents biases results.

- V_{soc} is the variance among socioeconomic strata. Cultural differences among families or social groups may be maintained by sociocultural inheritance leading to correlations between relatives that are very difficult to distinguish from those due to genetic determination. Such factors are important in comparisons among racial groups. The geographical isolation in different environments that has permitted the development of genetic differences among races has also created a parallel, but probably largely independent, development of cultural differences. Cavalli-Sforza and Feldman (1974) have formally analyzed the consequences of parent-offspring cultural transmission and have concluded that "cultural" inheritance is almost completely confounded with biological inheritance. The distinction between the two methods of transmission is not simple, and these workers conclude that only through careful studies of data on both adopted and biological offspring can the relative roles of cultural and biological inheritance be settled.

- V_{rac} is the variance in environmental conditions accompanying racial differences, included in which are sociocultural differences above. In some communities V_{rac} may be high, as are the differences between black and white Americans (Chapter 12).

- V_{GE} is the variance due to the genotype-environment interaction that occurs when given genotypes show different phenotypes in different environments (Section 6.2). It is difficult to give examples in man, but in addition to those cited in Section 6.2, it is instructive to look at an experiment on maze-learning ability in rats (Cooper and Zubek, 1958). Two lines were successfully selected by directional selection to be "bright" or "dull" at finding their way through a maze. In a normal laboratory environment for rats the difference in the mean number of errors between the maze-dull and maze-bright strains was about 50 (Fig. 6), but in a restricted environment both strains were equivalent, since the brights were reduced to the level of the dulls. This is a genotype-environment interaction such that maze-brights are much more affected in the restricted environment than the maze-dulls. Conversely, in a stimulating environment the relative improvement of the maze-dulls compared with the maze-brights was much greater: creating a better environment improved the maze-dulls relatively more than the maze-brights. Therefore, we have a complex genotype-environment interaction in rats that can be assessed because both genotype and environment are definable more precisely than is normally possible in man. Indeed, in man it is not normally possible to define the genotype or environment of a population. This means that the isolation of the interaction between the two components is intractable; hence, in man, separating V_{GE} from V_G and V_E is a problem not solved. As is shown in Chapter 12, this conclusion is of fundamental significance in issues such as the interpretation (but not the existence) of racial differences in IQ (and other behavioral traits).

Cattell (1965) has suggested an approach to the heredity-environment issue in man that uses several diverse family types and so has potentially more generality than the methods so far discussed. It is the multiple abstract variance analysis method (MAVA). The method entails a major difficulty of time and cost, since Cattell estimates that some 2500 pairs of children are needed for a comprehensive analysis. This is prohibitive from the practical point of view, so it is not surprising that few analyses have been carried out. The main categories of family types are (1) identical twins reared together, (2) identical twins reared apart, (3) sibs reared together, (4) sibs reared apart, (5) half-sibs reared together, (6) half-sibs reared apart, (7) unrelated children reared in same family, and (8) unrelated children reared apart. From these, information on the correlation between heredity and environment can be obtained. For example, a correlation of +0.25 was obtained between hereditary and environmental effects on intelligence; this is close to the values of +0.22 to +0.30 repeatedly obtained for correlations between intelligence and social status.

7.6   Can random mating be assumed in man?

In the derivation of correlations between relatives discussed in Chapter 6, it was assumed that mating occurs at random. As soon as there are deviations from random mating, such as inbreeding or assortative mating, the formulae given in Section 6.8 are no longer strictly accurate, as indicated in Section 7.5. In man, as long ago as 1903, Pearson and Lee

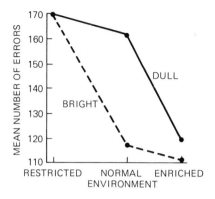

6. **Genotype-environment interaction.** Mean number of errors in a closed-field maze for bright and dull rats reared in enriched, normal, and restricted environments. (After Cooper and Zubek, 1958.)

found positive correlations between partners at marriage for physical traits such as stature and arm span. The correlation coefficients between mates were usually around +0.2. For example, Spuhler (1968) surveyed 105 physical traits in 40 human population samples. Correlation coefficients in the range of +0.1 to +0.2 are the most common for measurements of body size in Europeans and Americans of European descent, although coefficients smaller than +0.1 and in the range +0.2 to +0.3 are quite common. Correlation coefficients greater than +0.5 are rare. Data from Spuhler's paper are presented in Table 6. Studies of assortative mating in non-European populations are few. In two populations studied, the Rama-Navajo Indians and the Japanese, homogamy for body size was not found (Spuhler, 1968).

For behavioral traits the tendency toward strong positive assortative mating occurs in some cases (Spuhler, 1962). One estimate of intelli-

TABLE 6. **Correlation Coefficients Found in Various Studies for Various Physical Traits in a Human Population**

Trait	Correlation coefficients						Total no. of studies
	<0	0–0.1	0.1–0.2	0.2–0.3	0.3–0.4	>0.4	
Stature	1	6	8	7	4	1	27
Sitting height	1		3	3			7
Weight		1	2	3	1		7
Chest circumference		2	5				7
Head circumference	2	3	1	2			8
Cephalic index	2	12	5	3			22
Facial index	4	7	3		1		15
Nasal index	3	2	1	2			8
Hair color			2	2	1		5
Eye color	1	1	1	1		1	5

After Sphuler (1968).

gence, the Progressive Matrices Test of Raven, gave a correlation coefficient of +0.399 (P<0.01 for deviation from zero). A test on verbal meaning based on the selection of one of four words that best completes the meaning of each of 40 sentences gave $r = 0.305$ and 0.732 (P<0.01 in both cases) for the total number of right answers and the proportion of right answers, respectively. Spuhler also reported data showing substantial positive assortative mating for such psychological traits as association, neurotic tendency, and dominance, as did Beckman (1962) for musical ability. Generally, therefore, the tendency for assortative mating seems stronger for behavioral traits than for physical traits.

Although, as shown above, there are often positive correlations between mates, a positive correlation is frequently found between socioeconomic status and stature. Similarly, there is a strong correlation for socioeconomic status between husband and wife. Therefore, it can be argued that the positive correlations between mates with regard to stature may be partly due to a correlation with socioeconomic status. Furthermore, since stature is correlated with a number of other physical traits, such as weight and chest circumference, similar correlations are expected for them and are found. However, in the Rama-Navajo Indians no assortative mating for physical traits was found (Spuhler 1968), and it could be argued that the discrepancy compared with Caucasians is due to a different social structure.

Stature has been increasing in most western societies during this century. From data on Italian conscripts, Conterio and Cavalli-Sforza (1959) estimated the mean increase of stature as 0.1 cm per year or 3.0 cm per generation during this century. General improvement in living conditions, particularly with respect to nutrition and disease, is undoubtedly of great importance, as indicated by a significant positive correlation of stature and socioeconomic group in the Italian data. The relative importance of genetic factors is more difficult to assess, but heterosis as a result of the breakdown of isolated communities has been postulated. For age at marriage there is a correlation of 0.8 between mates. Hence, mates could show a correlation for stature merely because of a tendency to be born at the same time, if stature itself is increasing over the time period being considered. The literature contains other examples where positive correlations are removed when secular trends are taken into account. For example, Beckman (1962) found that the correlation between wives and husbands with respect to number of sibs disappeared when comparisons were restricted to single time periods.

For continuously varying traits, excluding nongenetic factors as discussed above, one of the main effects of positive assortative mating is to increase the additive genetic variance (V_A) compared with the random mating situation (see also Section 7.5). If V_A is the additive genetic variance under random mating, $\hat{V}_A$ is that under positive assortative mating, and r is the correlation coefficient between mates, then for a large number of genes (Crow and Felsenstein 1968):

$$\hat{V}_A \approx \frac{V_A}{1-r}$$

so that for $r>0$ then $\hat{V}_A > V_A$. For example, for $r = 0.2$, $\hat{V}_A = 1.25V_A$. Burt and Howard (1956) found the following correlation coefficients for IQ:

• Between mates (assortative mating), 0.388

• Between parent and child, 0.489

• Between sibs, 0.507

from which Cavalli-Sforza and Bodmer (1971) calculated the following relative contributions to variance components for IQ as:

• Additive, 51.2 percent

• Assortative mating, 19.2 percent

• Dominance, 23.4 percent

• Environmental, 6.1 percent

This shows that assortative mating may be important for quantitative traits in man, including behavioral traits.

Appendix 7.1   Intraclass correlation coefficient

If a twin has a measure x and another y, it is not possible, as in Equation 6.10, to decide which twin is x and which is y. In this case, the intraclass correlation coefficient (Equation 7.7) is used, which treats pairs symmetrically:

$$r = \frac{2\Sigma(x_i - \bar{x})(x_i' - \bar{x})}{\Sigma(x_i - \bar{x})^2 + \Sigma(x_i' - \bar{x})^2}$$

where x_i, x_i' are the pairs of measurements taken in a completely arbitrary order. Because of this, the mean must be defined over all measurements thus:

$$\bar{x} = \tfrac{1}{2}n[\Sigma x_i + \Sigma x_i']$$

There is a simple relation between the intraclass correlation above and the variance of the differences between pairs of twins, which can be derived as follows. We first evaluate the corrected sum of squares of the differences:

$$\Sigma(x_i - x_i')^2 = \Sigma[(x_i - \bar{x}) - (x_i' - \bar{x})]^2$$

This makes no difference to the variability, since all that is being done is to change the origin on which x and x' are measured by subtracting $\bar{x}$ in each case. From this expanded expression we obtain:

$$\Sigma(x_i - \bar{x})^2 + \Sigma(x_i' - \bar{x})^2 - 2\Sigma(x_i - \bar{x})(x_i' - \bar{x})$$

If we now evaluate $1-r$ or

$$1 - \frac{2\Sigma(x_i - \bar{x})(x_i' - \bar{x})}{\Sigma(x_i - \bar{x})^2 + \Sigma(x_i' - \bar{x})^2}$$

then from Equation 7.7

$$\frac{\Sigma(x_i - x_i')^2}{\Sigma(x_i - \bar{x})^2 + \Sigma(x_i' - \bar{x})^2} = 1-r \qquad (7.16)$$

In other words, there is a relation between the variance of the differences and the intraclass correlation that we can simply write:

$$r = 1 - \frac{V}{S^2} \qquad (7.17)$$

where V = mean square of differences between members of a twin pair and S^2 = total variance over all observations.

General readings

Cavalli-Sforza, L. L., and W. F. Bodmer. 1971. *The Genetics of Human Populations.* San Francisco: Freeman. This excellent text includes an advanced chapter on the genetic analysis of quantitative traits in man.

Mittler, P. 1971. *The Study of Twins.* London: Penguin. A very readable account of the place of twin studies in behavior-genetics research.

Shields, J. 1962. *Monozygotic Twins Brought Up Together. and Apart.* London: Oxford University Press. One of the few classic analyses of twins brought up together and apart.

The Genetics of Behavior: Drosophila

8.1 Single-gene effects

This chapter initiates the third section of this book (see Chapter 1 for its plan). To date we have concentrated on principles, but in this and the next four chapters we look at behavior phylogenetically, beginning in the present chapter with Drosophila, followed by Chapter 9 on rodents, and Chapter 10 on various other organisms ranging from bacteria and protozoa to assorted mammals. In Chapters 11 and 12, we proceed exclusively to man. The switch to specific organisms is made so that we can specify those behaviors that have been and can be studied in certain experimental subjects. As we saw in Chapter 1, the types of behavior that can be studied depend on the organism. In some species of Drosophila very sophisticated genetic techniques have been applied to many areas of study, including behavior genetics. Furthermore, the ease of rearing large numbers of individuals from a single, genetically controlled laboratory strain means that elegant statistical methods can be employed as an aid to drawing conclusions. Table 1 gives a full summary of the virtues of Drosophila as an experimental animal (see also Ashburner and Novitski, 1975; Wright and Ashburner, 1976). (We shall not endeavor to be comprehensive in this chapter — probably a hopeless

TABLE 1. Reasons for Drosophila's Value as an Experimental Organism for Genetic Research

● Short generation time. Many species develop from egg to adult in less than 2 weeks

● Ease of rearing. Even novices can breed and rear fruit flies with rewarding success

● Inexpensiveness. Fruit flies can be raised cheaply in large numbers; they eat fermenting fruit and yeasts

● Small size. Many fruit flies can grow in a small space, often a few jars

● Large numbers of progeny. A single gravid Drosophila female can produce hundreds of offspring

● Harmlessness. Drosophila carry no known disease that affects human beings; adults possess no biting or piercing mouth parts

● Sex ratio. Most species produce approximately equal numbers of each sex among their progeny

● Parthenogenesis. At least one species, *D. parthenogenetica*, is made up almost entirely of females

● Many species. There are an estimated 2000 species within the family Drosophilidae

● Wide distribution. *Drosophila* species are found throughout the world, from the arctic to the tropics

● Ease of collection. It is easy to trap Drosophila and to bring captured flies back to the laboratory in good condition

● Low chromosome number. Drosophila possess low numbers of easily identifiable chromosomes; some species have only six chromosomes or three pairs

● Larval salivary gland chromosomes. The large size of these giant, many-stranded chromosomes permits the researcher to identify even small sections of a single chromosome, much as one recognizes the face of a friend

● Hybrids. The large number of closely related member species permits the breeding of hybrid fruit flies in the laboratory

● Races and/or subspecies. A variety of races provides research material for those interested in the evolutionary process by which new species arise — speciation

● Isolating mechanisms. Drosophila exhibit several isolating mechanisms (such as sexual isolation and hybrid sterility) that prevent gene exchange between species

● Mutants. Because of their response to such mutagens as X-rays and chemicals, Drosophila mutants can be produced easily in the laboratory; mutation can alter the size, color, number, and/or structure of almost all parts of the fly's body

● Behavior. Drosophila behave consistently in courting, in response to gravity, and in response to light; these patterns lend themselves to genetic analysis and modification through selection

● Symbionts. Several *Drosophila* species carry a variety of microorganisms, allowing researchers to study symbiotic relations; since some of these microorganisms are "inherited," i.e., transmitted from parents to offspring, geneticists have a special interest in the process

● Cytoplasmic inheritance. Drosophila can in some instances transmit extrachromosomal units of heredity to progeny

From Ehrman (1971).

goal in the face of the great mass of genetic studies on Drosophila.) Many investigations appropriate to this chapter are discussed in earlier chapters, and we shall refer to them. Some of the evolutionary aspects are dealt with in Chapter 13. A review of the literature to mid-1971 appears in Parsons, 1973.

Besides the well known *D. melanogaster*, there are now more than 2000 formally recognized *Drosophila* species. Some six of these are recognized as cosmopolitan: *D. melanogaster, D. simulans, D. immigrans, D. hydei, D. funebris,* and *D. ananassae.* They live in six life zones: Nearctic, Neotropical, Palearctic, Ethiopian, Oriental, and Australian. (To be punctilious, we might also include two more in our list of cosmopolitan species: *D. putrida* and *D. busckii.*)

In deference to the Queen Mother of all *Drosophila* species, we begin with those *D. melanogaster* mutants that affect behavior (see Chapter 2, Fig. 3, a linkage map). The diversity of behaviors are controlled by single genes situated at loci scattered throughout the fly's genetic constitution. *D. melanogaster* has by far the best known *genome* (one haploid set of genes and chromosomes) among the *eukaryotes* (organisms made up of cells with nuclei bounded by nuclear membranes, undergoing meiosis); this makes it particularly valuable in the analysis of new subdivisions of genetics. Behavior genetics is no exception to this rule.

As a specific example, Figure 1 presents data on the electroretinogram (ERG, a record of the mass electrical response of appropriate cells to illumination) of wild-type, normal-visioned *D. melanogaster* and of flies with photoreceptor mutants known to be sex-linked. Flies were dark-adapted for at least 15 minutes before electroretinography in response to 0.5-second white light stimulus. Only intact, living flies were employed. The recording electrode was placed into the retina via a minute corneal hole. The independently occurring mutations either result in abnormal electroretinograms or delete them altogether, producing diminished visual acuity in either instance. (See Alawi et al., 1972, on mutation in *D. melanogaster* affecting phototransduction in insect vision, a mechanism by which the reception of sensory stimuli is associated with ionic events in receptor membranes.)

Benzer and collaborators have studied, among other point mutations affecting locomotor, visual, sexual, stress response, and neuromuscular behaviors, three single-gene mutations altering the normal circadian (about 24 hours) rhythm exhibited by *D. melanogaster* (Konopka and Benzer, 1971; Benzer, 1971, and references therein; Benzer, 1973). Flies of this species normally emerge (eclose) "around and about" dawn, when the presence of dew and conditions of temporarily higher humidity satisfy their omnipresent requirements for moisture; this indeed is the derivation of the noun Drosophila; "dew lover." Then, for most *Drosophila* species observed, a morning period of activity follows. It is over by midday and is followed by an evening spurt of activity.

Compelling evidence for the genetic control of this biological clock (Pittendrigh, 1958) is presented by flies with the *arrythmic* mutant, which eclose ad lib all day long; flies with the *short-period* mutant, which display a consistent 19- (not 24-) hour cycle; and flies with the *long-period* mutant, which have a 28-hour cycle (Fig. 2).

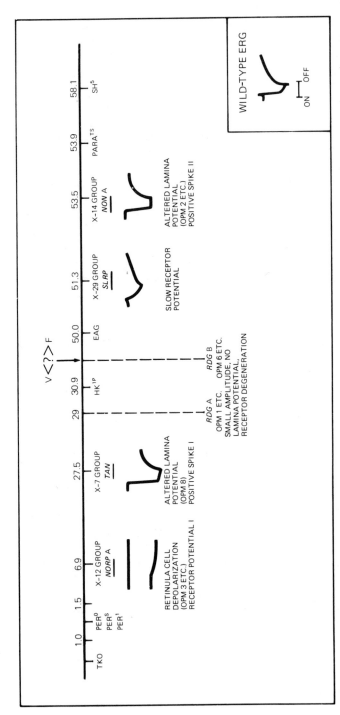

1. Neurological mutants on X chromosome of *D. melanogaster.* Map positions (not to scale) of mutants are indicated above line. Symbols for loci are shown below line together with a representative trace of the phenotypes of the electroretinogram (ERG) mutants. Two such traces are shown for X-12 group of mutants to demonstrate that some alleles at this locus have no response to a light stimulus, while other alleles show a small depolarization. An ERG on wild-type flies, with the duration of light stimulus indicated below, is presented for comparison. A phrase beneath each ERG provides a brief description of each mutant. Terms in parentheses show alternate designations of same lesions. Traces for mutant groups opm 3 and opm 6 are not shown. (From Grossfield, 1975.)

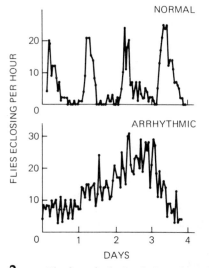
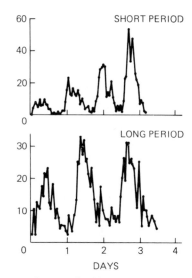

2.   **Rhythm of eclosion in** D. *melanogaster* from populations of wild-type and mutant pupae kept in constant darkness. (From Konopka and Benzer, 1971.)

Genetic Drosophila mosaics can be generated — as opposed to passively awaited for exploitation when they occur spontaneously, usually at very low rates (see Section 10.7 on wasp mosaics) — by depending upon the loss of a reliably unstable ring X chromosome. Ring chromosomes do not regularly successfully migrate to either pole in the terminal phases of either mitosis or meiosis, and so are omitted or destroyed in telophase, the final separatory stage of these cycles; the extent of male tissue depends upon the time during development when the loss occurs. At that point, two embryonic cell clones are established, one with a single X and the other with two X chromosomes. Subsequently, the X0 imago sections are male (unlike the X0 sterile human Turner female discussed in Section 4.3), while the XX sections are female. Given this basic technique plus the extreme genetic control extant with (only) D. *melanogaster*, two-dimensional embryonic *fate maps* can be constructed. These correlate precise anatomical sites with abnormalities affecting behavior (Hotta and Benzer, 1972; see Fig. 4).

Consider one of several detailed fate maps that has been constructed for the sex-linked *Hyperkinetic* (*Hk*) gene (Ikeda and Kaplan, 1970a,b): Homozygous females and hemizygous males, and to a lesser degree heterozygous females, display a rhythmic leg-shaking action during ether anesthetization. Figure 3 presents nine different gynandromorphic patterns, mosaics of female and male parts, with respect to this mutant. Six hundred such mosaics were collected, scored for surface landmarks

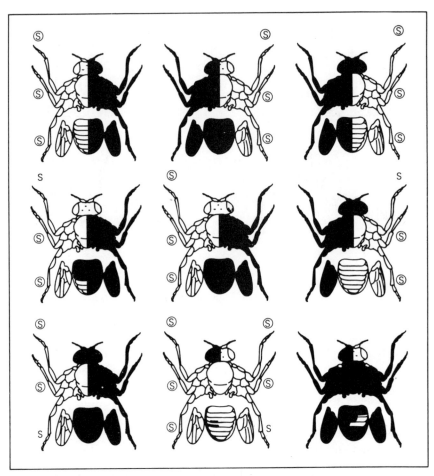

3.   **Hyperkinetic** *D. melanogaster* **genetic gynandromorphs.** *Black,* female tissue; *white,* male tissue; *S,* shaking leg; *encircled S,* rhythmic bursts from motor regions associated with the leg. (From Ikeda and Kaplan, 1970a,b.)

(male surfaces have different colors, eye and body, and differently shaped bristles), and examined when anesthetized for shaking. Ikeda, Kaplan, Hotta, and Benzer all agree that the shaking or control of each of the six legs is independent of all other legs and that there is strong likelihood of identity between the shakiness of a leg and the genotype of its cuticle (noncellular, waterproof external covering). Figure 4, shaped like one side of the ovoid Drosophila blastoderm (a layer of cells surrounding all the yolk in an insect's fertilized egg), depicts three separate foci, the structures in the fly which, when mutant, cause mutant behavior. There is one focus for every leg on each side of a fly, and all lie within the blastodermal region associated with the origin of the insect's ventral nervous system — ventral to blastodermal cuticular regions. This last is so despite the oft-observed sameness of the leg genotype and the cuticular genotype. There exists electrophysiological evidence of abnormally behaving thoracic ganglia in *Hk* individuals; expression of *Hk* is genetically autonomous, and the left and right sides of each thoracic ganglion are independent of each other. We may then cogently speculate about the location of parts of the thoracic ganglia on this fate map, e.g., perhaps in the shaded area imposed upon Figure 4.

8.2   Quantitative traits involving sexual behavior

We have seen elsewhere in this book that in Drosophila many quantitative traits are under genetic control and can be studied using biometrical methods and selection experiments. A list of the types of traits studied is presented in Section 5.1. Geotaxis is discussed in Section 5.2 to illustrate the selection experiment approach and its application in obtaining information on the genetic architecture of a trait. Diallel crosses for mating speed and duration of copulation are considered in Section 6.5, and in Section 6.11 locomotor activity is discussed as an example of the regression approach in the analysis of quantitative traits.

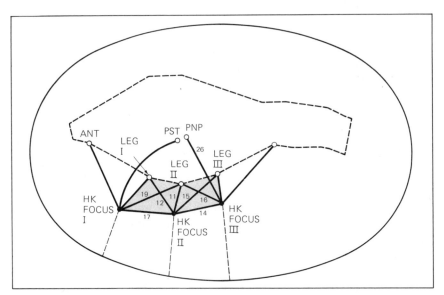

4.   Fate map for *Hk* gene in *D. melanogaster.* Three foci related to mutation are shown on a background representing one side of the embryonic blastoderm. **ANT,** antennae; **PNP,** postnotopleural bristle; **PST,** presutural bristle. Mutation at these sites is related to mutant behavior. (After Hotta and Benzer, 1972.)

We are now ready to consider those aspects of quantitative behavior in Drosophila whose hereditary bases are known and which we have not discussed in detail previously. In this section we consider traits involving sexual behavior.

Manning (1961) selected for fast and slow mating speeds in *D. melanogaster*, basing his selection upon the performance of 50 pairs of flies collected before mating could occur and placed together in a mating chamber. The 10 fastest and 10 slowest pairs were identified, selected, and used to initiate fast and slow selection lines. The response to this selection rapidly produced two fast and two slow lines; an unselected control line was maintained. After 25 generations, the mean mating speed averaged 3 minutes in the fast lines and 80 minutes in the slow lines. The difference between the lines is shown in Figure 5. Considerable variations in speed owing to environmental fluctuations occurred during selection, but the fluctuations were generally similar in all lines for a given generation. An approximate realized heritability of 0.30 was computed from the rate at which the selection lines diverged during the first few generations. Although no further genetic analyses were carried out, Manning analyzed in some detail how selection affected behavior. Hybridizing the fast and slow lines in both directions (reciprocal crosses) gave intermediate F_1 mating speeds, while intercrossing the two fast and slow lines themselves reciprocally gave fast and slow speeds, respectively. These results indicate that both sexes were affected by selection. Confirmation of this came from testing mating speeds against an unselected stock of flies: both sexes of the selected lines gave altered mating speeds in the expected directions. Activity differences between lines were measured by admitting flies to an arena where the number of squares entered by a fly in a given time period was scored. The slow lines exhibited much more of this type of activity than did the fast ones. Experiments using unselected females with selected males showed that the lag before courtship was much smaller for the fast than the slow lines; similarly, the frequency of licking (contact between the male proboscis and the female genitalia, see Section 3.2) was higher in the fast than in the slow lines. Thus, the fast lines have a high level of "sexual activity" and a low level of "general activity," and the slow lines have the opposite. Under natural conditions these two components are presumably coordinated at an optimum, since overresponsiveness or underresponsiveness would be obviously undesirable.

Selection for this trait operates on both sexes and there could be rather different genes controlling the response within the two sexes. Manning (1963) attempted to look into this by selecting for mating speed based on the behavior of one sex only. There was no response to selection for fast-mating males or slow-mating females, and a fast-mating female line

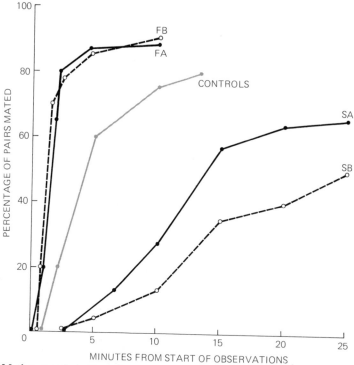

5.   **Mating speeds in** *D. melanogaster* in two lines selected for fast speeds (FA, FB), two selected for slow speeds (SA, SB), and controls at the eighteenth generation of selection. (From Manning, 1961.)

was not set up. One wonders if natural selection had not already fixed these genes for rapid mating in males, since it so clearly enhances their fitness. Indeed, as discussed in Section 6.5, Fulker (1966) found just such evidence for directional selection for fast mating. The general importance of rapid mating speed as a component of fitness in Drosophila is further emphasized by Parsons (1974). There was, however, a response in male lines selected for slow mating. In these lines, the mating speed of females was unaffected in early generations but somewhat reduced in later generations. Behaviorally both sexes showed lower general activity and the males showed reduced courtship activity, which contrasts with earlier experiments. Manning was unable, however, to come to any definite conclusions concerning possible differences between sexes in the genetic control of mating behavior.

Genes affecting general activity and sexual activity must therefore be wholly or partly controlled by separate genetic systems that can be modified independently. Ewing (1963), a student of Manning, selected for spontaneous activity and found that his inactive lines displayed greater sexual activity, as would be expected from Manning's observations. However, Ewing's technique involved placing 50 flies of one sex into the initial tube of a line of interconnected tubes and selecting the first 10 flies (active) and the last 10 (inactive) to reach the opposite end. This procedure separated flies that moved through rapidly from ones that did not, but when the two lines so formed were tested by introducing single flies into Manning's arena there was no significant difference between them; thus the two types of behavior seem to differ. (Here a caveat: this is but one example of an often observed and often referred to *apparatus effect* on behavior.) Manning was measuring spontaneous activity; Ewing was measuring dispersal activity or the reactivity of flies with one another.

In a study of the control of receptivity, Manning (1966, 1967) found female acceptance of a courting male to depend on two processes. The first determines whether a *D. melanogaster* female is accessible to the courtship of males. Young females are unresponsive for some 36 hours after eclosion, and then quite suddenly they become receptive and accept a male after a few minutes of courtship. The evidence suggests that this rapid change of receptivity occurs when the concentration of circulating juvenile hormone rises with the reactivation of the corpus allatum (one of the insect's endocrine organs), and the ovaries show a growth cycle parallel to that of receptivity. The second process can be called *courtship summation* and involves the addition of all the heterogeneous stimuli provided by a courting male that induce a female to allow him to mount once a critical level of stimulation is reached.

The evidence advanced indicates that the two processes are distinct and that the change from the unreceptive state to the receptive is an all-or-nothing process. Females are completely unresponsive to courtship or they accept within the normal time range for receptive females (about 95 percent of females accepting males within 15 minutes of introduction to them). There is no evidence that females become gradually more receptive by requiring less and less courtship before accepting. Virgin females remain receptive for many days, but after the first week of adult life an increasing proportion become unreceptive, and the switch-off seems to be a rapid all-or-nothing event like the switch-on. Old females which have mated and used up their stored sperm (by depositing many fertilized eggs) are more often receptive than are virgins of the same age. It is suggested that this is because their corpora allata are more active and the juvenile hormone concentration is kept above the critical level longer.

Manning (1968) selected successively for slow mating speed in *D. simulans*, a sibling species of *D. melanogaster*, in which the behavior of

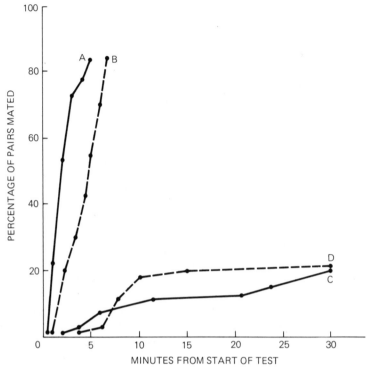

6. **Mating speeds in** *D. simulans* for (*A*) selected males × control females, (*B*) control males × control females, (*C*) selected males × selected females, (*D*) control males × selected females. (From Manning, 1968.)

males was not affected but in which there were marked effects in females (Fig. 6), in contrast to the mating speed selection experiments in *D. melanogaster*, in which both sexes were affected. (*D. melanogaster* and *D. simulans* are very closely related and morphologically similar species — females more so than males; they are therefore referred to as sibling species; see Section 4.2.) Most of the slow-mating-line females failed to become receptive on the second day after eclosion, as do normal flies. The females in fact performed the most vigorous repelling movements, by extruding the ovipositor and twisting or lifting the abdomen beyond the reach of the courting male. These movements normally are performed in these species by elderly virgins which have become unreceptive or by fertilized females whose receptivity is inhibited by the presence of already stored sperm in their seminal receptacles. However, the females in Manning's slow-mating line had normal ovarian growth and a corpus allatum complex that, when implanted into normal hosts, was capable of producing precocious receptivity. The experiments suggest that the females have a normal supply of juvenile hormones and that the genetic change involves one or more links in the chain of neural "target organs" on which the juvenile hormone acts. As Manning (1968) pointed out, this situation is comparable to that in some mammals, such as guinea pigs (Valenstein, Riss, and Young, 1955).

Kessler (1968, 1969) selected for fast and slow mating speeds in *D. pseudoobscura*, crossing three wild-type strains (from British Columbia, California, and Guatemala), and using a technique essentially similar to Manning's. After the twelfth generation of selection, tests were made on all possible combinations of fast, slow, and control lines as a 3 × 3 diallel cross in which observations were carried out for 30 minutes using 50 pairs in one container, copulating pairs being removed with an aspirator. It was found that slow-mating females reduced mating whenever they were involved and that fast-mating females were not significantly faster than the controls. Fast-mating males speeded up all matings in which they were involved, but slow-mating males were not significantly different from the controls. An analysis of variance showed females to account for more of the total variance than did males. This contrasts with data of Kaul and Parsons (1965) for ST/ST, ST/CH, and CH/CH karyotypes (ST is the Standard and CH the Chiricahua third chromosome gene arrangements in *D. pseudoobscura*), in which male determination was very strong. Note two things, however: first, Kessler was dealing, not with known karyotypes but with selection lines, and second, he was dealing with a mating chamber with 50 pairs of flies, while Kaul and Parsons examined single-pair matings. Here we see, as we shall a number of times in this text, evidence of gross differences in outcomes resulting from less than major differences in research techniques and apparatuses. In the two species of Drosophila mentioned — *D. pseudoobscura* and *D. melanogaster* — and in some not mentioned previously in this chapter — *D. persimilis*, *D. gaucha*, and *D. robusta* — male mating speed or male virility is usually an important, perhaps the most important, component of fitness. There are good data from a variety of sources to the effect that:

- Male mating speed is subject to selection for the rapidity of copulation
- Within a given species, rapid matings tend to be controlled by the genotype of the male involved, while the genotype of the female may assume importance in slower matings
- Mating speed is associated with fertility and with number of progeny
- Where studied in relation to other components of fitness encompassing the entire life cycle, mating speed is a most important component in drosophilids

8.3 Quantitative traits involving movement

Hirsch and Boudreau (1958) studied phototaxis in *D. melanogaster* in an apparatus consisting of a Y-maze (part of Fig. 7C), one arm of which was exposed to light during tests. Quite rapid responses to selection were found for both positive and negative phototaxis, which is reasonable as a heritability in the broad sense was estimated in excess of 0.5. Phototaxis has been studied with mazes having a design similar to those used for geotaxis, and responses to selection have been found in *D. melanogaster* (Hadler, 1964) and in *D. pseudoobscura* (Dobzhansky and Spassky, 1969). In both cases positive and negative selection lines were rapidly established. Upon relaxation of selection in *D. pseudoobscura*, convergence occurred almost as rapidly as the divergence under selection, showing that the photocentrality of natural populations is a trait subject to genetic homeostasis, the capacity for self-regulation and adjustment (Lerner, 1968). In fact, Dobzhansky and Spassky (1962) made a similar observation about geotaxis, and Spassky and Dobzhansky (1967) found that geographical strains of *D. pseudoobscura* and *D. persimilis* respond differently in tests of phototaxis. Later, Rockwell and Seiger (1973b), using laboratory populations, demonstrated the existence of a large amount of variation within populations of each species. A store of variability is apparently available in natural populations for phototaxis, as is the case for almost every quantitative trait — morphological, physiological, and behavioral (Parsons, 1974). For just one instance, isofemale strains, each derived from a single gravid *D. melanogaster* female caught in the state of Victoria, Australia, were individually found to be genetically different for three quantitative traits, two of them behavioral: scutellar chaetae (bristles) number, mating speed, and duration of copulation (Parsons, Hosgood, and Lee, 1967). In *D. melanogaster*, Médioni (1962) found variations between wild strains collected at different localities in the northern hemisphere: greater positive phototaxis was shown in flies of more northern origin. These variations are presumably under genetic control, but their adaptive significance is not known.

Phototaxis, as a behavioral trait, also illustrates a rather subtle environmental problem, in that under the usual conditions of handling in the laboratory, *D. pseudoobscura* shows positive phototaxis. Pittendrigh (1958), however, found the flies to be negatively phototactic in contrast to general observations. Lewontin (1959) carried out a series of experiments showing that *D. pseudoobscura* is negatively phototactic under conditions of low excitement, but if flies are forced to walk rapidly, or fly, they lose their negative phototaxis and become strongly attracted to light. So phototaxis is confounded with escape reactions. Hadler (1964) did indeed list numerous environmental variables affecting phototaxis in

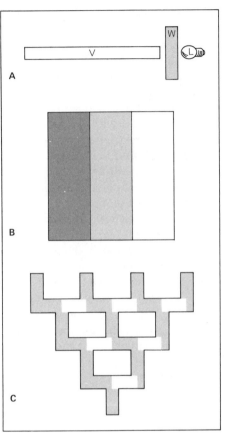

7. Experimental designs for phototaxis analysis in Drosophila. **A.** Measurement of rate at which flies approach light source. *V,* vessel containing flies; *W,* water-filled heat absorber; *L,* light source. **B.** Measurement of distribution of flies in lighted area. Shaded areas correspond to areas of different intensities of overhead light. **C.** Measurement of movement of flies in partially illuminated maze. Unshaded areas represent possible light choices. Lateral movement across several intersections is prevented by one-way cone inserted in each arm. Illumination comes from above. (**A,** after Carpenter, 1905; **B,** after Rockwell and Seiger, 1973b; **C,** after Spassky and Dobzhansky, 1967.)

addition to those just mentioned — temperature, hour of day when tested, time since anesthesia, rearing conditons, time since feeding, energy and wavelength of light, state of dark adaptation, number of observations or trials per individual, age, and sex. A phototactic response is, therefore, a product of particular stimulus-environment variables as well as genotypes. It is quite clear that any form of accurate genetic analysis must be based on precisely defined environments, as has been repeatedly stressed.

Another complication is the method used in studying phototaxis. Three different experimental designs (Fig. 7) have been employed (Hadler, 1964; Rockwell and Seiger, 1973a): (1) measurement of the rate at which flies approach a light source at the far end of a tube (e.g., Carpenter, 1905; Scott, 1943); (2) recording, after a specific period, of the distribution of flies in a field with a directed or undirected light source (e.g., Wolken, Mellom, and Contis, 1957; Koch, 1967), and (3) analysis of movement in mazes (Rockwell and Seiger, 1973b). Hadler considers that one of the major difficulties in comparing studies of phototaxis originating from different laboratories results from interlaboratory differences in experimental technique. They may be measuring slightly different behaviors; for one example, the first method confounds phototaxis with photokinesis.

Rockwell and Seiger (1973a) concur that the measurement of phototaxis is operationally defined. They further discuss how the various general designs differentially accentuate components of the chain of events composing the totality measured as phototaxis, and indicate that the designs thus differ in their utility for research directed at diverse aspects of the response. They warn, for those interested in the potential adaptive significance and evolution of the behavior, that the operational nature of the measurement must be taken into account in any generalizations, since there is no certainty that the response measured in the laboratory is necessarily the same as that occurring in nature. This, of course, is a crucial problem relevant to every laboratory-analyzed behavior discussed in this book.

While it is clear that all environmental factors influencing either the sign or intensity of the response must be carefully controlled and taken into account in any comparisons, Rockwell and Seiger (1973a) consider that the flexibility of the response relative to environmental changes may be an important adaptive qualification of the response allowing increased survival in a heterogeneous environment. For this reason they, among others, propose that in studies concerning the adaptive significance and evolution of phototaxis and of other behaviors, responses relative to several values of various environmental parameters found in the behaving animal's habitat are as important as the response to a certain set value of *each* parameter (Perttunen, 1963). For example, Parsons (1973, 1974) maintains that the only valid way of assessing the importance of mating behavior in natural populations is to study it in all the environments to which the population under surveillance is likely to be exposed — unfortunately, an impossible charge.

Optomotor response in *D. melanogaster* is another trait for which evidence of genetic determination has come from selection experiments (Siegel, 1967). The optomotor response to an illuminated moving striped plate was measured, and each fly was given 10 opportunities to respond. Scores ranged from 0 (no response) to 10. Selective breeding procedures based on low, middle, and high scores were instituted. This led to the development of three strains differing with respect to optomotor response.

Becker (1970) has made a start on the genetics of chemotaxis of *D. melanogaster* with a Y-unit maze of the type used for geotaxis and phototaxis. Selection over 12 generations yielded two lines insensitive to insect repellents, and the appropriate crosses indicated that the genes responsible for insensitivity were at least in part dominant. Compared with geotaxis and phototaxis, chemotaxis seems to have the likely advantage of allowing researchers to relate the stimulating molecule to the specificity of the receptor, and further attempts to select genetic variants for chemotaxis are awaited with interest. This point applies not only to Drosophila but also to bacteria, protozoa, nematodes, and a variety of organisms, some of which are discussed in Chapter 10.

A number of other traits appear amenable to genetic analyses, especially with selection techniques, but have not yet been fully exploited. One such trait is preening or cleaning behavior, which has been described in terms of a number of separate behavioral elements (Connolly, 1968). The various preening movements serve to keep the insect clean and to free the sensory surfaces from contaminants. The presence of other flies increased the amount of preening behavior, even if no additional physical contact between these flies was permitted. Further descriptions of this type of behavior are given by Szebenyi (1969), whose approach is analogous to Bastock's (1956) detailed study of mating behavior of flies bearing the *yellow* mutant and of wild-type flies (Section 3.2). Szebenyi divided preening or cleaning behavior into a series of component behaviors, then judged preening to be a good trait for behavior-genetics analysis. Hay (1972) agrees and has indicated this to be so both for preening and for more general activity, using sophisticated biometrical methods.

8.4 Frequency-dependent mating

Random mating is discussed in Chapter 2 in relation to the Hardy-Weinberg equilibrium, and tests for the randomness of mating are treated there. In this chapter we consider frequency-dependent mating, which is known best but not exclusively in Drosophila. Frequency-dependent mating occurs when the proportion of matings among different genotypes depends on the proportion (frequency) of individual genotypes represented in the mating population. In particular it has been found that the genotype that is rare in a mating situation tends to be favored at the expense of the common type. In a series of multiple-choice mating experiments with *D. melanogaster*, Petit (1958) showed that the genotype of females is unimportant relative to that of the males; hence, this is sometimes known as "rare male mating behavior." Analyzing the sex-linked dominant gene *Bar* (producing a narrower eye than that of the nonmutant wild type), she found that *Bar* males were disadvantaged relative to wild-type males in competition for mates and that this disadvantage was most pronounced when *Bar* males were commoner than wild type. Analyzing the sex-linked recessive gene *white* (altering the normal red eye color to white), she found that *white* males were at a disadvantage to wild-type males when they represented 40 to 80 percent of the males, but were at an advantage outside these limits.

A number of recent reports have suggested that frequency-dependent mating occurs quite often (see Petit and Ehrman, 1969, for a review). The genotype present in a minority mates more frequently than would be expected if matings took place at random. Spiess (Appendix to Ehrman, 1966) used *D. persimilis* homozygous for two autosomal inversions — Klamath (KL) and Whitney (WT). By varying the ratio of WT to KL males, he showed that a clear advantage was awarded to the minority homokaryotype. Ehrman et al. (1965) and Ehrman (1968; review in 1972) observed *D. pseudoobscura*, a sibling species of *D. persimilis*, using Elens-Wattiaux mating chambers (Chapter 3, Fig. 3). They showed that frequency-dependent mating took place between strains with different karyotypes, different geographical origins, mutants versus wild type, strains selected for positive and negative geotaxis, and even flies of the same genotype raised at different temperatures. Frequency-dependent sexual selection has thus far been reported in seven *Drosophila* species: *D. melanogaster, D. pseudoobscura, D. persimilis, D. willistoni, D. tropicalis, D. equinoxialis,* and *D. funebris* (Spiess, 1968; Spiess and Spiess, 1969; Petit and Ehrman, 1969; Borisov, 1970). It has also been reported in the flour beetle, *Tribolium*, by Sinnock (1969, 1970) and to a lesser degree in *Mormoniella*, a wasp (Grant, Snyder, and Glessner, 1974). Also see Neel and Schull (1968) and Mayr (1970).

The generality of the rare male mating advantage in Drosophila can hardly be doubted, so studies have recently been made of possible mechanisms by which it is mediated. By some means the female must recognize each male genotype and its frequency relative to other male genotypes present. Ehrman (1969) has explored the use of olfactory stimulation as the basis of this discrimination. Using specially constructed cages (Chapter 3, Fig. 3), she was able to show that the minority advantage disappears in the front part of the cage if a current of air is introduced into it that has first passed through the rear section containing courting and copulating couples of the "rare" type. Since direct tactile cues are excluded, it appears that the commonness of the olfactory stimulus derived from the rare males overrides and obscures their actual rarity. Recent work by Shorey and Bartell (1970) is suggestive in this context. They found that a volatile sex pheromone produced by *D. melanogaster* females stimulates the males to courtship. They concluded that the pheromone increases the probability that the male will initiate courtship with a nearby female. Male courtship behavior is also stimulated by odors released by other males; however, the male odor appears to be less than one tenth as effective as that of the female. (Individual odor differences and their social function in insects have been profitably considered by Barrows, Bell, and Michener, 1975.)

Recent work has shown that the olfactory cues involved in the rare male behavior between *D. pseudoobscura* Arrowhead and Chiricahua strains (differing in autosomal inversions) are susceptible to chemical purification and analysis. Fractions obtained by extracting whole flies in organic solvents can produce nonrandom mating even when the two types of males are equally abundant in the mating population. Interestingly, different solvents must be used for the two strains (Ehrman, 1972); the Arrowhead acetone extract and the Chiricahua petroleum ether extract have each been fractioned by chromatographic techniques to yield highly active substances (Leonard, Ehrman, and Pruzan, 1974; see Leonard, Ehrman, and Schorsch, 1974, for description of the bioassay).

From the viewpoint of population genetics, the rare male advantage should lead to an enhancement of the frequency of the initially rare genotype in the gene pool, provided there are no other selective forces acting against it. As the rare type becomes commoner, its advantage diminishes, leading to equilibrium. Table 2 shows an experimental demonstration of this. It seems likely that there are a number of gene and chromosome polymorphisms in Drosophila that are maintained by such frequency-dependent equilibria. These are polymorphisms for which minimal fitness differentials between the component genotypes are expected at equilibrium, leading to a different sort of selection from that in the heterozygote advantage model (Dobzhansky, 1970) discussed in Section 4.2. Thus frequency-dependence may represent a way of maintaining a high level of genetic variability without obviously associated fitness differentials. This may be of considerable evolutionary significance, since it has been argued that there is a limit to the number of polymorphisms a population can maintain under the classic heterozygote fitness advantage model (Sved, Reed, and Bodmer, 1967).

In the frequency-dependence studies so far cited, direct observation techniques for determining the number and nature of matings between flies confined to a small space have been used. Ehrman (1970a) carried out experiments in which *D. pseudoobscura* individuals were allowed to mate in mass cultures and the proportions of the two types in each generation were determined by their mating success in the previous generation. From an initial 80:20 ratio, the proportion of the two types converged to approximate equality because of the advantage of the rare type in securing mates (Table 2). Even more significant are experiments carried out (Ehrman, 1970b) in a room with a volume of approximately 75 m³. Two *D. pseudoobscura* strains were used: the wild type and flies homozygous for the recessive orange-eyed mutation, *or*. (This vigorous and easily identified mutation is found in nature in a heterozygous condition.) An 80:20 ratio and about 2000 flies were used in each of the two experiments, one in which *or* was rare and one in which the wild type was less abundant. In both cases an advantage was demonstrated for the rare type, although it varied in magnitude depending upon which genotype was rare. This is the closest approximation to a natural population yet studied (but see Borisov, 1970, who observed *D. funebris* in nature utilizing chromosomal inversions as marker genes), and it suggests that the phenomenon, if widespread, may serve an important evolutionary role.

TABLE 2. Distribution of Matings in Mass Cultures of *D. pseudoobscura* Where *or* and *pr* Females Had a Choice of *or* and *pr* Males

Genera-tion	Pairs		Matings				Matings observed with rare ♂♂	Matings expected with rare ♂♂
	or	*pr*	*or* ♀ × *or* ♂	*or* ♀ × *pr* ♂	*pr* ♀ × *or* ♂	*pr* ♀ × *pr* ♂		
1	80	20	22	17	9	4	21	10.4
2	60	40	25	8	14	10	18	22.8
3	68	32	23	17	6	6	23	16.6
4	56	44	9	12	6	21	33	21.1
5	31	69	7	5	17	9	14	26.2
6	63	37	34	17	8	9	26	25.2
7	62	38	30	21	16	10	31	29.3
8	60	40	19	17	10	12	29	23.2
9	50	50	7	18	17	10	28	26.0
10	46	54	20	24	33	22	46	53.5
1	20	80	8	7	12	43	20	14.0
2	29	71	12	16	12	24	24	18.6
3	38	62	9	17	16	29	25	27.0
4	35	65	9	12	17	25	26	22.1
5	41	59	12	7	8	13	20	16.4
6	50	50	13	10	18	19	31	30.0
7	52	48	21	25	21	17	42	43.6
8	50	50	18	25	19	22	37	42.0
9	44	56	18	22	18	19	36	33.9
10	47	53	10	8	5	11	15	16.0

Orange *(or)* and purple *(pr)* are two unlinked autosomal recessive mutations altering eye color.

$\frac{or\ pr^+}{or\ pr}$ or $\frac{or\ pr^+}{or\ pr^+}$ have orange eyes; $\frac{or^+\ pr}{or\ pr}$ or $\frac{or^+\ pr}{or^+\ pr}$ have purple eyes; $\frac{or^+\ pr^+}{or\ pr}$ or

$\frac{or^+\ pr^+}{or^+\ pr^+}$ or $\frac{or^+\ pr^+}{or\ pr^+}$ or $\frac{or^+\ pr^+}{or^+\ pr}$ have wild type red eyes; and $\frac{or\ pr}{or\ pr}$ have white eyes.

From Ehrman (1970a).

TABLE 3.   Number of Matings of Female *D. pseudoobscura* (AR Homokaryotype) with Various Kinds of Males

Ratio AR♂♂: or♂♂	Measures		4-day virgins	11-day virgins	Mating experience AR♂♂	Exposure to AR♂♂	Exposure to AR♀♀ + AR♂♂	Mating experience or♂♂	Exposure to or♂♂	Exposure to AR♀♀ + or♂♂
						Condition of females				
1:1	Replicates		3	4	10	3	4	15	4	4
	Percent ♀♀ mating		90	70	28	91	77	25	63	77
	No. of matings with AR♂♂	Observed	37	45	40	37	35	18	23	35
		Expected	27.0	28.0	28.5	27.5	27.5	33.0	25.0	31.0
	No. of matings with or♂♂	Observed	17	11	17	18	20	48	27	27
		Expected	27.0	28.0	28.5	27.5	27.5	33.0	25.0	31.0
	χ_1^2		7.41[a]	20.64[a]	9.28[a]	6.56[a]	4.09[a]	13.64[b]	0.32	1.03
4:1	Replicates		3	4	7	4	4	8	4	4
	Percent ♀♀ mating		90	68	33	73	73	35	71	68
	No. of matings with AR♂♂	Observed	35	41	39	46	46	34	49	43
		Expected	43.2	44.0	36.0	46.4	46.4	45.6	45.6	43.2
	No. of matings with or♂♂	Observed	19	14	6	12	12	23	8	11
		Expected	10.8	11.0	9.0	11.6	11.6	11.4	11.4	10.8
	χ_1^2		7.78[b]	1.02	1.25	0.02	0.02	14.75[b]	1.27	0.00

1:4								
Replicates	3	6	7	4	4	9	4	4
Percent ♀♀ mating	79	57	28	64	68	36	68	65
No. of matings with AR♂♂								
Observed	20	28	22	16	19	18	19	17
Expected	9.0	13.6	8.0	10.2	11.0	13.0	11.0	10.4
No. of matings with *No. or* ♂♂								
Observed	25	40	18	35 6	47	36	35	
Expected	36.0	54.4	32	40.8	44.0	52.0	44.0	41.6
χ^2_1	16.81	19.06[a]	30.63[a]	4.12[a]	7.27[a]	2.40	7.27[a]	6.28[a]

Females varied according to age and previous experiences; males varied according to genotype, karyotype, and proportions present.
AR, Arrowhead; *or*, orange-eyed homozygous orange (autosomal recessive) genotype, ST karyotype.
$\chi^2 = 3.84, P < 0.05; \chi^2 = 6.65, P < 0.01.$
Expected, expected number of matings given panmixia; *Observed*, observed number of matings. Total number of observed matings, 1324.
[a] Significant in favor of AR males.
[b] Significant in favor of *or* or males.

TABLE 4. Number of Matings of Female *D. pseudoobscura* (CH Homokaryotype) with Various Kinds of Males

Ratio CH♂♂: or♂♂	Measures	Condition of females							
		4-day virgins	11-day virgins	Mating experience CH♂♂	Exposure to CH♂♂	Exposure to CH♀♀ + CH♂♂	Mating experience or♂♂	Exposure to or♂♂	Exposure to CH♀♀ + or♂♂
1:1	Replicates	3	5	9	4	4	15	4	4
	Percent ♀♀ mating	85	60	28	63	72	24	68	65
	No. of matings with CH♂♂								
	Observed	36	47	38	35	39	27	28	25
	Expected	25.5	30.0	25.5	25.0	29.0	36.5	27	26
	No. of matings with or♂♂								
	Observed	15	13	13	15	19	46	26	27
	Expected	25.5	30.0	25.5	25.0	29.0	36.5	27	26
	χ_1^2	8.65[a]	19.27[a]	12.25[a]	8.00[a]	6.90[a]	4.95[b]	0.07	0.08
4:1	Replicates	4	8	11	4	4	15	4	4
	Percent ♀♀ mating	86	53	25	72	75	23	63	63
	No. of matings with CH♂♂								
	Observed	48	77	48	45	44	45	42	40
	Expected	55.2	68.0	43.2	46.4	48.0	55.2	40	40.8
	No. of matings with or♂♂								
	Observed	21	8	6	13	16	24	8	11
	Expected	13.8	17.0	10.8	11.6	12.0	13.8	10	10.2
	χ_1^2	4.70[b]	5.96[a]	2.67	0.21	1.66	9.42[b]	0.50	0.08

1:4

Replicates	3	6	15	4	5	17	4	4
Percent ♀♀ mating	90	49	17	74	61	16	84	80
No. of matings with CH♂♂								
Observed	19	19	18	13	18	14	12	15
Expected	10.8	11.6	10.4	11.8	12.2	11.2	13.4	12.8
No. of matings with or♂♂								
Observed	35	39	34	46	43	42	55	49
Expected	43.2	46.4	41.6	47.2	48.8	44.8	53.6	51.2
χ_1^2	7.78[a]	5.90[a]	6.94[a]	6.94[a]	3.45	0.88	0.18	0.47

Females varied according to age and previous experiences; males varied according to genotype, karyotype, and proportions present.
CH, Chiricahua; or, orange-eyed homozygous orange (autosomal recessive) genotype, ST karyotype.
$\chi_1^2 = 3.84$, $P < 0.05$; $\chi_1^2 = 6.65$, $P < 0.01$.
Expected, expected number of matings given panmixia; *Observed*, observed number of matings. Total number of matings, 1416.
[a] Significant in favor of CH males.
[b] Significant in favor of or males.

Recently the phenomenon of frequency-dependent sexual selection has been studied from a psychological viewpoint by Pruzan. She was interested in the effect on females of various experiences such as exposure to copulating Drosophila couples, exposure to males only, or actual mating experience as well as the effect of age on subsequent choice of mates (Pruzan and Ehrman, 1974; Pruzan, 1975). Direct observation tests were conducted employing D. *pseudoobscura* females of the previously described Arrowhead (AR) and Chiricahua (CH) homokaryotypes. Four-day-old virgins conferred a rare male advantage on all tested minority males — *or*, AR, and CH — confirming previously published results. A micromesh divider, which allows the passage of light, airborne stimuli such as odor and vibration, and licking but prevents copulation, exposed test females to either copulating couples or males only. Results of these experiences were not clear; in some cases mere exposure shifted the mating disadvantage of particular males employed in a female-choice test toward random mating. Exciting and consistently replicable results were obtained when females were inseminated by males of one karyotype initially, allowed to deposit their eggs, and then retested for male preference. These experienced females awarded a rare male advantage only when the rare male was of the same karyotype as the first mate; otherwise mating was random (Tables 3 and 4). Such females, then, were showing a change in behavior as a result of previous experience; using a basic definition, one can make a case for demonstration of learning (Le Francois, 1972). (For elegant studies demonstrating learned behavior in D. *melanogaster*, the reader is referred to a 1974 study by Quinn, Harris, and Benzer and recently Hay, 1975.) Equivalently aged AR virgins (11 days old) mated significantly more frequently than expected with minority males when those males were of their same karyotype; otherwise, matings were nearly random. Thus, frequency-dependent mating itself is dependent on age and experience, leading to a complex interaction of these variables.

The demonstrated existence of genetic variability for behavioral and nonbehavioral traits is the raw material of research efforts, especially those aimed at dissection of both the chain of events composing the total response and the genetic bases of such responses. Dobzhansky's *Genetics of Natural Populations*, a lengthy series of articles pertinent to the point of this paragraph, should be consulted about these matters (see bibliography in Levene, Ehrman, and Richmond, 1970).

Benzer (1971) remarked that "experience thus far with the fly as a model system for unraveling the path from gene to behavior [and for the reverse, the genetic consequences of behavior] is encouraging. In any case, it is fun."

The Genetics of Behavior: Rodents

In this chapter we consider the behavior of rodents, with reference particularly to mice, and to a lesser extent to rats and guinea pigs. The aspects of rodent behavior discussed in previous chapters can be summarized as follows:

- The behavior-affecting mutant gene *fidget* is considered early in Chapter 2 in connection with the basic principles of Mendelian genetics, and in Section 2.4 comments are made on the behavioral effects of the mutant gene *yellow* in mice.

- In Section 3.3 mating success as determined by single genes is discussed as an example of sexual selection.

- In Chapter 4, where sex chromosome aberrations are considered, mention is made of the need to study chromosomally aberrant mice in greater detail.

- Chapter 5 begins detailed considerations of many genes and behavior. Commonly studied rodent behaviors under such genetic control are listed in Section 5.1, and in Section 5.5 one of the classic behavior-genetics selection experiments concerning defecation in rats is described.

- In Chapter 6 the genetic analysis of quantitative traits is reviewed in experimental animals, expecially rodents. An example of the effect of early experience on the time required to reach food is presented as illustrative of the complex genotype-environment interactions often occurring in mammals (Section 6.1). Defecation, ambulation, and escape-avoidance training in rats illustrate methods for estimating components of variance and the heritability of traits (Section 6.5). Also discussed are maternal effects (Section 6.6), repeatability (Section 6.10), selection experiments (Section 6.11), and correlated responses to selection (Section 6.12), all illustrated by rodent examples.

Tallying all examples, we note that much information on the genetics of behavior in rodents has been presented prior to this chapter. The discussion to follow derives from our selection of certain examples from which conclusions supplemental to those already presented emerge.

9.1   Mice: Single-gene effects

It is clear that a diversity of single genes affects behavior in rodents. Many genes with obvious behavioral effects have been mapped in mice. In addition, a number of studies have shown that genes identified via morphology, usually coat color, display subtle behavioral effects (Thiessen, Owen, and Whitsett, 1970). The work on mice exemplifies the point made in Section 8.1 in relation to Drosophila: no genes affect behavior per se. Rather, the behavioral changes result from genetic effects on enzyme level, hormone level, tissue sensitivity, membrane permeability, and other functions. The influence of heredity on behavior is indirect.

Over 300 mutant genes occupying over 250 loci are listed for the mouse (Green, 1966). By mid-1965 92 mutant genes associated with neurological defects were known in the mouse (Sidman, Appel, and Fuller, 1965), affecting almost every conceivable biological function. They were classified into defects of regional development, structural defects of individual cells that fail to make some special product, and functional defects requiring biochemical study. Most of the known mutant disorders affect the nervous system during its development, and several mutant genes are known specifically to affect the cerebellum. A number of mutants affect the inner ear and related structures, leading to a primary disorder in the early embryonic central nervous system and interfering with the subsequent induction of peripheral structures. Some representative mutant genes that affect the nervous system are listed in Table 1; they range from those that produce severe defects of the central nervous system to those responsible for more minor disturbances. Behavioral changes are associated with various morphological, physiological, or biochemical effects, just as in Drosophila (Chapter 8). Detailed studies on various mutants associated with behavioral effects are summarized in Wilcock (1969). Included are the mutant genes *short ear, hairless, furless, pintail, looptail, tailless, waltzer, wobbler lethal, jerker, quaking, ducky, twirler, dystrophic, reeler,* and *jumpy.* For example, jerkers, waltzing mice, and quaking mice are unable to walk normally and cannot balance well. Wobblers develop more slowly than normals in open-field activity level, ability to rear up on the hindlegs, and ability to climb an incline. Looptail mice perform fewer face-cleaning acts than normals, and on histological examination are found to have enlarged ventricles and deformation of extrapyramidal motor systems in the forebrain.

Assessment of the work on nervous-system-affecting mutants is difficult because the effects of these mutant genes on behavior are gross. To date these studies have contributed little information about the relation of genes to normal behavior, since the behavior associated with the mutant is so abnormal as to be outside the usual variation found in behavioral phenotypes. However, as Wilcock (1969) has suggested, developmental studies of such mutants could have important clinical implications.

TABLE 1. Mutant Genes Affecting Nervous System, and Consequently Behavior, in the Mouse

Gene name	Gene symbol	Linkage group	Biological* phenotype	Behavioral phenotype
Absent corpus callosum	ac		Absence of all or part of corpus callosum	None now known
Cerebral degeneration	cb		Degeneration of cerebral hemispheres and olfactory lobes	Diffuse; progressive overall degeneration of behavior
Dancer	Dc		Absence of macula of utriculus	Circular; failure to swim
Deafness	dn		Degeneration of Deiters' cell in organ of Corti	Accentuation of sniffling
Dilute lethal	d	II	Myelin degeneration; lowered phenylalanine hydroxylase activity	Convulsions
Eyeless	ey		Absence of eye and optic tract	Visual incapacity
Muted	mu		Absence of otoliths in one or both ears	Postural defects
Pirouette	pi	XVII	Degeneration of organ of Corti	Circular
Wobbler lethal	wl	III	Myelin degeneration; elevated succinic dehydrogenase levels	Locomotor difficulties

* By *biological* we mean morphological, physiological, or biochemical effects produced directly by the mutant gene and in turn responsible for the abnormal behavior.
  After Thiessen, Owen, and Whitsett (1970).

Thiessen, Owen, and Whitsett (1970) have pointed out that only a few single-gene substitutions with more "normal" functions have been studied in terms of their behavioral effects (Table 2). These involve mainly alterations in coat color. Failure to observe behavioral effects due to single genes has been reported, but this does not constitute proof that behavioral effects are lacking, since in no case can the complete spectrum of possible behaviors exhibited by an organism ever have been tested. Consider, for example, the *albino* allele. The double recessive *cc* at the *C* locus on linkage group I blocks the enzymatic synthesis of tyrosinase, which is necessary for the conversion of tyrosine to dopa and finally to melanin:

TABLE 2. Influence on Behavior of Single-Gene Substitutions in the Mouse

Gene name	Gene symbol	Linkage group	Biological phenotype	Behavioral phenotype
Albino	c	I	Absence of pigment in fur and eyes	Decreased nibbling; delayed audiogenic seizure; decreased water-escape performance; decreased active avoidance, increased passive avoidance; decreased activity; competitive advantage in sex partners; decreased alcohol preference; decreased water-escape performance and open-field activity; poor black-white discrimination
Brown	b	VIII	Brown fur instead of black pigment	Increased grooming
Dilute	d	II	Blue-gray fur	Decreased activity
Misty	m	VIII	Dilute coat color, tail and belly spots	Decreased nibbling
Pink eye dilution	p	I	Pink eyes, reduced or brown pigment	Decreased staring at examiner, less paw lifting, more grooming and shaking
Pintail	Pt	VIII	Short tail	Fast avoidance extinction
Short ear	se	II	Reduced cartilaginous skeleton	Impaired avoidance learning to sound
Yellow	A^y	V	Yellow or orange fur and black eyes	Decreased short and long-term activity; failure of male to synchronize mating in groups of females

After Thiessen, Owen, and Whitsett (1970).

tyrosine
$\downarrow$ (tyrosinase)
dopa
$\downarrow$ (dopa-oxidase)
dopa-quinone
$\downarrow$
melanin

Extensive behavioral changes are noted in *cc* mice, and the problem is to elucidate the pathway from gene to behavior. There is some indication (DeFries, Hegmann, and Weir, 1966) that photophobic reactions related to a loss of eye pigment mediate the extreme hesitancy exhibited by albino mice.

To assess the generality of behavioral effects associated with traits lacking obvious behavioral effects, a number of coat-color mutants were investigated (Thiessen, Owen, and Whitsett, 1970). The pigment in rodent fur is melanin of two kinds, pheomelanin and eumelanin; the former is always yellow; the latter may be brown or black. Therefore, the wide variety of coat colors in mice is the result of genetic influences on only two kinds of pigments. For reference, the most common inbred strains of mice encountered in behavior-genetics research are listed in Table 3 and the genetic determinants of their coat colors are specified. The primary coat-color loci are *agouti, black, albino,* and *dilute,* which specify the distribution of black, brown, and yellow pigments in rodent hairs; *pink eye* and *piebald* are secondary loci that govern not only the amount of different pigments but also the shape, size, and distribution of individual pigment granules. Generally, at each locus three or more alleles are present. Dominance and epistatic relations are often complex.

TABLE 3. Genetic Determinants of Mouse Coat Color at Common Loci

Strain	Locus					
	Agouti	*Black*	*Albino*	*Dilute*	*Pink eye*	*Piebald*
C57BL	aa	BB	CC	DD	PP	SS
C3H/2	AA	BB	CC	DD	PP	SS
DBA/2	aa	bb	CC	dd	PP	SS
I	aa	bb	CC	dd	pp	ss
BALB/c	AA	bb	cc	DD	PP	SS
A	aa	bb	cc	DD	PP	SS
R III	AA	BB	cc	DD	PP	SS
Linkage group	V	VIII	I	II	I	III

Double recessive phenotypes are aa nonagouti, bb black, cc albino, dd dilute, pp pink eye, ss piebald.

In an F_2 derived from intercrossing F_1s from the inbred strains AKR/J (*aaBBccDD*) and DBA/2J (*aabbCCdd*) (see Table 3), albinos were compared with nonalbinos using a battery of 11 tasks with the apparatus illustrated in Figure 1. The tests are described briefly below to illustrate the type of test battery the behavior geneticist interested in rodents can use. It is important to follow detailed experimental procedures in any test to minimize the degree of subjectivity. Reference to the original publications involved is needed to assess and appreciate this.

The testing equipment illustrated in Figure 1 includes:

- An open field (*A*), used to assess open-field activity, a measure of general exploration, described by the number of lines crossed. This apparatus used as an inclined plane allows an estimate of geotaxis.

- A brightness plane (*B*), used to measure the degree to which an animal prefers a light or a dark environment.

- A tactual plane (*C*), with floor divided into smooth and rough halves, used to measure tactual preference.

- A visual cliff (*D*), used to measure depth perception. The tendency of the animal to go to the shallow or the deep side is assessed after placing the animal on the centerboard.

- An arena (*E*), used to measure general activity as assessed by the number of lines crossed in a 2-minute period.

- An olfactory alley (*F*), in which olfactory sensitivity to a noxious stimulus is assessed by placing ammonia at one end of the runway and water at the other. After placing an animal in the middle of the alley the experimenter records the amount of time spent on the side of the alley containing water during a 5-minute test.

- An activity wheel (*G*), whereby the number of rotations can be counted over long periods to provide measures of long-term activity.

- A water-escape apparatus (*H*), in which learning and escape performances are assessed. It consists of an aquarium filled with water at a temperature of approximately 25°C. A small restraining box with a trap door is positioned above the water at one end, and at the opposite end a wire-mesh ramp extending into the water provides an exit. A trial consists of placing a mouse in the restraining box and orienting it toward the exit ramp; the trap door is opened and the mouse dropped into the water. Latency or tendency to swim to the ramp over five trials is measured.

- A temperature gradient (*I*), ranging from 10°C to 51°C, used to determine mouse temperature preferences.

- An auditory alley (*J*) in which preference for or aversion to an auditory stimulus is measured. At the end of each arm of the apparatus is a speaker into which a 14,000-Hz sine wave is fed by an audiogenerator and an audiostimulator. An animal is placed in the center of the alley and allowed

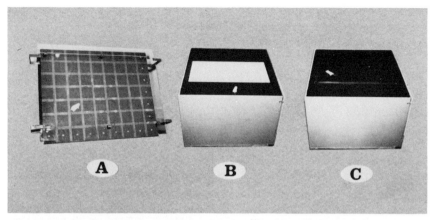

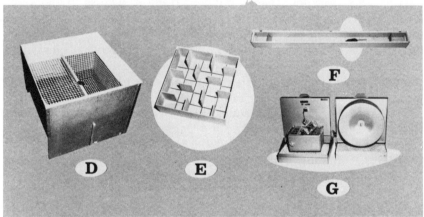

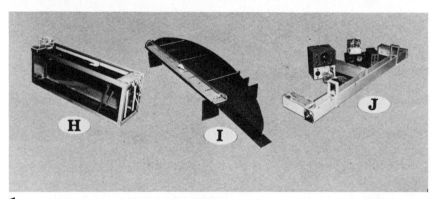

1. **Apparatus for testing mouse behaviors. A.** Open field in horizontal and inclined plane. **B.** Brightness plane. **C.** Tactual plane. **D.** Visual cliff. **E.** Arena. **F.** Olfactory alley. **G.** Activity wheels. **H.** Water-escape setup. **I.** Temperature gradient. **J.** Auditory alley. (From Thiessen, Owen, and Whitsett, 1970.)

to explore for 5 minutes. Then sound is delivered at one end of the alley and the time spent by the animal at either end is recorded. Scores range from primarily positive, indicating a preference for the sound, to primarily negative, indicating aversion to the sound.

Because the *albino* allele was studied in F_2 animals, results may not be entirely free of linkage effects, although Thiessen, Owen and Whitsett (1970) consider such effects relatively unimportant. A considerable array of behaviors was found to differ between albino and nonalbino mice. The *albino* gene lowered the normal sensitivity to change in environmental incline (the inclined plane), reduced activity under white-light conditions (arena, water-escape apparatus, and visual cliff) but not red-light conditions for the open field or when activity was measured primarily at night on an activity wheel. Albinos stayed in a lighted environment or on a rough surface longer than pigmented mice. Even so, both classes tended to avoid light and remain on a rough surface. Albinos responded toward a sound source while nonalbinos moved away, and the albinos tended to avoid the olfactory stimuli more. Photophobia, mentioned above, seems the basis of the light reaction, but explaining the whole complex of behaviors in an integrated way is clearly difficult. The one generalization that can be made is that the albinos hesitate in their reactions to environmental changes; further evidence in agreement with this is presented in Section 9.3.

Fourteen other genotypes involving coat-color variations (but not albinos) were studied in the C57BL/6J strain. Since for practical purposes the only gene allowed to vary was the one of interest, any behavioral effects can be ascribed to that gene. Four tests were used: open-field activity, geotaxis, water-escape behavior, and activity wheel revolutions. The results showed that of the 14 genotypes tested, 71 percent modified some aspect of behavior; in other words, it is not difficult to find effects. Further, the larger the number of behaviors sampled, the more likely it is that a gene substitution affecting behavior will be recognized. Thus 14 percent of genotypes affected open-field behavior; 36 percent open-field behavior and geotaxis; 57 percent open-field behavior, geotaxis, and water-escape behavior; and 71 percent these three behaviors plus activity-wheel revolutions. Clearly, a number of behavioral effects are associated with the ordinary coat-color alleles, suggesting that almost any mutant may have behavioral effects if the battery of tests used is adequately comprehensive.

9.2   Mice: Quantitative traits involving open-field behavior

Compared with the research for major genes for behavioral traits, described in Section 9.1, much more work has been done on quantitative traits, usually without the identification of the specific loci or chromosomes involved. Studies of open-field behavior date from Hall's (1951) and Broadhurst's (1960, 1967a) work in rats. An enclosure is used that provides a strange open-field situation, and in this both a measure of emotionality, as assessed by defecation and urination, and a measure of activity, as assessed by the number of squares crossed in a given time, can be obtained. As shown in Section 5.5, selective breeding affects emotionality, and negative correlations between ambulation and defecation in the open-field have been found (Hall, 1951; Broadhurst, 1967a). Such negative correlations seem fairly general, although the association is affected somewhat by environmental variables such as light and noise (Archer, 1973). The relation also to some extent depends on the species, strain, sex, sample size, and early experience of the subjects.

A number of studies indicate heritable differences in activity for mice. Thompson (1953) found striking differences among 15 strains when individuals were tested in a square arena (30 × 30 inches). The floor was divided into 25 squares, and barriers were placed at the base of every other square. The number of squares traversed by a subject during a 10-minute test was used as the activity score. Of the 15 strains found to differ widely in this behavior, 5 were tested later for activity level in a Y-maze and the arena (Thompson, 1956). For arena activity, the same rank order of the strains was observed, and with one exception the rank order of the strains in the Y-maze corresponded to that in the arena. These parallel results reflect the degree of situational generality in these activity data — an important consideration.

These and many other studies (see Fuller and Thompson, 1960) show clearly that the observed individual differences in level of activity in an unfamiliar situation are a function of heritable differences. The same conclusion applies to defecation. Open-field behavior itself can be objectively and efficiently measured, permitting the large sample size that is essential for detailed genetic analyses. DeFries and Hegmann (1970) carried out a detailed analysis of the differences in open-field behavior for two inbred strains of mice and their derived generations. The two parental inbred strains, BALB/cJ and C57BL/6J, were known to differ widely in open-field behavior. The field used was a square (36 × 36 in.) of white-painted Plexiglas divided into 36 squares, each 6 × 6 in. Testing was carried out on mice 40±5 days old. Activity was measured as the total number of light beams (used to demarcate the squares) that were interrupted during two 3-minute tests. The total number of fecal boluses dropped was also recorded. Data were obtained on the inbred strains P_1 and P_2, their F_1, backcross, F_2, and F_3 generations, and five generations of selection in both directions, i.e., for high and low open-field activity.

Heritabilities were estimated based on the data for the parental, F_1, BC_1 ($P_1 × F_1$), BC_2 ($P_2 × F_1$), and F_2 generations after applying the square root transformation to both activity and defecation scores. These transformations were carried out in an attempt to fulfill scaling criteria, some of which are discussed in Section 6.3. The heritabilities so estimated were appreciable — of the order of 0.4 (Table 4). The heritabilities in the narrow sense were generally a little lower than the degrees of genetic determination, indicating that the additive genetic variance (V_A) was greater than the dominance variance (V_D). In other words, most of the genetic variability was due to the additive genetic variance. Heritabilities were also estimated from the regression of offspring on midparent (Section 6.9) and from half sib correlations (Table 4). Immediately there is a problem, since although heritabilities are generally greater than zero (except for females in half sib correlations), they vary highly according to the breeding procedure. This could argue for the effects of inadequate scale, genotype-environment interactions, or other difficulties in the data. The same problem occurs for defecation. Heritabilities and degrees of genetic determination for defecation are rather lower than for activity, indicating a higher environmental component in determination of this behavior. In terms of the genotypic variance, $V_A > V_D$; this can be seen by

TABLE 4. Heritabilities in the Narrow Sense and Degrees of Genetic Determination of Activity and Defecation Scores of Mice in Open-Field Behavior

A. Heritability from parental, F_1, BC_1, BC_2, and F_2 generations

	Males	Females
Activity	0.58 ± 0.06	0.28 ± 0.04
Defecation	0.42 ± 0.07	0.36 ± 0.06

B. Heritability from regression of offspring on midparent

	Males on midparent	Females on midparent
Activity	0.24 ± 0.12	0.19 ± 0.12
Defecation	0.04 ± 0.09	0.17 ± 0.08

C. Heritability from half-sib correlations

	Males	Females
Activity	0.50 ± 0.32	−0.25 ± 0.31
Defecation	0.30 ± 0.32	−0.29 ± 0.31

D. Degrees of genetic determination from parental, F_1, BC_1, BC_2, and F_2 generations

	Males	Females
Activity	0.63 ± 0.06	0.49 ± 0.06
Defecation	0.39 ± 0.06	0.38 ± 0.06

After DeFries and Hegmann (1970).

comparing the estimates in Sections A and D of Table 4. Heritability for the half-sib correlation in females is negative (but not significantly less than zero), while the remaining heritabilities are positive, some being significantly greater than zero. For both traits we can thus ask which value of heritability forms the best estimate? A possible way of approaching this is to assess the realized heritability (h^2) from a directional selection experiment (Section 6.11).

Such an experiment was done for open-field activity beginning with a foundation population of 40 litters chosen at random from the F_3 generation of the animals used in the preliminary analyses (with the restriction that each litter must contain at least two males and two females). The most active male and female and the least active male and female from each of 10 litters were selected. The resulting 10 high-active males and 10 high-active females were then mated at random at approximately 60 days of age so as to produce progeny representing the first selected generation (S_1) of one high-active line (H_1). Similarly, the 10 low-active males and females were mated at random to produce the S_1 generation of a low-active line (L_1). (The parents of lines H_1 and L_1 were littermates.) In addition, high-active and low-active males and females were selected from each of 10 other litters and mated at random within their level of activity; their offspring represented generation S_1 of lines H_2 and L_2. Two control lines, C_1 and C_2, were set up. (See DeFries and Hegmann, 1970, for further details.)

Over five generations of selection there were clear and consistent responses to selection, from which realized heritabilities can be computed. After five generations of selection, the realized heritability — as measured by the response to selection (R) divided by the selection differential (S or R/S) (Equation 6.25) — was 0.31 ± 0.04 for H_1 versus L_1, and 0.19 ± 0.07 for H_2 versus L_2, with a pooled value of 0.26 ± 0.03. This is in good agreement with the single-generation regression of offspring on midparent (Table 4). Therefore, under the conditions of this study, the regression of offspring on midparent can be argued to be a good predictor, but there is no theoretical reason why this should always be true, in view of the complications and assumptions that have to be accommodated in these various estimates. This difficult-to-interpret example is presented as an instance of the problem of interpretive assessment in behavior-genetics analysis; we chronicle it here precisely because of these difficulties.

Realized additive genetic correlations (r_A) (Section 6.12) computed between activity (the trait selected for over five generations) and defecation and body weight (Table 5) showed the expected negative correlation between activity and defecation. A positive but not significant correlation was found between activity and body weight. In situations where more than one trait is measured, the computation of such correlations is often informative.

One remaining facet of this study merits mention. The above analysis shows that open-field activity is a quantitative trait, presumably under multifactorial additive control. However, a major gene effect on activity has been found, as discussed in Section 9.1, since albino mice have lower activity and higher defecation scores than pigmented animals. The relative importance of this single-gene effect was measured by assessing its contribution to the additive genetic variance and covariance associated with these behaviors. Segregation at the C locus was estimated to account for 12 percent of the additive genetic variance of open-field activity, 26 percent of the additive genetic variance of defecation, and 21 percent of the additive genetic covariance between them. Therefore, even though there is an important major gene effect, a relatively large proportion of the genetic variance of the genotypes tested remains and is due to segregation at an unknown number of unidentified loci.

9.3   Mice: Quantitative traits: the behavioral phenotype

The observation from several sets of data in mice (and in rats) of a negative correlation between activity and emotionality in the open field under a variety of genetic situations (such as comparisons between major genes, different strains, and in lines selected for high and low activity and for high and low defecation scores) indicates the complexity of the total behavioral phenotype. Furthermore, as mentioned in Section 6.11, in Broadhurst's (1960) rat lines selected for high and low defecation scores, correlated responses for many traits, some behavioral and some physiological, were found to agree with what is expected from the dichotomy of emotionality occurring in the reactive and nonreactive lines. Furthermore, Blizard (1971) found that reactive rats had higher heart rates after handling than nonreactive rats. This leads us to ask whether there is generally a "behavioral phenotype" corresponding to a particular genotype. In other words, does a particular genotype lead to a

TABLE 5.   Realized Genetic Correlations from Results of Five Generations of Selection in Mice

Traits	H_1 vs. L_1	H_2 vs. L_2	Pooled estimate
Activity and defecation	-0.88 ± 0.15	-0.80 ± 0.26	-0.86 ± 0.14
Activity and weight	0.33 ± 0.23	0.35 ± 0.44	0.34 ± 0.22

After DeFries and Hegmann (1970).

set of behaviors, as suggested for the *albino* locus? It is difficult to answer this completely, but the evidence favors it as a working hypothesis.

The proposal of a complex behavioral phenotype corresponding to a particular genotype is supported by Parsons (1972a, 1974) for activity (open-field and exploratory), emotionality, and weight in studies on three inbred strains of mice, C57BL, BALB/c, and C3H (Table 6). C57BL, the lightest strain, was the most active, had the highest exploratory activity, and was the least emotional, while BALB/c was the complete reverse, with C3H intermediate but often quite close to C57BL.

By the use of inbred strains and mutant stocks, it can be shown that much of the variation in skeletal morphology among strains is genetic (Grüneberg, 1963), and in fact Grüneberg and others have suggested that many if not most minor skeletal variants are expressions of generalized or localized size variations. For this reason, Howe and Parsons (1967) classified skeletons of mice in the three strains for the presence or absence of 25 minor skeletal variants — 15 affecting the skull, 8 the vertebral column, and 2 the appendicular skeleton. From the percentage of incidences of each variant in the strains, a mean measure of divergence among strains can be obtained, as given in Berry (1963). Data on skeletal divergence show BALB/c > C3H > C57BL (Table 7). Weight differences and the pattern of skeletal divergence are thus associated, and the incidence of many skeletal variants may indeed be associated with body weight, as suggested by Grüneberg (1963). Even though the number of strains examined is limited, these results allow

TABLE 6. Order of Three Inbred Strains of Mice for Morphological, Physiological, and Behavioral Traits

Trait	Order*
Open-field activity	C57 > C3H > BA
Exploratory activity	C57 > C3H > BA
Open-field emotionality	BA > C3H > C57
Weight	BA > C3H > C57
Skeletal divergence	BA > C3H > C57
Percentage of no-shock jumps†	C3H > C57 > BA
Temperature preference‡	C57 > C3H ≫ BA
Body temperature‡	BA ≫ C3H ≈ C57
Abdominal fur density‡	C57 > C3H ≫ BA
Tail length‡	
15–18 days at measurement	BA ≈ C3H > C57
55–58 days at measurement	BA ≈ C3H ≈ C57

* C57 = C57BL; BA = BALB/c.
† See Table 8.
‡ See Table 9.
Data of Howe and Parsons (1967), Rose and Parsons (1970), Silcock and Parsons (1973).

one to argue for an association between genotype, skeletal morphology, weight, and various behavioral parameters. This argument is reasonable since skeletal variants are presumably associated with variants of the muscular, nervous, and vascular systems, and such variants presumably have consequences at the behavioral level. Therefore, even if the argument cannot be generalized fully, it seems worth considering in the study of any quantitative behavioral trait.

In the same three strains of mice, a measure of learning was assessed by testing conditioned avoidance (Rose and Parsons, 1970). The apparatus used consisted of a see-through Perspex box with a grid floor. The floor was divided into two equal parts by a low central barrier. Shock could be applied to either side and to the central barrier. The latter was shocked to prevent the mouse from "sitting on the fence." The mouse was placed into this apparatus for 1 minute; then a light source above the apparatus was switched on. Two seconds later a shock was administered to the mouse's feet through the grid floor, and times were recorded from the commencement of the light signal until the mouse had jumped the central barrier to the safe side. The times recorded for the first jump in the apparatus were used as a measure of "initial reaction to shock." The mouse was then removed from the apparatus, allowed to rest for 1 minute, and returned for another shock test. Altogether the mouse was given 10 shock trials in the following sequence:

- 4 trials, each 1 minute apart
- 1 hour rest
- 3 additional trials, each 1 minute apart
- 24 hours rest
- a final 3 trials, each 1 minute apart

Table 8 shows the percentage of no-shock jumps (jumps to the safe side of the apparatus after the light signal was switched on but before shock was applied). Trials 2 to 10 were used to determine this parameter. Trial 1 was omitted because an initial crossing of the barrier cannot be regarded as a conditioned avoidance response. The highest percentages of no-shock jumps occurred for trials 4, 7, and 10 at the end of each

TABLE 7. Mean Measures of Skeletal Divergence (and Standard Deviations) Among Three Inbred Strains of Mice

	BALB/c	C3H
C57BL	1.326 ± 0.221	1.012 ± 0.275
BALB/c		0.348 ± 0.156

set of trials, and low percentages occurred for the first trial in each series after rest, as expected. The order of superiority of the inbred strains is C3H >> C57BL > BALB/c, which does not correspond to the sequence obtained for activity and emotionality. Therefore, the association between morphology and the behavioral phenotype does not hold. In this case, the link between genes and the behavioral output can be regarded as less direct than for the various simpler forms of behavior discussed earlier, because of the assumed importance of the learned component. The data show some variability according to trial number, since up to trial 5, BALB/c > C57BL but thereafter the situation is reversed (C57BL > BALB/c). This represents a genotype-environment interaction among trials in that C57BL mice take longer for a score to be registered, but even so C57BL mice end up with a higher score in later trials. The experiments were repeated using a similar technique but with a buzzer rather than a light as the signal (Rose and Parsons, unpublished data). Sequences of 10 trials were used and after the first sequence C57BL > BALB/c > C3H, but after the second sequence, C3H and C57BL were very similar but superior to BALB/c. Differences occur according to the technique of assessment.

Of more ecological significance are the remaining traits listed in Table 6. Temperature preference was assessed in a cage with a temperature gradient (Fig. 2) along the floor ranging from 23°C to 43°C over a distance of about 120 cm (Silcock and Parsons, 1973). Mice could be seen actually choosing a preferred temperature. The behavioral process consisted of a mouse lowering its abdomen onto the floor of the cage as it moved about in a certain section, eventually settling in the position that presumably was the most comfortable temperature for it. The C3H and

TABLE 8.   Percentage of No-Shock Jumps in Trials 2 to 10 Recorded for Male Mice of Three Inbred Strains

Trial No.	BALB/c	C3H	C57BL	BALB/c × C3H	BALB/c × C57BL	C3H × C57BL
2						1.3
3	0.7	14.8		3.1	2.2	9.3
4	2.8	22.2	2.0	10.2	9.6	22.7
5	1.4	16.0	1.0	2.0	7.4	12.0
6	0.7	20.0	3.2	19.4	14.8	17.3
7	3.6	20.0	7.4	20.4	29.6	32.0
8	1.6	4.5	6.9	12.7	18.9	12.0
9	4.0	13.6	9.7	11.4	23.6	24.0
10	9.5	31.8	12.5	16.5	37.8	29.3
All trials	2.6	15.8	4.3	10.4	15.8	17.8

From Rose and Parsons (1970).

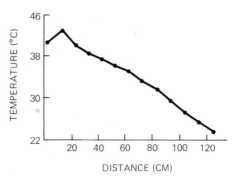

2. Temperature gradient for testing mouse preferences. Temperatures along floor of test cage vary over distance of 120 cm. Mice of different strains were introduced at the midpoint and watched until a preferred temperature was selected, as described in the text. (From Silcock and Parsons, 1973.)

BALB/c mice often slept in these positions. The BALB/c strain preferred the lowest temperature, especially mice between 55 and 58 days old (Table 9). Associated with this was a high body temperature (as measured anally within 30 seconds of death with a quick-read thermometer) and low abdominal fur density (the abdomen was chosen because of its apparent importance in selecting optimal temperature). There is a positive association between fur density and temperature selected by the various strains, associated with a negative relation between these variables and body weight and body temperature. Because the larger BALB/c mice have a low surface area/volume ratio, they would be favored in a cooler environment and so are expected to select it. In a cooler environment the high metabolic rate may be adaptive and this may be indicated in these mice by the higher body temperature. Based on the assumption of these adaptations, high fur density may not be of great importance to them. Conversely, C57BL and C3H mice have a lower body temperature and a higher fur density, and are lighter in weight. In a warm environment, heat is clearly of less significance than in a cold one; hence, under these circumstances little selection for high body temperature is expected. Furthermore, thicker fur may be adaptive in a warm environment, reducing water loss from the skin to a minimum.

Thus, temperature preference is associated with morphological and physiological traits. The behavior patterns observed during the process of choosing a preferred temperature show a direct selection indicating that the behavior is essentially innate. In the wild, therefore, we expect animals to prefer temperatures closely resembling those of their natal homes, as described in various species of *Peromyscus* (Ogilvie and Stinson, 1966).

There are discussions in the literature on the role of the tail as a thermoregulatory organ. In Table 9 data for the inbred strains show no significant differences in tail length among strains. Even though the tail

TABLE 9. Temperature Preference, Mean Weight, Body Temperature, Abdominal Fur Density, and Tail Length of Three Inbred Strains of Mice and Their Hybrids

Strain	Temperature preference (°C)		Mean weight (gm)		Body temperature (°C)	Abdominal fur density hairs/mm²	Tail length (cm)	
	Males	Females	Males	Females			Males	Females
Age: 15 to 18 days								
BALB/c	26.60	26.66	9.1	9.4			5.3	5.0
C3H	37.74	38.06	6.8	7.4			5.0	5.0
C57BL	38.71	38.79	7.1	7.0			4.8	4.9
C57BL × C3H	37.85	37.46	7.2	7.2			4.9	4.8
BALB/c × C3H	35.37	33.96	8.2	7.6			5.4	5.2
BALB/c × C57BL	35.68	37.29	7.7	7.7			5.1	5.1
Age: 55 to 58 days								
BALB/c	25.67	26.30	24.0	20.8	38.03	31	8.3	8.1
C3H	36.78	35.92	21.5	19.0	35.95	64	8.4	8.2
C57BL	34.30	37.47	21.1	18.1	35.55	59	8.0	8.0
C57BL × C3H	30.94	37.95	22.5	19.8	37.05	63	9.1	8.9
BALB/c × C3H	30.00	30.65	22.8	19.3	37.29	64	9.0	9.0
BALB/c × C57BL	33.10	37.25	23.6	20.6	36.98	64	8.9	8.8

After Silcock and Parsons (1973).

is regarded as having a thermoregulatory function (Harrison, Morton, and Weiner, 1959), the data of Silcock and Parsons (1973) and other published data allow the conclusion that tail length is less important in temperature preference selection than body weight. There are, however, tail-length differences in nature such that mice in cool environments often, but not always, have longer tails than mice in warm environments (Berry, 1970).

The study of inbred strains suggests that a given genotype can be thought of as controlling a series of interrelated traits, giving an overall phenotype, since the ranking is the same for all traits except learning and tail length.

The exceptional nature of the learning and tail-length data is reinforced when data from hybrids are considered (Tables 8 and 9). Heterosis occurs for the learning measure between the two hybrid pairs, BALB/c × C57BL and C3H × C57BL, and for all hybrids for tail length. None of the other traits listed in Table 6 shows heterosis. Thus the two exceptional traits based on the ordering of the three inbred strains are exceptional in showing heterosis (and inbreeding depression). It has been argued that such traits are subject partially or wholly to directional selection in the direction of the hybrids (Mather, 1966) and are traits with a fairly direct relation to fitness. Traits not showing heterosis or inbreeding depression have been argued to be relatively peripheral to overall fitness and to be subject to stabilizing selection. Therefore, for the traits subject to stabilizing selection, there is a fairly simple connection from genotype to morphology and physiology and then to behavior, and hence a complex behavioral phenotype associated with a given genotype.

In the wild situation, for a trait important in choice of habitat, stabilizing selection rather than directional selection toward some undefined optimum is expected to be the rule. Such normalizing selection would keep traits within fairly narrow limits, and animals not exhibiting behavior close to the norm of a certain population would be less likely to breed with other members. As already pointed out, such traits are expected to show little or no dominance, and in fact, temperature preference and the associated variables of weight, body temperature, and fur density accord with this.

Considerable other published data compare different strains of mice. Rodgers and McClearn (1962) found when giving mice a choice of alcohol of various concentrations that the ranking of the three strains on the basis of mean daily consumption was C57BL > C3H > BALB/c. However, for learning performance, McClearn (1972) concluded from a literature survey that C3H mice are generally poorer performers than C57BL and BALB/c in various experiments, but they performed relatively well in a water-escape situation and a shuttle box avoidance apparatus. In any case, as already mentioned, according to how learning was assessed, Rose and Parsons (1970, unpublished data) obtained variable results. It is generally difficult to accommodate learning into the behavioral phenotype.

Erlenmeyer-Kimling (1972) surveyed a number of strains in relation to early experience. The C57BL strain responded to all treatments more frequently than other strains, verifying observations made by Ginsburg (1967), Henderson (1968), and others on the particular lability of the C57BL strain to environmental variables. BALB/c showed generally low responsiveness to the treatments considered, and C3H seemed intermediate. The BALB/c result is not surprising, since the mice are albino; the hesitancy of albinos in reactions to environmental changes is discussed in Section 9.1. For strains C57BL and C3H, general background variables such as isolation, environmental enrichment, or cage illumination may be less critical than more specific and possibly traumatic events, such as handling, shock, and noxious noise. The opposite may be true for BALB/c. Therefore, for lability to early experience effects we can probably write C57BL > C3H > BALB/c, and this then can tentatively be regarded as a further component of the behavioral phenotype. However, further work paying particular attention to those traits for which learning is involved seems necessary.

The behavioral measures most likely to reflect the influences of certain treatments also tend to differ among the various strains. In the highly responsive C57BL strain, most behavioral measures are affected by early experience, but defecation scores are not greatly affected. C3H, which showed treatment effects for only about 40 percent of the behavioral measures tested, seems especially unresponsive when the measures involve learning (Erlenmeyer-Kimling, 1972). Similarly, the C3H data in Table 8 show a much smaller relative change between trials 3 and 10 than do the C57BL data.

Staats (1966) and others have presented tables showing the origin of the various inbred strains of mice and the relations among them. As is clear from the papers in Lindzey and Thiessen (1970), there is a vast amount of information on the various strains, much of it involving behavioral traits. Unfortunately, although many strains and behaviors have been studied, comprehensive studies of many behaviors in many strains are rarer. The three strains under discussion provide a reasonable basis for such comparisons, and from them a behavioral phenotype corresponding to a given genotype emerges. This correspondence seems restricted to traits under stabilizing selection, for which the ordering of the three strains is almost invariant. This association was not found for learning or tail length, both of which are probably more subject to directional selection as indicated by the presence of heterosis for these traits. It seems reasonable to postulate that learning may be subject to fairly intense directional selection for high learning ability, as learning no doubt has a fairly direct relation to fitness. Similarly, Collins (1964) has presented conditioned avoidance data showing heterosis. To argue that tail length is under continuous directional selection does not seem realistic. However, it may be a trait subject to directional and stabilizing selection simultaneously, a possibility indicated as likely for metrical traits in certain plant populations (Allard, Jain, and Workman, 1968). This does not exclude the possibility of heterosis for the other types of traits for which it has not been reported in this chapter. For example, Bruell (1964a,b) found heterosis for exploratory behavior and wheel running. Such results could be explained by differing experimental techniques such that the traits have higher learning components than those under discussion, by differing genotypes, or by differing environments.

Additional published work exists for the traits discussed above, as well as for many others (see Parsons, 1967b, and Lindzey and Thiessen, 1970, for references).

9.4   Mice: Male sexual behavior

In this section we turn to a genetic analysis of male sexual behavior in the mouse. McGill (1970) describes the behavior of the males after the females are brought into estrus by hormone injections:

> In a typical series a male first encountering an estrus female carefully investigates her, concentrating his attention on the anogenital regions. If sufficient sexual arousal occurs, the male mounts the female palpating her sides with his forepaws while simultaneously executing a series of rapid, pelvic thrusts. Quite frequently the first attempt at gaining intromission fails and the male dismounts and engages in genital cleaning. When a male is successful in gaining intromission, the rate of pelvic thrusting is greatly reduced while the amplitude is increased. The thrusts of intromission average about one-half second and are easily counted. During intromission, the male keeps one hindfoot on the floor and rests the other on the hindquarters of the female. The number of thrusts in each intromission varies from only a few to 300 or more. After an intromission, both animals generally engage in genital cleaning. This behavioral sequence of mounts, intromissions, and genital cleaning usually continues until the male ejaculates. During the ejaculatory intromission, the speed of pelvic thrusting increases, and, finally, the male quivers strongly while maintaining deep penetration of the female. At this point, he raises the hindfoot which has been resting on the floor and clutches the female with all four limbs. Most frequently, this results in both animals falling to one side. Following the male's ejaculation, both male and female again engage in genital cleaning.

Preliminary data on male sexual behavior using the inbred mouse strains C57BL/6J and DBA/2J and the F_1 are given in Table 10 for the 14 measures defined therein. Considerable differences among inbred strains are shown for the various components of male sexual behavior. The inheritance of sexual behavior is clearly not simple, since the data suggest three different modes of inheritance: (1) dominance of one parental genotype or the other (measures 1 to 4 and 5 to 7), (2) where the F_1 is between the parents (measures 8 to 10), and (3) overdominance or heterosis, where the F_1 is superior to both parents (measures 11 to 13). In other words, for this spectrum of behaviors, all associated with male

TABLE 10. Median Scores and Significance Levels of the Three Possible Comparisons for 14 Measures of Male Sexual Behavior in Three Inbred Strains of Mice

Measure*	Median scores			Significance levels		
	C57BL/6J	DBA/2J	F_1	C57 vs. DBA	C57 vs. F_1	DBA vs. F_1
1	42	85	42	0.02		0.002
2	400	129	546	0.002		0.002
3	17	5	18	0.02		0.02
4	0	20	0	0.02		0.001
5	23	17	19	0.02	0.02	
6	15	20	19	0.02	0.01	
7	2	0.5	0	0.01	0.001	
8	28	137	42	0.002	0.002	0.002
9	2	7	3	0.002	0.001	0.02
10	1	4	2	0.002	0.02	0.002
11	18	16	7		0.02	0.05
12	16	20	25		0.02	
13	107	179	93			0.02
14	1252	1376	1091			

* Definitions of measures
1. Mount latency (number of seconds from introduction of female until male mounts).
2. Total number of thrusts with intromission preceding ejaculation.
3. Number of intromissions preceding ejaculation.
4. Percentage of times male bites female following ejaculation.
5. Ejaculation duration (number of seconds male spends clutching female and maintaining vaginal contact, following ejaculation).
6. Time of intromission (number of seconds from beginning of mount with intromission until male dismounts).
7. Number of head mounts during a series.
8. Interintromission interval (number of seconds from end of one mount with intromission until beginning of next).
9. Time of mount (length of mount without intromission in seconds).
10. Preintromission mount duration (number of seconds from beginning of mount with intromission until male's penis is inserted into female's vagina and first thrust of intromission occurs).
11. Number of mounts without intromission per series.
12. Number of thrusts making up each intromission.
13. Intromission latency (number of seconds from introduction of female until male gains intromission).
14. Ejaculation latency (number of seconds from beginning of first intromission until beginning of ejaculation).
After McGill (1970).

mating, there is a complete spectrum of modes of inheritance. Furthermore, when a different cross (DBA/2J × AKR/J) was carried out, differing results were obtained for many of the traits. Therefore, the modes of inheritance found are specific to the particular strains studied. More generality would be possible if many strains were studied, as can be done for a diallel cross or the simplified triple test cross (Section 6.5).

McGill (1970) has commented on the importance of environment, since the matings described in Table 10 were observed and scored while pairs were housed in plastic cylinders under normal room illumination. Under these circumstances, a C57BL/6J male mated with three females during 10 nights of testing. Using a cage placed in the dark, which would approximate more closely the situation normally experienced by the animals, this figure increased above five. This demonstrates a point we have repeatedly made — the results of any experiment are specific to the environment in which it is carried out, and generalizations on inheritance should be made only for experiments carried out under a wide range of conditions.

A full biometrical analysis was carried out for the two parental strains C57BL(P_1) and DBA(P_2), and the F_1, F_2, $BC_1(P_1 \times F_1)$, and $BC_2(P_2 \times F_2)$ generations from which components of variation can be estimated. It was found from reciprocal crosses that sex linkage or maternal effects could be largely ruled out.

As an example, consider intromission latency (measure 13, Table 10). The mean values in seconds were:

C57BL(P_1)	DBA(P_2)	F_1	F_2	BC_1	BC_2
151.91	171.02	115.40	123.48	127.87	136.03

Some heterosis is evident, as in Table 10. Biometrical analysis showed that a logarithmic transformation provided the best scale. The components of variation then came out to be: $V_E = 0.045$, $V_A = 0.008$, and $V_D = -0.002$, giving $h^2 = 0.154$. Thus in the F_2, about 15 percent of the variance is due to additive genetic causes, and the dominance component is unimportant. McGill concluded that this trait is controlled by additive genes coupled with a large environmental variance. He concluded from repeated testing that the large environmental variance was due primarily to specialized nonlocalized variation occurring from test to test within individual animals, since the repeatability as defined in Section 6.10 was only 0.16 in the raw score data.

For ejaculation latency (measure 14), the mean times in seconds were:

C57BL(P_1)	DBA(P_2)	F_1	F_2	BC_1	BC_2
1368.91	1977.27	1189.82	1204.73	1354.35	1316.94

Again some heterosis is evident, as in Table 10. An attempt to find a suitable scale was unsuccessful; therefore, the analysis was carried out on the raw data. This gave a heritability in the region of 0.15 to 0.25, which is similar to that for intromission latency. Again there was major variation due to within-animal variation from test to test. For both intromission and ejaculation latencies it may be reasonable to attribute much of the variation to differences in the behavior of the females, even though efforts were made to control this.

The population variance (Section 6.10) can be subdivided into additive genetic variance (V_A); dominance variance (V_D); special environmental variance (V_{Es}) due to within-individual variance arising from temporary or localized circumstances (in this case presumably differences between females is a major variable); and general environmental variance (V_{Eg}). Since V_{Es} is 91 percent and 73 percent of V_E for the two traits, respectively, repeat measurements are desirable. It is also important that any environmental source of variation be precisely controlled and minimized for the most accurate interpretations. Possible environmental factors include barometric pressure, recency of feeding or drinking, time of day at test (circadian rhythms), and interaction with other males prior to test — in addition to differences between females. In both the examples discussed, $V_A > V_D$; therefore, if directional selection were carried out, a positive response would be expected. The response would probably be slow because of the large environmental variances.

9.5   Mice: Traits with detectable physiological basis

Audiogenic seizures are a series of psychomotor reactions to intense acoustic stimuli. One complete syndrome (Schlesinger and Griek, 1970) consists of (1) a latency period of variable duration during which the mouse may huddle while appearing to be attending to the stimulus, or appear to ignore the stimulus while washing and grooming excessively; (2) a wild running phase characterized by frenzied running along the boundary of the container; (3) a clonic convulsion during which the animal falls on its side while drawing up its rear legs toward its chin; (4) a tonic seizure during which all four legs are extended caudally; and (5) death due to respiratory failure. Variations of this pattern have been observed. The duration of the latency period ranges considerably. The wild running phase, which clearly differentiates audiogenic seizures from other convulsive patterns, may be accompanied by a change of gait appearing as a series of stiff-legged bounds. It may then terminate without a subsequent clonic seizure. Tonic seizures may or may not be fatal, and death may or may not be prevented by artificial resuscitation.

Even so, the five discrete phases itemized above are characteristic of an audiogenic seizure.

Interest in audiogenic seizures has resulted in the publication of an enormous research literature over the last 40 years. Information of interest to the behavior geneticist is reviewed in Fuller and Thompson (1960) and Schlesinger and Griek (1970). Prior to 1947 most of the literature was in psychological journals, but this emphasis has now disappeared. The other concomitant change as mentioned in Chapter 1 is a shift away from an almost total reliance on the laboratory rat as an experimental animal. Indeed much of the recent work is with mice. Schlesinger and Griek (1970) consider that studies of audiogenic seizures have now become truly interdisciplinary, involving biochemistry, genetics, pharmacology, psychiatry, and psychology.

Coleman (1960) studied *dilute* mice (*dd* and *d'd'*) which have a light coat color provided the background genotype is such that the phenotypic effect can be seen. They have a lower audiogenic seizure threshold than normal mice. Furthermore, *dilute* mice have only 14 to 50 percent as much phenylalanine hydroxylase as nondilute wild-type mice. The particular interest in this enzyme is that it converts the essential amino acid phenylalanine to tyrosine and is the enzyme missing in human phenylketonuria. However, there are some differences between the enzyme defect in *dilute* mice and in humans. First, in *dilute* mice, the enzyme deficiency is only partial and the remaining activity is sufficient to metabolize dietary phenylalanine adequately. No excess phenylalanine can be demonstrated in *dilute* mice maintained on standard laboratory diets. Second, the defect in *dilute* mice does not appear to be due to a failure to produce the enzyme, since enzyme activity is normal in the supernatant fraction of liver homogenates after centrifugation. The implication is that in *dilute* mice an endogenous inhibitor of phenylalanine hydroxylase is associated with the mitochondrial fraction. When fed a diet containing excess phenylalanine, *dilute* mice excrete this amino acid more slowly than nondilute mice, and under normal dietary conditions, they excrete certain abnormal phenylalanine metabolites such as phenylacetic acid. Phenylacetic acid has been shown to inhibit decarboxylating reactions in a number of tissues, and it is possible that *dilute* mice are deficient in certain products of decarboxylating reactions. Specifically, it is possible that these animals are deficient in the neurotransmitter substances GABA (γ-aminobutyric acid), NE (norepinephrine), and 5HT (5-hydroxytryptamine, also called serotonin) in the brain, and that these deficiencies in turn account for the high proportion of seizures in *dilute* mice. As pointed out by Schlesinger and Griek (1970), several assumptions are needed if the above speculations are to form a working model.

It must be assumed that GABA, NE, and 5HT are inhibitory in their action on the central nervous system and that phenylacetic acid in the amount present in *dilute* mice inhibits decarboxylation. Problems remain to be dealt with, but Schlesinger and Griek regard the working model as reasonable. Irrespective of the final situation, there is clearly likely to be a fairly intimate association of genes, biochemical and physiological processes, and behavior. The full demonstration of this association will be scientifically very valuable.

The inbred strain DBA/2J is genetically *dilute*. Predictably DBA mice are sensitive to audiogenic seizures; C57BL/CJ mice are not. F_1 mice are intermediate, although phenotypically closer to the nonsusceptible parent. Age is a major factor in seizure susceptibility; in one survey (Schlesinger and Griek, 1970) DBA mice had a 90 percent risk at 21 days (as assessed by full clonic-tonic seizures) and 13 percent at 14 and 28 days, while C57BL mice were resistant to clonic-tonic seizures at all ages, and F_1 mice had a developmental pattern similar to that of DBA mice. These results parallel those reported by other investigators (e.g., Fuller and Thompson, 1960). The exact age of onset varies in different laboratories, indicating the importance of as yet unspecified environmental factors such as diet, housing conditions, temperature, and/or diurnal rhythms, all of which may interact with genetic factors to give slightly different developmental patterns.

These same two strains were also tested for seizures resulting from the drug Metrazol and for electroconvulsive seizures. In both cases the DBA strain was more susceptible, suggesting that this particular strain is simply more susceptible to seizures regardless of the inducing agent. In agreement with the metabolic hypothesis stated above was the finding of lower endogenous 5HT and NE levels in DBA mice. Furthermore, when 5HT or NE was depleted by drugs or by diet, susceptibility to audiogenic, Metrazol-induced, and electrically induced seizures was increased. Conversely increased levels of 5HT, NE, or GABA protected the animals. The general conclusion is that DBA mice have extremely excitable nervous systems. Therefore, we have evidence for a behavioral phenotype with physiological correlates.

The genetic situation needs further investigation, since Schlesinger and Griek (1970) have argued that the *dilute* locus may be not directly implicated but just closely linked. The evidence comes from Schlesinger, Elston, and Boggan (1966), who obtained single-gene mutations to full coat color in DBA/2J strains. In these mice, genetically *Dd* or *DD* on a DBA background, the dilute locus did not importantly contribute to seizure susceptibility. But Lindzey et al. (1971) consider that no single genetic hypothesis can account for these data, and evidently we must await further work.

As another example, we consider alcohol preference and aversion. A good recent review of this appears in Lindzey et al. (1971) with particular reference to mice, although studies have been carried out on rats and man. Like all the traits discussed, differences have been reported among strains of mice, in that given a series of choices of drinking fluids, some strains (e.g., C57BL/6J) prefer alcohol while others (e.g., DBA/2J) do not. In one series of experiments (Rodgers and McClearn, 1962) four inbred strains of mice were offered simultaneously ad lib choices of water and six alcohol solutions ranging in strength from 2.5 to 15 percent over a 3-week period. Table 11 shows the mean daily consumption of water and the various alcohol solutions, and the proportion of the liquid consumed weekly that was alcohol. This last figure provides a single index representative of the alcohol preference of each strain on a weekly basis. In each week the rate of alcohol consumption for the four strains was in the order C57BL > C3H/2 > BALB/c > A/3. For strains C57BL and C3H/2 the percentage of alcohol consumed increased over a 3-week period, a marked preference for 10 percent alcohol developing by the third week. In strains BALB/c and A/3, however, there was a progressive reduction in alcohol consumption and the development of an increasing preference for water. Thus in the strains tested, the tendency for alcohol preference to increase is positively correlated with initial preference. An analysis of variance of the proportion of liquid consumed that was alcohol revealed a highly significant effect due to different strains (genotypes). Clearly, alcohol preference is under genetic control, but also depends on the environment, in that alcohol preference varies according to the period of previous consumption. Nachman, Larue, and Le Magnen (1971) showed that the removal of the olfactory bulbs eliminated the aversion to alcohol in BALB/c but did not abolish the preference for alcohol in C57BL mice. This, along with the observation that BALB/c mice appear to avoid alcohol immediately, without prior experience, led to the hypothesis that BALB/c mice are more responsive than are C57BL mice to alcohol as a sensory stimulus.

Not a great deal more is known about the genetics of the trait, although there is some evidence that major gene effects may be important in differentiating among strains (Lindzey et al., 1971).

From the physiological point of view, there seems to be some association between differences in the liver enzyme alcohol dehydrogenase (ADH) and alcohol preference. ADH is involved in the first step in the metabolism of ethanol to acetaldehyde and so could be of critical importance. Most of the earlier research concentrated on ADH (Lindzey et al., 1971), but attention is now turning to the enzyme that oxidizes acetaldehyde — aldehyde dehydrogenase (ACDH). In combination, the two enzymes are responsible for the breakdown of ethanol through the citric

TABLE 11. Mean Daily Consumption of Various Alcohol Solutions and Proportion of Absolute Alcohol to Total Liquid Consumed Weekly in Four Inbred Strains of Mice

Strain	Week	Mean daily consumption of various alcohol solutions (ml)								Proportion absolute alcohol/ total liquid
		0.0%	2.5%	5.0%	7.5%	10.0%	12.5%	15.0%	Total	
C57BL	1	1.88	1.83	1.75	1.17	1.81	3.90	2.55	14.89	0.085
	2	1.19	0.78	1.69	3.24	1.38	4.33	2.90	15.51	0.093
	3	1.31	0.72	0.46	0.80	5.22	3.04	3.66	15.21	0.104
C3H/2	1	2.64	2.52	3.00	2.62	2.24	1.80	1.29	14.82	0.065
	2	2.44	2.29	3.96	2.34	2.84	1.36	1.32	16.55	0.066
	3	2.57	1.41	2.71	1.93	5.72	3.11	1.34	18.79	0.075
BALB/c	1	10.03	0.65	0.66	0.58	0.85	0.56	0.82	14.15	0.024
	2	9.76	0.42	0.42	0.39	0.38	0.48	0.42	12.27	0.019
	3	9.30	0.31	0.40	0.43	0.40	0.40	0.47	11.71	0.018
A/3	1	10.20	0.96	0.72	0.52	0.44	0.59	0.46	13.89	0.021
	2	11.16	0.53	0.67	0.41	0.36	0.43	0.40	13.96	0.016
	3	10.85	0.36	0.31	0.41	0.41	0.44	0.39	13.17	0.015

After Rodgers and McClearn (1962).

acid cycle. ACDH may be of considerable importance, since alcohol-drinking and non-drinking strains differ by as much as 300 percent for this enzyme. We await with interest the further linking of the genetic, biochemical, physiological, and behavioral components of alcohol preference.

9.6 Other rodents

Quite an amount of early work was carried out by psychologists on rats, in particular the Norway rat, *Rattus norvegicus*. In fact, as Beach (1950) pointed out (Chapter 1), the trend in American journals specializing in comparative psychology was to reduce the number of species studied over the period 1911 to 1948, with mammals displacing the invertebrates; and of the mammals, the Norway rat was by far predominant. Slightly more than 50 percent of the articles dealt with conditioning and learning, and about 15 percent to 20 percent on reflexes, simple reaction patterns, and sensory capacities. Other forms of behavior, such as reproductive behavior, emotional reactions, social behavior, feeding, and general habits were much less frequently studied. Therefore, it is not surprising that the proportion of this literature directly relevant to behavior genetics is minimal, since the behavior geneticist depends on comparisons within and among species and strains. In addition, the concentration on conditioning and learning restricts the use of the data for behavior geneticists, since these are the traits for which genotype-environment interactions are likely to assume a major role.

With these comments in mind, we can turn to some of the relevant experiments. Tolman (1924) reported the results of what was the first selection experiment for maze learning by rats. The foundation population consisted of 82 white rats of heterogeneous ancestry. From this population, nine "bright" and nine "dull" pairs were mated to obtain the first selected generation. A second selected generation was produced by further selection among the brights and dulls. Selection was successful in the first generation, but less so in the second, and Tolman suggested the discrepancy may have resided in extraneous environmental factors. The general problem, however, was not abandoned; Tryon (1942) published the results of 18 generations of selection for maze-learning ability over the period 1925 to 1940. He bred rats selectively on the basis of their error scores in learning in a multiple T-maze. The foundation population was a heterogeneous sample of rats. By the eighth generation the maze-brights and maze-dulls did not overlap and there was negligible response thereafter. Detailed biometrical genetic analyses were not carried out, although Broadhurst and Jinks (1963) later calculated a degree of genetic determination of about 40 percent using the F_1 and F_2 crosses between the selection lines. The data are somewhat unsatisfactory for biometrical analyses as there was a significant genotype-environment interaction that could not be scaled out. Further discussions of other maze-learning selection experiments differing in techniques are presented by Fuller and Thompson (1960), including some experiments with negative results.

Another form of behavior (already dealt with in mice), the conditioned avoidance response, which has a learning component, has been studied less in rats. Bignami (1965) carried out successful selection for the trait, indicating genetic control. Satinder (1971) found genotypic differences among four selectively bred strains of rats for escape-avoidance. Of particular interest was the finding that the strains responded differentially to the drugs d-amphetamine sulfate and caffeine — genotype-environment interactions of a special kind. This is an area that should be explored fully with the aim of testing the range of effects of drugs on behaviors of different genotypes. Related to this are experiments showing that rats can be selectively bred for susceptibility to morphine addiction (Nicholls and Hsiao, 1967).

In a few other traits studied in rats, there is evidence for genetic control either from crosses between inbred strains or from selection experiments. Broadhurst's work on defecation in rats as a measure of emotionality and its associated variables, especially activity score, is discussed above. Directional selection for activity has also been carried out, since Rundquist (1933) successfully selected for active and inactive strains based on activity in a revolving drum. Generally the data favor a multifactorial interpretation for the genetic basis of these traits. Finally, just as in mice, a number of major gene effects have been described in rats (reviewed by Wilcock, 1969), but no new principles emerge. Much of this work involves major pigmentation genes. Since the behaviors are similar to those described for mice, a detailed discussion is not given here.

An experiment in guinea pigs is relevant. Goy and Jakway (1959) and Jakway (1959) studied sexual behavior in two inbred strains and in the F_1, F_2, BC_1, and BC_2 generations. For females the response to the injection of female hormone was assessed in terms of four behavioral measures elicited experimentally by tests for lordosis (arching of the back) prior to copulation. For the males, females in heat were used as the stimulus and six behavioral measures taken. Broadhurst and Jinks (1963) found difficulties in scaling for all but two variables — the number of malelike mountings during estrus for females, and the number of penile intromissions for males. Estimates of heritability fell between about 0.5

and 0.6 for these traits, and there was a marked dominance component for both traits. Heterosis was found for the highest measures of activity in males — the intromission rate and number of ejaculations.

The basic conclusion is that in parallel with results in the mouse, most quantitative behavioral traits in other rodents are controlled multifactorially, as can be shown by differences among strains and the results of selection experiments. For this reason, a more detailed review of these results is not necessary. Mice have an advantage as experimental animals because they have a short generation time and a much better known chromosome map than other rodents. It is likely that behavior-genetics analyses of mice will develop rapidly even though they were initiated later than those with rats.

General reading

Lindzey, G., and D. D. Thiessen. 1970. *Contributions to Behavior-Genetic Analysis: The Mouse as a Prototype.* New York: Appleton. A collection of papers on various aspects of mouse behavior, considering genetic analysis, gene-environmental interplay, single-gene effects, gene-physiological determination, and evolutionary aspects.

The Genetics of Behavior:
Other Creatures

In 1968, in an eminently recommendable anthology entitled *Roots of Behavior* (Bliss, 1968), Dilger comments that "direct evidence of the genetic control of behavior in vertebrates is, unfortunately, even scarcer [than in invertebrates] and the precise identification of responsible genes is almost nonexistent."

Our purpose in this chapter is to avoid the erroneous impression that the genetics of behavior must be approached via the manipulation of the genes and the behavior of Drosophila, rodents, or man. We offer a survey of what we know to be tentatively successful approaches to studying other organisms. These creatures are not now as easy to manipulate or to know genetically as Drosophila, rodents, and man (about whom genetic knowledge has been accumulated despite the absence of planned crosses) because no equivalently profound degree of knowledge of their genomes exists, ready to be tapped. This, however, makes them all the more enticing as subjects, we believe and hope to convince our readers. Our examples are not all confined to this particular composite chapter; some are dispersed earlier in the book — the trisomic monkey whose retarded behavior is reported in Chapter 4, Dilger's lovebirds discussed in Chapter 5, the barkless dogs described in the same chapter, and Rothenbuhler's tidy bees mentioned in Chapter 3. These are utilized as examples of principles in the genetic analysis of behavior.

We present the material contained in this chapter in the form of isolated examples to indicate the range of organisms investigated and the scope of these often technically arduous investigations, individually considered because of the individual limitations presented by the experimental subjects.

10.1 Bacteria

"What is the repertoire of behavioral responses in an organism?" asked Adler, Hazelbauer, and Dahl (1973). When the organism is a peritrichously flagellated bacterium such as the ubiquitous *Escherichia coli*, one may inquire about the direction of these flagella in the presence of specific substances, as is done with Kung's ciliated *Paramecium* (Section 10.2). Motile bacteria are attracted to a variety of chemicals (see Pérez-Miravete, 1973, for a survey). Chemoreceptors detect specific chemicals without participating in their metabolism (Hazelbauer and Adler, 1971). How is the clustering depicted in Figure 1 accomplished? Adler et al., utilizing mutagens, have isolated dozens of true-breeding mutant clones that do not behave positively chemotactically toward a variety of substances, such as sugars, amino acids, and oxygen, attractive to wild-type, nonmutant *E. coli* (Mesibov and Adler, 1972). This is so even

1. **Attraction of *E. coli*** to aspartate contained in a capillary tube. (From Adler, 1969. Copyright 1969 by the American Association for the Advancement of Science.)

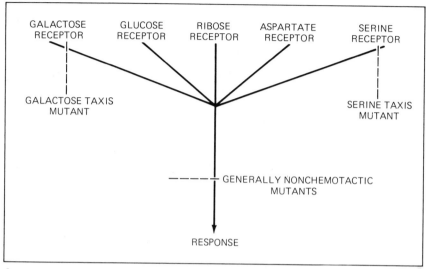

2. **Chemotaxis in *E. coli*.** Possible explanation of defects exhibited by mutants with impaired attraction toward specific amino acids or sugars. (From Adler, 1969. Copyright 1969 by the American Association for the Advancement of Science.)

though the mutant bacteria are perfectly motile with full complements of normal flagella. They can and do respond normally to all but the particular attractant to which they no longer possess the capacity to respond. This situation is schematically presented in Figure 2. Armstrong and Adler (1969) have located some genes for motility and chemotaxis on the *E. coli* genetic map, itemizing loci (genetically active spots on chromosomes) such as *curly* (altered flagellin protein, wavelength approximately half of normal, only rotational movement possible); *motile* (possess normal looking flagella but cannot move); *flagella* (no flagella present, nonmotile); *chemotaxis* (nonchemotactic, fully motile, three genes involved).

How many different chemoreceptors are involved? Nine exist for strong sugar (N-acetyl glucosamine, fructose, galactose, glucose, maltose, mannitol, ribose, sorbitol, and trehalose) attractants alone. Two (aspartate and serine) serve to detect amino acids. We can then assert the legitimacy of our anticipation of major contributions from the study of the behavior of *E. coli* and other bacteria to the science of genetics.

10.2 Paramecia

The new behavior genetics of *Paramecium aurelia* is aimed at a genetic dissection of the organism's outer, limiting, excitable membrane. The locomotor behavior of *Paramecium* is under the control of this surface structure (Eckert, 1972), and the direction and beating frequencies of the *cilia* (fine membrane-enclosed cytoplasmic hairlike threads projecting en masse outward from the surface of a cell) are correlated with cross-membrane electrical events, i.e., voltage-sensitive calcium conductance. Cilial reversal resulting in an alteration of swimming direction is called *avoiding*. Various stimuli elicit avoiding behavior in ciliated protozoa. These have been known since 1906, when Jennings described a paramecial avoiding reaction as an interruption in forward swimming due to the temporary reversal of the ciliary beat. This produces a short backward jerk or a sudden halt before forward propulsions begin anew in an altered direction. Only very recently, however, have the *Pawn* mutants of *P. aurelia* been isolated and subjected to behavioral, genetic, and electrophysiological analyses, i.e., the work of Kung and collaborators (Chang and Kung, 1973; Satow and Kung, 1974; Chang et al., 1974, and references therein; Pérez-Miravete, 1973).

Pawn, one of a series of genetic mutants, cannot swim backward as the wild type does; it is named after the chess piece which functions under similar limitations (Fig. 3). *Pawns* may be temperature-sensitive or temperature-independent; one of the former behaves normally at 23°C, moving backward, but fails to avoid strong stimuli (an assortment of salts in a toxic solution) at 35°C. Some of these mutations are induced with chemical mutagens such as versions of nitrosoguanidine (see Vogel and Röhrborn, 1970, and Hollaender, 1971, for general references on mutagenesis), and more than 100 *Pawns* are now available for study.

More specifically, the *Pawn* single-gene mutation is a functional defect in voltage-sensitive calcium cation conductance changes on the membrane bearing the animal's cilia. *Pawns* with detergent-disrupted membranes can swim backward when sufficient Ca^{++} and adenosine triphosphate are added to the medium. Thus, only a membrane defect causes the absence of backing movements; the ciliary motile apparatus is intact in *Pawns*. The temperature-sensitive *Pawns* are even more valuable in that they, as conditional mutants, allow the turning on and off of certain membrane processes at will. Four of the five known independently arising, temperature-sensitive *Pawns* were found to be allelic at a locus known also to generate temperature-independent *Pawn* mutations, and there are now a total of three *Pawn* loci, at least one unlinked with the other two. (*Paranoiacs*, another mutation, violently overreact to Na^+ but not to other stimuli; Satow and Kung, 1974.) Remember that the maintenance of all these lines is facilitated by the facultative sexuality of *P. aurelia*. The organism can be easily cloned and/or crossed (Sonneborn, 1970). We then have ample reason to recommend this organism and its ilk to behavior geneticists as a potentially fertile source material.

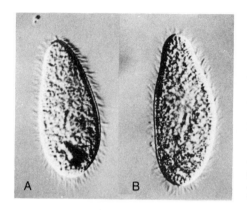

3.   **Locomotor behavior of P. aurelia.** Left: *Pawn* depicted during its resting position at the end of a stroke; its cilia are static and pointing toward the rear. Right: Wild type depicted swimming backward because salts were added to the medium. Photograph made with Nomarski interference-contrast illumination. X200. (From Kung and Naitoh, 1973. Copyright 1973 by the American Association for the Advancement of Science.)

10.3   Nematodes

The work of Brenner and of Ward (Ward, 1973, and references therein; Brenner, 1973) in manipulating the behavior of a nematode, *Caenorhabditis elegans*, must be judged pioneer, not because of the behavior investigated (chemotaxis among other behaviors), but because of the organism studied. Their efforts are unique in that this invertebrate has heretofore not been investigated genetically. Yet the organism possesses inherent advantages for a geneticist. It is a self-fertilizing hermaphrodite wherein sperm is first produced and stored. Developing eggs, some 300 of them per female, are subsequently deposited, and the adult-to-adult life cycle spans 3 to 4 days at 20°C. Such endogeny fosters homozygosity, but different induced mutations can be coupled in the same individual because a small number of males (0.1 percent) are routinely produced by meiotic nondisjunction. These may then be crossed with hermaphrodites, carrying genetic markers with them.

C. *elegans* has a cylindrical, threadlike, unsegmented body whose tracks can be quite easily seen in an agar-coated Petri dish. These visible grooves in agar can be made upon a gradient consisting of attractants such as chemical compounds (cyclic nucleotides), cations (Na^+, Li^+, K^+, Mg^{++}), or alkaline pH values. The animal's patterns of movement may reflect:

- Orientation. Movement up a concentration gradient, involving "lateral" motion of the worm's head

- Accumulation. Persistent location at a specific point in a gradient

- Habituation. Always occurring last since the subject has by then become accustomed to both the container and its contents. This involves a familiarity with both the gradient and the attractant. The behavior of the worm changes after it remains in a region of maximum attraction, i.e., it swims away only to repeat its cycle later.

The chemotactic behavior of wild-type, nonmutant nematodes was recorded and compared with tracks made by mutants with head or with tail blisters on their cuticles, with heads bent ventrally or dorsally, with defective head muscles, or with shortened heads. From such comparisons, we can conclude that the sensory receptors located in the head mediate orientation in a chemical gradient. Worms with distal tail blisters orient normally, but head blisters preclude this behavior. Bent-headed animals track complex spirals with the head bent toward the center. Defective head muscles or a shortened head reduces the efficiency of orientation compared with that of wild types. But why do these animals orient toward the common, biologically active cyclic nucleotides such as cyclic adenosine monophosphate at all? Perhaps because *C. elegans* normally eats soil bacteria which release such compounds into their environment. Juvenile and adult worms, as well as *dauer* (lasting) larvae (which accumulate when cultures are starved) all respond similarly to these attractants. The role of attraction to ions or to pH in this nematode's natural environment is not known at present.

On many levels then, here is a splendid, easily laboratory-reared organism for exploitation by the behavior geneticist. It has a haploid number of six chromosomes corresponding to six linkage groups. One more tempting fact — this worm has less than 300 neurons in its entire nervous system. This figure should be evaluated in light of the estimated 612 million to 9.2 billion neurons in the cortex of just one human cerebral hemisphere (Blinkov and Glezer, 1968), and of the 70,000 to 80,000 "brain neurons" of the crayfish *Procambus clarkii* (Wiersma, 1967). This last range represents the only count available on an arthropod. Note though that the tarsal (distal part of arthropod limb) contact chemoreceptors of the Diptera are known to function with but one sensory neuron.

10.4 *Ephestia*

In the flour moth *Ephestia kühniella*, a tropical and temperate zone cereal-consuming pest in flour mills, spinning behavior just prior to pupation has been found to have both genetic and environmental bases (Cotter, 1951; Caspari, 1958). They may interact to induce lengthened intervals between termination of larval feeding with emergence from the food mass and the initiation of pupation. The opposite, an abbreviated interval, results in what Caspari and Cotter have called *nonspinning*: full-grown larvae in their final instar (molt) stop feeding, leave their food by crawling up culture dish walls or stay on the surface of the food mass depending upon degrees of crowding, spin a cocoon, and pupate. This is the normal wild-type behavior, involving spinning only of a closed cocoon of which one tip is less firmly constructed and is directed upward so that this tip serves as the site of eclosion, with full-grown larvae pupating in the laboratory within 7 to 10 days.

The lengthened interval, terminating in pupation up to a month later, results in sheet or mat spinning, with the very occasional omission of a spun cocoon altogether (Fig. 4). But Ephestia able to spin mats will not do so if kept in the light; perhaps the negative phototaxis consistently displayed by these insects prevents them from leaving the food when cultured in light.

Caspari (1951) points out the difficulty of analyzing the genetics of spinning behavior in these subjects. Mats are spun by *populations*, not by *individuals*, and Cotter (personal communication) points out that there has been no adequate testing of differential spinning capacities; heavy spinners might be those individuals in the population with the greatest number of "spinning" alleles. (Much the same caveat must be applied to studies of phototaxis in other insects, as pointed out by Rockwell and Seiger, 1973a.) In an F_1 generation produced by crossing the variant forms, nonspinning seems to be dominant, if not perfectly so. The F_1 spins little silk. The F_2 progeny produce somewhat more silk than their immediate F_1 ancestors, and this observation is interpreted as indicative of genetic segregation. Furthermore, the backcross (F_1 × spinning strain) yields an amount of silk intermediate between amounts produced by the F_1 and by the spinning strain. The best explanation is Mendelian inheritance and segregation for two or more pairs of unlinked genes (Cotter, 1951; Caspari, 1955; Caspari and Gottlieb, 1959).

In addition, a nutritional factor is vital to the manifestation of mat-spinning behavior since sheets are spun *only* if three conditions — two environmental and one genetic — are met: if the food is cornmeal, not whole corn; if light is absent; and, of course, if the appropriate genes are present. Larval density within individual culture dishes also must be considered as an environmental parameter. Keeping in mind that the fabrication of sheets lengthens the entire developmental time and the generational intervals, Caspari (1958) argues:

> It could be imagined that under circumstances the spinning of sheets rather than of tunnels would constitute a selective advantage for the species. That this possibility exists is suggested by the fact that some close relatives of *Ephestia kühniella* actually spin mats rather than tunnels. This advantage would result in selective pressure in favor of the genes necessary for the spinning of sheets. At the same time, it would cause the organism to be adversely affected by whole corn food, but favored by cornmeal. The choice of foods for the resulting organisms would therefore be severely restricted. This may serve to suggest the origin of food specialism in insects, where it is frequently found that a certain species is restricted to a specific kind of food plant.

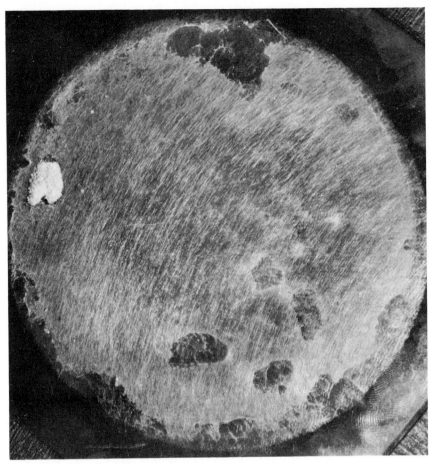

4.   **Mat silk spinning in** *E. kühniella,* the flour moth. Two cocoons are located outside the food mass. (Courtesy of William Cotter.)

This argument is cogent and is generally true of animals other than Ephestia — e.g., *Drosophila flavopilosa* (Brncic, 1966, 1968) and *D. pachea* (Heed and Kircher, 1965). (See also Section 13.2 on habitat selection in Drosophila.) The very act of spinning silk outside a feeding tunnel in the food mass implies larval exposure to surface-feeding predators during that extended period of time; but if the food is additionally manipulated after harvesting, the subsequent mode of storage — whether in bags, boxes, or granaries — tends to exclude a very high percentage of potential predators. The mat itself presumably acts (1) to protect the subsurface-feeding larvae (the larger percentage of the population) from common parasitic wasps such as *Habrobracon juglandis,* and (2), perhaps less important, to function as a site for the deposition of eggs by the adult female.

10.5 *Musca domestica:* The housefly

Female houseflies produce (z)-9-tricosene, a pheromone that attracts conspecific males and induces courtship and mating behavior in them (Voaden et al., 1972; Rogoff et al., 1973). The compound has been appropriately called *muscalure*.

Pseudoflies were constructed of knots of black shoelaces either impregnated with benzene extracts of female flies (containing their pheromone) or with benzene alone (controls). The quantity and quality of the pheromone in solvent, the female flies from which the chemical was extracted, light, and temperature were all controlled, and 347 males were individually scored on their responses. These are sketched in Figure 5. Differences between flies exposed to treated and to control pseudoflies were due to two components of behavior: (1) pheromone-mediated attraction to treated pseudoflies and (2) activity of individual flies, i.e., number of mating strikes (from flight to mount) per male.

This latter category, responsiveness, was found to be a heritable characteristic. Selective breeding (breeding from the two males which each made the most or the fewest mating strikes during two separate tests with virgin females) for high (most mating strikes) and for low (fewest mating strikes) male responsiveness was practiced through the F_4 generation with two high and two low lines. Interestingly and unfortunately, one low line was lost due to failure to breed, and the remaining

5. **Mating behavior in *M. domestica*.** Transition from flight to mating position (a strike) in male housefly with a conspecific female and with a pseudofly (black material soaked in pheromonal extract). (From Cowan and Rogoff, 1968.)

low line averaged 6.24 ± 3.6 mating strikes per hour with a range of 0 to 15.6 strikes per hour by the F_4 generation. The two F_4 high lines averaged 21.72 ± 8.7 and 20.34 ± 9.8 mating strikes per hour with ranges of 0 to 90.0 and 0 to 41.7, respectively. Such results resemble those acquired when selection is practiced for fast and slow mating speeds in Drosophila (the work of Manning, 1961, 1963, discussed in Chapter 8).

10.6   Mosquitoes

Much begins to be known about the genetics of mosquitoes (Craig, 1965; Wright and Pal, 1967). The bulk of the information now available concerns formal genetics — the location of loci and knowledge of the chromosomes involved (Craig and VandeHey, 1962). *Aedes atropalpus* breeds in streambed rockpools and occurs in two behavioral forms, autogenous and anautogenous. Autogenous reproduction involves egg maturation without an exogenous source of protein, such as a blood meal. Females fed only sugar and raisins deposit eggs that hatch. The resulting F_1 progeny can again be reared without blood meals. This is not true of anautogenous forms, the common ones, which must feed once on blood before egg maturation. Autogeny is controlled by an autosomal dominant gene.

Gwadz (1970) picked two *A. atropalpus* strains carefully for his study of the genetics of this mosquito behavior. One was homozygous for the dominant gene for autogeny and was designated GP (for Gunpowder Falls, Maryland, its site of origin). The other was homozygous for the recessive gene for anautogeny and was denoted TEX (for Austin, Texas). These subjects were raised under the optimal conditions for each strain, including the avoidance of crowding. Among other details, these conditions involved the control of temperature (27 ± 1°C), relative humidity (80 ± 5 percent), day length (16 hours), and age (females hatched within 30 minutes of each other). Males came from the same population as the females being tested and were older than these females. Figure 6 represents data acquired from dissections in physiological saline followed by microscopic examination for the presence or absence of stored sperm in the spermathecae of the females — evidence of mating and insemination.

The results are clear: GP mosquitoes mate sooner after eclosion than do TEX mosquitoes. Their hybrids in either direction are intermediate but closer to the GP parent. The details on mean insemination time after exposure to males are: GP, 38 hours; F_1GP/TEX or F_1TEX/GP, 54 hours; and TEX, 120 hours (5 days). The long refractory period TEX displays is not unexpected, since TEX females must fly and search and locate a blood meal before they can get down to the serious business of maturing hatchable eggs, sexual activity aside.

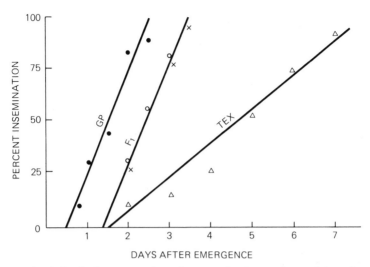

6. **Mating behavior in** *A. atropalpus,* the mosquito. Onset of receptivity to insemination for two parental and two hybrid populations. **Circles,** GP/TEX (GP female × TEX male) hybrid; **crosses,** reciprocal TEX/GP (TEX female × GP male) hybrid. Each point represents a minimum of 200 females. (From Gwadz, 1970.)

The times of onset of sexual receptivity for the four (all the possible) backcross generations were then ascertained and are graphed in Figure 7. If one assumes genetic control exercised by means of a semidominant autosomal gene, then the broken line represents expected rates of insemination.

Put another way, if R is the (autosomal) gene for "rapid" receptivity and R' is its semidominant allele for delayed receptivity to insemination, then the GP strain is RR', intermediate. (This intermediacy of the heterozygote is referred to as semidominance at a locus.) Then

GPTEX/GP = TEXGP/GP
          = RR'(F_1 hybrid) × RR(GP male)
          = 1 RR (early) : 1 RR' (intermediate)
GPTEX/TEX = TEXGP/TEX
          = RR'(F_1 hybrid) × $R'R'$(TEX male)
          = 1 $R'R'$(late) : 1 RR'(intermediate)

And progeny of both backcrosses to GP males should, on the average, allow insemination before either backcross to TEX males. They do so with appropriate overlap between the upper end of the two GP male backcross times and the lower end of the two TEX male backcross times. The performance of the backcross progeny fits expected results well. Note the impressive numbers of individuals observed and scored in every category.

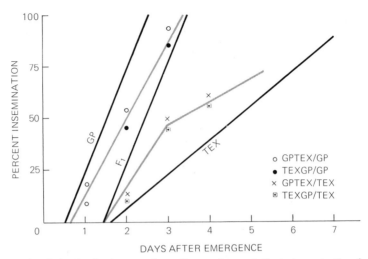

7. **Mating behavior in *A. atropalpus*.** Onset of receptivity to insemination for four backcrosses of F_1 hybrid females to parental males. **Gray lines,** expected percentages if a single-gene hypothesis is postulated. Each point represents a minimum of 200 females. The four backcrosses are GPTEX/GP (GP female × TEX male), F_1 female × GP male; TEXGP/TEX (TEX female × GP male), F_1 female × GP male; GPTEX/TEX (GP female × TEX male), F_1 female × TEX male; and TEXGP/TEX (TEX female × GP male), F_1 female × TEX male. (From Gwadz, 1970.)

10.7 A parasitic wasp

During the course of his lifelong investigations into the genetics of wasps (1934, 1939), Whiting obtained a number of gynandromorphs of the parasitic (on caterpillars) wasp, *Habrobracon juglandis*. Figure 8 depicts one type of gynandromorph (the female-male form, a sexual mosaic) and the normal female and male of the species.

Like other members of the order Hymenoptera (bees, ants, hornets, etc.), wasp males emanate from unfertilized eggs. Unmated females can parthenogenetically (without a male partner) produce all male broods. The mother-to-son transmission of hereditary traits is direct and not shared. A mated female, however, produces fatherless sons again and biparental daughters. If a female is homozygous for a recessive trait and is mated to a homozygous dominant male, her female offspring are heterozygous dominant and her male offspring hemizygous, since they are haploid, and recessive.

Gynandromorphs come from abnormal eggs with two nuclei, only one of which is fertilized. Female parts develop from the diploid part, male from the unfertilized portion.

Consider, as one example, an individual in which the gynandromorphic parts could be easily identified as to sex because it was produced by

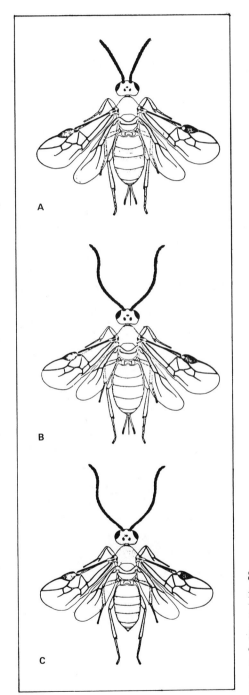

8. *H. juglandis,* **a parasitic wasp.**
A. Normal female. Note the relatively longer wings, shorter antennae, and sting sheathed by a pair of sensory appendages, the gonapophyses, at the tip of the abdomen. **B.** Gynandromorph. **C.** Normal male. Note the relatively shorter wings, longer antennae, and sexually dimorphic abdominal tip. (From Whiting, 1932.)

a cross between a female recessive for orange eyes (*oo*) and defective wing venation (*dd*) and a wild-type male. Such a cross produced 82 wild-type (*OoDd*) females, 17 orange-eyed males with defective wing veins (*od*), and 1 gynandromorph. These mosaics are rare, having an incidence ranging from 1 in 1000 to 1 in 10,000. Figure 9 shows the head of the gynandromorph resulting from the cross described. It had an orange (*o*) right eye, which was male; part of its left eye was orange, the rest black (*Oo*) and female. The right antenna was longer (male) than the left (female), and the two right wings had defective veins (*d*) and were shorter (male) than the left ones (female). Additional secondary sexual dimorphisms and pigmentation make it possible to identify other bodily parts. In this individual, the left side was female; the right, male.

Table 1 summarizes data on the behavior of 50 sexual mosaics; reactions, for the most part, are normal for one or the other sex. Though each body is physically a sexual mixture, behavior is not. Clearly, it is the sex of the head upon which responses of the insect depend. For example, normal female responses to caterpillars (here, those of the Mediterranean flour moth, *E. kühniella*; see Section 10.4) involve thrusting the abdomen forward and downward to bring the sting into protruding position and the antennae straight forward. Then she advances slowly and inserts her sting, with no specific region on her victim preferred for insertion. Antennae are passed over the caterpillar's body surfaces during stinging. After wriggling ceases, the sting is withdrawn and the female applies her mouth to the now quiet caterpillar to suck fluids from it. Afterward, a fold in the skin is selected for egg laying.

Males ignore caterpillars, even avoid them. Upon introduction to

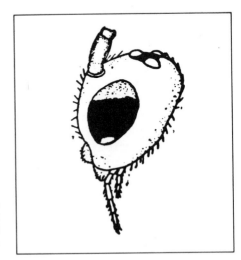

9. **Head of H. juglandis gynandromorph.** Lateral view. This left eye is part male (lighter sections) and part female (darker section). (From Whiting, 1932.)

wasp females, they become agitated and attempt to mount as soon as a female is located. A male may copulate with a single female more than once or with several successively, beating wings and antennae while copulating. The male grasps and shoves aside the female's wings while mounting and he may try to mount other males. Copulation lasts up to 2 minutes.

The behavior of other insect gynandromorphs has been observed — e.g., the wasp *Habrobracon brevicornis* and the bee *Megachile gemula* (Mitchell, 1929) — but more recently and with greater precision, *D. melanogaster* mosaics have been created and analyzed by Hotta and Benzer (1973), discussed in Chapter 8. This continues, with greater genetic control and varieties, and incorporating the deliberate production of mosaics via chemical mutagenesis, the work of Sturtevant, Morgan, and Bridges (Morgan and Bridges, 1919).

10.8 Some acoustical insects

I think it will be admitted by naturalists, without my entering on details, that secondary sexual characters are highly variable. It will also be admitted that species of the same group differ from each other more widely in their secondary sexual characters, than in other parts of their organisation; . . . The cause of the original variability of these characters is not manifest; but we can see why they should not have been rendered as constant and

TABLE 1. Behavior of *H. juglandis* Gynandromorphs Sexually (Toward Females) and Parasitically (Toward Caterpillars), According to Sex of Head and Abdomen

Head	Abdomen		Reactions toward females		Reactions toward caterpillars	
			Positive	Indifferent	Positive	Indifferent
Male	Mixed	9	9			9
	Female	20	20			15
Female	Male	1		1	1	
	Mixed	3		3	3	
Mixed	Male	2		2	2	
		1	1			
	Mixed	3	3			3
		3	3			
		1		1	1	
		1			1	
	Female	2	2			2
		1	1			
		3			3	
Total		50	39	7	11	29

From Whiting (1932).

uniform as others, for they are accumulated by sexual selection, which is less rigid in its action than ordinary selection, as it does not entail death, but only gives fewer offspring to the less favoured males. What ever the cause may be of the variability of secondary sexual characters, as they are highly variable, sexual selection will have had a wide scope for action, and may thus have succeeded in giving to the species of the same group a greater amount of difference in these than in other respects.

Charles Darwin, 1859

Part of *The Descent of Man and Selection in Relation to Sex* deals with species other than our own; Darwin (1871) apparently found this necessary to explain and then to defend his newly introduced theory of behavioral sexual selection. One chapter (Chapter 10 of Volume 1) deals with the secondary sexual characteristics of insects. Darwin included the illustration reproduced here as Figure 10, showing the stridulatory apparatus of a male field cricket. He commented upon the prodigious volume and duration of nocturnal cricket choruses and the fact that, "all observers agree that the sounds serve either to call or excite the mute females."

In this section we are mainly concerned with insects of the family Gryllidae, the common field cricket. Reproductive pairing is brought about by long-range acoustical signals. A sexually mature male produces sound pulses by moving a wing that bears friction mechanisms. Each closing stroke of the forewings produces a sound pulse, and the wings are reset for another cycle by a silent opening of these structures. In this way, sound is produced by a cyclic opening and closing of the wings in which sound occurs only during the closing stroke. Each movement of the wings is precisely timed by the concerted contraction of two sets of wing muscles that operate antagonistically. Contraction is initiated by neural discharge in the motor neurons of each set of antagonists. Thus the wing movements that produce sound are said to be neurogenically controlled: contraction and motor neuron discharge are related causally in a one-to-one way (Bentley and Kutsch, 1966). This means that observing or recording the behavior (a procession of sound pulses) also provides a direct way to monitor precisely the activity of the motor components of the nervous system that underlies the behavior. This provides the neurobiologist with a simplified behavior for neural analysis and makes the system appealing for neurogeneticists, one of whose goals is to relate neural activity to genotype.

Let us now return to the behavior and its function. The male produces a calling song that attracts conspecific females to him, presumably over distances ranging from several meters to tens of meters. The effective radius of calling has not been carefully studied and is likely to be a complex matter affected by temperature, humidity, terrain, and wind conditions. Whatever the true effective limits of acoustical communica-

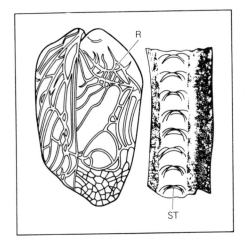

10. **Stridulatory apparatus of male** *Gryllus campestris,* the field cricket. **Right.** Highly magnified underside of part of the wing nervure showing the teeth (ST). **Left.** Upper surface of the wing cover, with the projecting, smooth nervure (R) across which the teeth (ST) are scraped. (From Darwin, 1871.)

tion, it is certain that acoustical signals play the major, if not sole, role in attracting females. Most crickets are acoustically active at night, which eliminates the possibility of a role for vision. Chemical factors may play a minor part (this issue has not been investigated), but dating from the experiments of Regen in 1914, it has been known that acoustical signals are sufficient to attract females to the source of the sound in the "absence" of visual, chemical, or tactile environmental stimuli. Once a male has attracted a female to within a few centimeters of him, and tactile interaction between them occurs, he usually sings a courtship song that is obviously (even to human ears) different from the calling song. One might wonder about conferring species specificity on courtship song rather than on calling song. That this is not the case can be rationalized in terms of saving the female's time and energy (two factors known to be important in an animal's reproductive success). If many different species of crickets sang in close sympatry, a female would risk making many "wasted trips" if she did not know whether the caller was a conspecific male until she was within a few centimeters of him. In terms of a female's time and energy this would be a wasteful way of forming conspecific pairs for mating.

How does a species encode specificity into a calling song given the mechanism described above? The "song" can be varied in the physical domains of time and frequency. There is evidence that frequency, in this case the fundamental frequency, of the song plays some, but not the major, role in species specificity. Cricket calls are relatively pure in harmonic content, consisting of a fundamental frequency and a variable number of harmonics. Experiments with artificial songs have demonstrated that the fundamental frequency coupled with temporal pattern of the species are sufficient to mediate phonotactic behavior. Although sympatric cricket species may differ in the value of the fundamental frequency, the range of variance is not great within the sympatric group. Furthermore, frequency is not modulated in cricket songs as it is in bird songs. The most striking difference in the songs of different species is in the temporal pattern of the procession of pulses that constitutes calling song. These parameters of rhythm are remarkably stereotyped among individuals of a local population or species, and markedly different from species to species. It is a simple matter to tape-record a cricket's calling song and subsequently display the pulse trains on an oscilloscope. Then one can film the pulse train displayed on the scope, measure precisely the interval (in millimeters) between pulses, and translate the measurement back to time (in seconds). This provides exact information about the temporal structure of the call. As mentioned above, it is possible to infer the activity of the motor neurons involved in song generation by looking at the pulse train. Thus, knowledge of the temporal structure of the song not only enables us to characterize the behavior, but also provides a convenient "window" into the nervous system as it produces the behavior being studied. For purposes of providing species specificity, the rhythmic structure of song is the most characteristic aspect of the acoustical signal. Most workers in the field feel that the species specificity resides in the call rhythm; while this may contribute to species information, it does not define the species (Hoy, personal communication).

GENETIC CONTROL OF CALLING SONG IN MALE CRICKETS

Earlier studies (Bigelow, 1960; Leroy, 1964) established the fact that hybrid F_1 crickets can be produced in the laboratory, even though hybrids are not commonly found in the field (Alexander, 1968; Hill, Loftus-Hills, and Gartside, 1972). In these earlier studies the focus was on measuring the pulse rate of hybrid calls compared with parental calls, or comparing the "phrase structure" of hybrid and parental calls. Bentley and Hoy (1972) analyzed the F_1 Teleogryllus hybrids (studied earlier by Leroy, 1964) with the specific aim of making exhaustive measurements of interval classes in the song by sampling calling songs from numerous individuals. Hybrids from the Australian field crickets T. commodus and T. oceanicus were produced in the laboratory. The calling songs of these species are complex because each phrase is composed of two pulse types (Fig. 11). This gives rise to numerous interval classes whose stereotype provides behavioral units that can be followed in genetic experiments. It was also possible to produce reciprocal cross F_1 hybrids. These were denoted T-1 and T-2 by these researchers: T-1 was

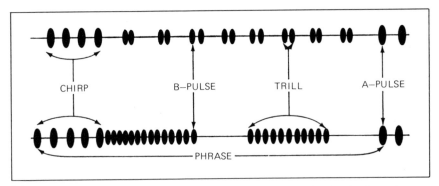

11. **Phrase structure of calling song of** *Teleogryllus,* a cricket. Each phrase is composed of two pulse types: A-pulses contained in the chirp portion of the phrase, and B-pulses contained in the trill (warble). *T. oceanicus* is shown above and *T. commodus* below. (From Bentley and Hoy, 1972.)

the hybrid from *T. oceanicus* female × *T. commodus* male; T-2 was the hybrid from *T. commodus* female × *T. oceanicus* male.

Figure 12 shows oscillograms (produced by playing a section of a tape-recorded calling song and displaying it on an oscilloscope, where it can be filmed) of the calling songs of *T. commodus, T. oceanicus,* and the hybrids. While visual examination of the oscillograms reveals differences in the F_1 song compared with the parental song, many interpulse intervals must be compiled before definite statements can be made about the genetic control of the calling song. From such measurements it is possible to construct interpulse interval frequency histograms, as shown in Figure 13.

It is clear that calling songs of the F_1 hybrids are distinctly different from either parental call. The intrachirp and intratrill intervals in the hybrid call are intermediate between corresponding parental intervals. This excludes a simple dominant-recessive genetic control of these rhythmic parameters. In fact, there is no evidence for simple dominance in any parameter of song rhythm. Intermediate inheritance is usually interpreted in terms of polygenic inheritance, a mechanism that has been invoked in earlier studies of song genetics (including Teleogryllus by Leroy, 1964) in crickets. Intermediate inheritance can also be explained by monofactorial control in which there is incomplete penetrance. Backcross hybrids can be examined to determine which mechanism is actually responsible, and in Teleogryllus four backcross classes can be produced. The backcross hybrids themselves reveal intermediate inheritance, so that the polygenic hypothesis is supported (Bentley, 1971). This finding confirms the same conclusion reached by Leroy (1964) in her studies of Teleogryllus.

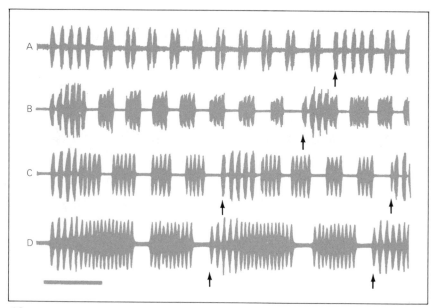

12.   **Oscillograms of calling songs** of *T. oceanicus* **(A)**, *T. commodus* **(D)**, and their reciprocal hybrids, *T. oceanicus* female × *T. commodus* male **(B)** and *T. commodus* female × *T. oceanicus* male **(C)**. **Arrow** marks the beginning of the next phase. **Solid bar** is a 0.5-second time mark. (From Bentley and Hoy, 1972.)

The study of Bentley and Hoy demonstrated that sex-linked factors affect the calling song. The two reciprocal hybrids, T-1 and T-2, differed from each other in the duration and variability of the intertrill interval and in phrase repetition rate. The arrows in the frequency histograms of interpulse intervals (Fig. 13) point to the intertrill interval that is prominent in the song of *T. oceanicus* and absent from the song of *T. commodus*. The intertrill interval is also present in hybrid T-1 songs and is absent from the songs of T-2. Crickets have X0 sex determination; the male receives his single X chromosome from his mother. Thus, all T-1 males receive *T. oceanicus* X chromosomes; similarly T-2 males receive the X chromosome from *T. commodus*. The presence or absence of an intertrill interval in the calling song is associated with the origin of the X chromosome. We can, therefore, infer the presence of X-chromosome-linked factors that affect the rhythmic structure of calling song.

The genetic control of song production can be summarized as follows:

- Polygenic inheritance of the call rhythm is supported by the finding that intrachirp and intratrill intervals are intermediate in the F_1 between the parental values and are similar among individuals of both reciprocal hybrid classes. Hybrid calls are distinctly different from either parental call.

- A few aspects of call rhythm, such as the intertrill interval, are influenced by sex-linked factors.

• Genetic control is distributed among an unknown number of autosomes and the X chromosome. Thus, genetic control of song is multichromosomally distributed as well as polygenic.

GENETIC CONTROL OF FEMALE RESPONSE TO CALLING SONG

Whereas the male produces a stereotyped calling song that serves to identify both his location and his species, the intended recipient of his call, a conspecific female, cannot call back. Females are mute and re-

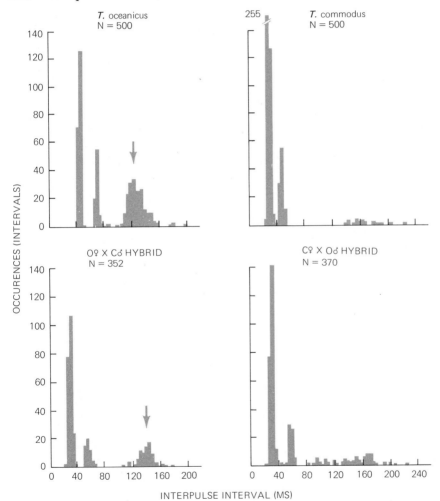

13. **Interpulse interval frequency histograms of calling songs** of *Teleogryllus* species and their F₁ hybrids. Each histogram represents the analysis of one individual cricket song. N, number of intervals measured. The intervals form three modes: intratrill, intrachirp, and intertrill intervals. *Arrows* denote well-defined intertrill intervals. (From Bentley and Hoy, 1972.)

spond to the call by walking to its source. It is obviously important that a female cricket be somehow attuned to the call of her species, presumably as a result of genetic programming. This can be tested by measuring the ability of a female to discriminate the call of her species from hetero-specific calls. Females can and must discriminate and be attracted to the conspecific males' calling song.

We previously have described the genetics of calling song in the cricket Teleogryllus. Hybrid males were found to sing their own calls, quite distinctive from either parental call. What is the responsiveness of the female hybrid to the call of its parental species and particularly that of its male siblings? The relative attractiveness of parental and hybrid calling songs for a female cricket can be measured by placing a tethered female on a Y-maze in a directional sound field. The tethered female walks on a Styrofoam maze in midair; when she walks, the maze is propelled backward beneath her (Fig. 14). The maze consists of three straight paths, 10.5 cm long, interconnected by two Y-shaped choice points (at 120° angles). At each Y, the female must choose either the right or the left arm of the maze. Once having made the choice she is back on one of the straight paths, followed by another Y junction, and so on. Analysis of this behavior enabled Hoy and Paul (1973) to measure a female's responsiveness to calling songs under controlled conditions. Virgin females from each species as well as the hybrid T-1 were placed on the Styrofoam Y-maze. They were then presented the recorded mating calls from one of two loudspeakers, placed on the right and left

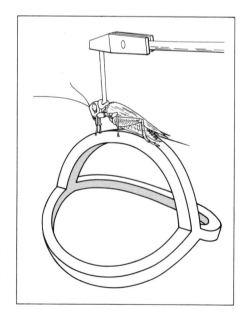

14. Tethered female field cricket on a Y-maze. Her movements on the maze reflect her preferences for certain male calling songs. (Courtesy of R. Hoy.)

of the maze. Choice behavior in a directional sound field (termed the *phonomotor response*) provided a measure of the relative attractiveness of a given song. The phonomotor response confirmed the locomotor behavior of a free-walking cricket in a directional sound field. Hill, Loftus-Hills, and Gartside (1972) found that free-walking *T. oceanicus* and *T. commodus* make species-specific discriminations. Hoy and Paul found that on the Y-maze, *T. commodus* and *T. oceanicus* also perform species-specific discrimination. The surprising result involved the hybrids. T-1 hybrid females preferred the song of sibling T-1 males over the song of either parental type (Table 2). Hoy and Paul (1973) arbitrarily defined a phonomotor response as meeting a "strong" criterion if at least 15 of 20 choices were made in the direction of the sound source in *each* direction. Calling sound was played through only one speaker at a time; when 20 choices were made, the same song was played through the other speaker for another 20 choices. With *T. oceanicus* on the maze, 14 out of 22 females met the established criterion when they were played the song of *T. commodus*. Similarly, *T. commodus* females on the maze preferred their homospecific calling song to that of *T. oceanicus*. This is precisely the behavior found by Hill, Loftus-Hills, and Gartside (1972) in the free-walking situation. Finally, Hoy and Paul found that 75 percent of hybrid T-1 females met criterion when played the hybrid calling song, but only 38 percent met criterion for the *T. commodus* song and only 27 percent met criterion for the *T. oceanicus* song.

Hybrid male Telogryllus produce unique calling songs that are easily

TABLE 2. Phonomotor Responses of 147 Female Crickets to Conspecific and Heterospecific Calling Songs

Calling song	Females at criterion/ total tested	Females at criterion (%)
	T. oceanicus on Maze	
T. oceanicus	14/22	63.6
T. commodus	4/22	18.0
	T. commodus on Maze	
T. oceanicus	3/15	20.0
T. commodus	21/28	75.0
	Hybrid (T-1) on Maze	
T. oceanicus	3/11	27.3
T. commodus	8/21	38.0
Hybrid (T-1)	21/28	75.0

Criterion was 15/20 correct choices in direction both for sound played from the right and for sound played from the left.

From Hoy and Paul (1973). Copyright 1973 by the American Association for the Advancement of Science.

distinguishable from either parental song. Hybrid female Teleogryllus find the calling song of their hybrid siblings more attractive than the songs of either parental species. Presumably, both phenomena have a genetic basis. The implication is that the production of the signal in the male and its reception and translation into locomotor behavior in the female are somehow genetically coupled. Perhaps the receptive apparatus of the hybrid female is somehow attuned to the song of the hybrid male. It is tempting to propose neurological models by which this genetic coupling might lead to neurophysiological coupling, but it is too early to speculate on precise mechanisms. Clearly, however, this would be an appealing way to "design" a communication system. Evolutionary biologists have pointed out the attractiveness of coupling transmitters and receivers in species-specific communication (Alexander, 1962).

CONCLUSION

The study of behavior genetics in crickets offers a novel approach to the study of polygenically controlled behavior. The calling song provides measurable units of behavior that can be followed in hybridization experiments. The fact that acoustical behavior is the basis of species-specific communication provides the opportunity for applying genetic analyses that have implications beyond traits in an individual animal. Even though very little is known about the formal genetics of crickets (e.g., no chromosomal mapping has been done), a genetic approach to acoustical behavior promises to illuminate more general problems of behavior genetics and the evolution of behavior (Hoy, 1974).

10.9 Fish

Records of sexual behavior in the platyfish, *Xiphophorus* (*Platypoecilus*) *masculatus*, and in the swordtail, *X. helleri*, and in the males of their F_1, F_2, and backcross hybrids have been taken from a series of over 1000 10-minute-long observations made during experiments designed to elucidate the mechanisms of insemination in these freshwater fishes. (Our account follows closely that of the authors, Clark, Aronson, and Gordon, 1954.) Some features are:

- Gonopodial (appendages used for reproductive purposes) thrusting that can be distinguished behaviorally from copulation. As determined by a gonaductal smear technique (the observation of living sperm extracted with a pipette under a microscope), thrusts alone never result in insemination of the female.

- The gonopodium is a holdfast organ in which the tip is modified to form an effective device for attachment. In the absence of this holdfast mechanism, copulations do not occur and males so mutilated do not inseminate females.

- The pelvic fin on the side to which the gonopodium is swung also moves forward, and this is an integral part of the copulatory mechanism. In the absence of both pelvic fins, the ability to transfer sperm to the female is greatly reduced.

- Although courtship patterns are similar in platyfish and swordtail, a number of qualitative and quantitative differences are revealed. Several behaviors — swinging, sidling, quivering, nipping, thrusting, and copulation — are observed during courtship in both platyfish and swordtails. Two behavior patterns typically shown during courtship by male platyfish — pecking and retiring — are not observed during swordtail courtship. Male swordtails, on the other hand, perform types of courtship behavior referred to as exaggerated backing and nibbling that are not seen in platyfish.

The most striking quantitative differences in the sexual behavior among platyfish, swordtails, and their various hybrid combinations were associated with copulation. The mean duration of copulation was longer for swordtails (2.39 seconds) than for platyfish (1.36 seconds), and swordtail pairs mated sooner (averaging 1 minute) during the 10-minute observation period than did platyfish (averaging 5 minutes). However, platyfish mated more frequently (in 26.7 percent of observations compared with 13.4 percent in swordtails), and the number of inseminations resulting from copulations was higher in platyfish (86.0 percent) than in swordtails (39.4 percent). In F_1 hybrids the frequency of copulation was slightly higher than in platyfish (29.0 percent), and the number inseminated after copulation was intermediate between parental types (64.3 percent). In the F_2 and backcross hybrids these values were much lower. In general, copulatory behavior in F_1 hybrids was either intermediate or more like that of the swordtails. Some features of male sexual behavior in these fishes apparently are under genetic control, but there is no simple mode of inheritance to account for the data obtained; polygenic inheritance should be considered.

Studies on interspecies fish groups reveal almost complete reproductive isolation between platyfish and swordtails when a choice of mates is offered, even though some heterospecific courtship activities may be observed. When no choice of mates is available, heterogamic copulations occur with a relatively low percentage of inseminations (18.2 percent) (Fig. 15).

Effective reproductive isolation between the swordtail and the platyfish appears to depend on an array of partially isolating mechanisms. None alone is sufficient to ensure isolation, but acting in concert, these factors so reduce the probability of hybridization that under natural conditions reproductive isolation is apparently complete. It is unfortunate that genetic analyses of the sophistication attainable with closely related populations of Drosophila (Section 5.3) are not possible here (but see Franck, 1970). For example, within the *D. paulistorum* species com-

plex and with *D. pseudoobscura* × *D. persimilis* (Tan, 1946), reproductive isolation was shown to be under polygenic control; the same may be so for this pair of fish species. Their array of isolating factors is as follows:

- Ecological and geographical isolation
  Partial segregation of swordtails in swift headwaters and platyfish in the more slowly flowing lowland streams

- Physiological isolation
  Differences in sensory apparatus
  Differences in threshold for sexual responses
  Differences in sexual behavior
  Differing responses to stimuli

- Gametic isolation
  Sperm less viable in oviduct of heterospecific female
  Competition between homospecific and heterospecific sperm

- Genetic isolation
  Hybrid inferiority
  Partial hybrid sterility

10.10 Quail

An autosomal recessive gene *sg*, when homozygous, causes star gazing in the Japanese quail, *Coturnix coturnix japonica* (Fig. 16). This abnormal behavior involves primarily the withdrawal of the head caudally. The motion occurs when the birds are suddenly confined or if an opaque object is placed above them. It becomes more pronounced as the birds age, rarely being detected until at least 3 weeks of age. By then, the exaggerated motion is obvious, sometimes resulting in a circular movement or squatting on the hocks with the head resting on the floor. This does not appear to affect a bird's capacity for the acquisition of nourishment, and affected birds grow normally. Males and females homozygous for the *sg* gene are fertile.

10.11 Chickens

Courting in chickens has a dual role that is both agonistic and sexual (Kruijt and Hogan, 1967; Siegel, 1972). Careful genetic analyses of courtship and aggressive mating behavior have been undertaken in chickens (Cook, Siegel, and Hinkelmann, 1972; Siegel, 1972). Mating behavior

---

15. **Female platyfish and male swordtail hybrids. Top:** Male (below) and female (above) F₁ hybrids. **Middle:** Male and two female F₂ hybrids. The male (uppermost of the three fish) is double recessive for nonspotting. **Bottom:** Male (above) and female (below) F₂ hybrids. The male carries the dominant gene for spotting. (From Clark, Aronson, and Gordon, 1954.)

16.   Star gazing Japanese quail at 1 month of age. (From Savage and Collins, 1972.)

was recorded at a mature age (30 to 35 weeks) by placing a lone male with a flock of eight hens for eight 10-minute observation periods. Related behaviors such as treading (where the male covers the female), mounting, courting, and relative aggressiveness as indicated by paired

TABLE 3.   **Sample Size, Means, and Standard Errors of Mating Behavior, Aggressiveness, Body Weight, and Reproductive Performances in *Gallus gallus domesticus*, the Chicken**

Mating combi-nation*	N	Behavior				Aggres-siveness (male)
		Completed matings	Treads	Mounts	Courts	
HH	115	14.3 ± 0.9	14.9 ± 1.0	16.6 ± 1.1	88 ± 4	42 ± 2
HC	48	11.7 ± 1.2	12.1 ± 1.2	13.5 ± 1.4	79 ± 5	57 ± 3
CH	47	10.2 ± 1.2	10.6 ± 1.2	11.2 ± 1.3	93 ± 6	55 ± 4
HL	21	7.5 ± 1.5	7.8 ± 1.5	9.0 ± 1.6	76 ± 9	61 ± 4
CC	75	5.9 ± 0.7	6.3 ± 0.7	7.0 ± 0.8	81 ± 6	49 ± 3
LC	20	5.8 ± 1.4	6.0 ± 1.4	6.2 ± 1.5	59 ± 7	37 ± 5
LH	53	4.8 ± 0.7	5.2 ± 0.7	5.6 ± 0.7	70 ± 4	52 ± 4
LL	55	3.3 ± 0.6	3.5 ± 0.6	4.0 ± 0.7	66 ± 4	43 ± 3
CL	5	0.8 ± 0.4	1.4 ± 0.5	2.6 ± 1.2	60 ± 20	40 ± 9

* H, high-mating line; L, low-mating line; C, unselected control line.
From Cook, Siegel, and Hinkelmann (1972).

encounters were measured. Table 3 reports, for males, sums of completed copulations and of three intimately related behaviors — treads, mounts, and courtings — plus relative aggressiveness and correlated body weights at two ages. Females were scored for body weight and egg-laying capacity. After 10 generations of selection for high and low mating, reciprocal crosses were made among three lines (the third being a randomly bred control line). Parents used to produce the cross progeny were taken at random from the selected and the control lines.

The genetic control of high levels of treading, mounting, and completing mounts (with insemination) is autosomal, polygenic, and additive (see Chapter 6, especially Sections 6.3 and 6.4). No major loci controlling these behaviors were located, no one chromosome or chromosomal section or arm bore linked loci of any greater importance than any other in these regards, and genetic material on the sex chromosomes (in birds, males are the homogametic sex, bearing two like sex chromosomes, as do mammalian females) was not involved. Reciprocal crosses — high-line females × low-line males and low-line females × high-line males — revealed no maternal effects (genetic consequences fostered by the contents of the egg cytoplasm; see Chapter 6) for those behaviors; such effects would initially be detected by significantly different trends in reciprocal crosses.

After examining Table 3, do you detect any relations between male and/or female body weight, female sexual maturity and/or productivity, and mating behavior and success? Other than general impressions, how would you test for them statistically? If you were required to design a

TABLE 3. *(Continued)*

| Body weight (male) | | | Body weight (female) | | | Reproductive performance (female) | | |
N	4-week	8-week	N	4-week	8-week	N	Age at first egg (days)†	Hen-day egg production
186	289 ± 2	826 ± 6	181	246 ± 2	655 ± 6	112	180 ± 2	62 ± 2
81	294 ± 4	844 ± 9	58	256 ± 4	684 ± 10	36	181 ± 4	58 ± 3
69	297 ± 4	854 ± 14	62	272 ± 4	716 ± 10	37	166 ± 3	62 ± 4
35	320 ± 6	873 ± 19	24	266 ± 9	667 ± 21	15	178 ± 4	63 ± 4
126	300 ± 4	875 ± 8	107	266 ± 4	713 ± 7	55	188 ± 3	53 ± 3
37	288 ± 5	858 ± 17	31	257 ± 6	731 ± 14	21	174 ± 5	64 ± 5
80	306 ± 4	906 ± 10	58	273 ± 4	753 ± 10	34	174 ± 4	62 ± 4
99	278 ± 3	844 ± 10	105	250 ± 3	702 ± 8	59	175 ± 4	61 ± 3
15	266 ± 1	823 ± 32	16	248 ± 8	688 ± 23	7	172 ± 11	62 ± 8

† Age at first egg indicates sexual maturity.

breeding plan to maximize body weight and productivity, the former in both sexes and the latter in females, how would you employ the data presented in Table 3 and the selected lines from which the data were extracted? (We recommend a set of papers reporting experimental investigations of genetic selection for social dominance in chickens and considering inbreeding and social dominance: Craig and Baruth, 1965; Craig, Ortman, and Guhl, 1965.)

Before attempting answers to these commercially motivated questions, consider an interesting but preliminary experimental design — Wood-Gush's (1972) study of behavior that precedes egg laying. The white strain employed was white leghorn, the brown strain was of Rhode Island red and light Sussex origin, and both whites and browns were like those strains available on the open market. Remember that sitting time (when eggs are finally deposited) is more important to a poultryman than pacing time!

Tables 4 and 5 show results of tests on individual birds. The mean pacing score for whites is 767.6, and for browns 414.8, paces in an hour of activity. The mean time to nest reentry after pacing is 8.59 minutes for whites and 5.10 minutes for browns. Note the considerable variability from bird to bird and that the between-bird, not between-strain, differences are significant. Whites pace more but they take longer to get back to their nests than browns.

Tables 6 and 7 show results of tests of eight birds from each of the two strains and supply information on the influences of external conditions. Does environment influence pacing and sitting times, both pre-egg-laying behaviors? Browns consistently sat more than whites, regardless of the hour, but they sat more during the prelaying hours. Light made the brown females pace more before oviposition. Whites again paced more than browns, and they also paced most during pre-egg-laying times. As expected, there is an inverse correlation between enhanced pacing and decreased sitting. Enclosed (solid, nonreflective walls on

TABLE 4.   Mean Number of Paces Taken by Each Hen During Two Half-Hour Prelaying Periods in Pens with Nests

Whites		Browns	
Bird	Mean pacing	Bird	Mean pacing
1	218.5	A	108.5
2	377.5	B	317.0
3	475.0	C	363.0
4	622.5	D	377.0
5	1412.0	E	522.0
6	1500.5	F	801.5

From Wood-Gush (1972).

TABLE 5.   Mean Time for Each Hen to Reenter Nest after Removal from It

	Whites			Browns	
	Mean time of second entry			Mean time of second entry	
Bird	Min.	Sec.	Bird	Min.	Sec.
4	1	38.5	G	2	13.5
1	2	9.0	H	2	46.5
5	3	46.6	I	4	51.5
2	8	5.0	C	6	46.0
6	13	31.0	D	8	51.0
3	22	21.5			

From Wood-Gush (1972).

three sides) or open (wire) cages did not statistically significantly alter the tendency to pace on the part of white birds. This tendency was always emphasized during the prelaying hour. Brown hens again sat more, but did so especially in enclosed cages.

Elsewhere, Wood-Gush has investigated nesting behavior (1954, 1963) and the effects of different lighting and cage types (1969, 1972) on the behavior of the domestic fowl. He has speculated whether the essential sitting component of nesting and prelaying behavior depends upon timed key stimuli and whether the nonsitting white strain has lost, via genetic selection over many generations of domestication, the ability to respond to particular stimuli by sitting, while the brown strain generalizes even suboptimal stimuli and sits. One wonders how the ancestral chicken behaved or behaves. Was this ancestor the red jungle fowl *Gallus spadiceus* (depicted in Weisz, 1969)?

What indeed does domestication imply to the behavior geneticist? For

TABLE 6.   Responses of Females of Brown and White Strains in Prelaying Period in Cages Under Dark and Light Conditions

	No. of paces		Sitting time (min.)	
Strain	Prelaying hour	Nonlaying hour (control)	Prelaying hour	Nonlaying hour (control)
		Dark		
Brown	119	104	18.8	1.6
White	704	82	0.5	2.1
		Light		
Brown	248	113	16.2	3.9
White	684	127	0.3	1.3

From Wood-Gush (1972).

his own benefit, man controls the reproduction of his domesticated plants and animals, always breeding from his ideal, always assuming a high positive correlation between genotype and phenotype. These populations then represent only a portion of the total possible range of variation, and this portion is consistently restricted. *For the sake of the acquisition of high frequencies and/or magnitudes of desirable traits, man may be obligated to include associated traits, sometimes behavioral ones.* He often cannot exclude these latter without sacrificing at least a part of his primary goal. An obvious example is the extreme nervousness encompassing tendencies to self-mutilation found in lines of thoroughbred race horses. Another may be the flighty behavior of the otherwise commercially valuable white leghorns. In just about any selection program correlated responses to selection are likely (Sections 5.5 and 6.12). Often the correlated response leads to lower fitness, and the desired response may then cease altogether.

"Therefore," Wood-Gush (1972) concludes, "from the agricultural point of view, it would be desirable to continue to experiment . . . to produce stimuli that will elicit sitting in these birds."

In "The Nuns' Priest's Tale," part of Chaucer's *Canterbury Tales*, is what may be the earliest allusion to nonrandom mating in chickens:

> And in the yard a cock called Chanticleer.
> In all the land, for crowing, he'd no peer.
> His voice was merrier than the organ gay
> On Mass days, which in church begins to play;
> More regular was his crowing in his lodge
> Than is a clock or abbey horologe.
> By instinct he's marked each ascension down
> Of equinoctial value in that town;
> For when fifteen degrees had been ascended,

TABLE 7. Responses of Females of Brown and White Strains During Prelaying Period in Enclosed and Open Cages

	No. of paces		Sitting time (min.)	
Strain	Prelaying hour	Nonlaying hour (control)	Prelaying hour	Nonlaying hour (control)
		Open		
Brown	144	75	32.66	7.41
White	1035	104	9.28	6.87
		Enclosed		
Brown	67	124	48.59	9.09
White	939	78	17.12	15.16

From Wood-Gush (1972).

Then crew he so it might not be amended.
His comb was redder than a fine coral,
And battlemented like a castle wall.
His bill was black and just like jet it shone;
Like azure were his legs and toes, each one;
His spurs were whiter than the lily flower;
And plumage of the burnished gold his dower.
This noble cock had in his governance
Seven hens to give him pride and all pleasance,
Which were his sisters and his paramours
And wondrously like him as to colours.

See "Genetics and Sexual Selection" (Ehrman, 1972), where Lill and collaborators' work on the role of vision and homogamy in avian mating behavior is reviewed. We also recommend Selander (1972).

10.12 Geese

In the Canadian Arctic, the lesser snow goose occurs in two color phases, blue and white. This plumage dimorphism is determined by a single autosomal gene with the blue form dominant over the white one (though traces of white on a blue animal indicate the heterozygote). Although the blue goose and the lesser snow goose are still sometimes incorrectly classified as different species, they are both *Anser cerulescens*. The rare instances where blue goslings hatch from nests belonging to two white parents are due to dumping, the deposition of eggs in a nest by a female not incubating on that particular nest (Cooke and Mirsky, 1972).

Positive assortative mating (more matings between like phenotypes than expected assuming random mating; see Chapter 2) occurs in these birds so that of 3480 lesser snow goose families surveyed between 1968 and 1970 in Manitoba, Canada, the following was found:

Parents	Observed no. of families	Expected no. of families assuming random mating
White female × white male	3099	3036.5
White female × blue male	195	257.5
Blue female × white male	109	171.5
Blue female × blue male	77	14.5

Note the excess number of matings between likes and the corresponding paucity of matings between unlikes. This pattern can be explained if birds select their mates according to the color of their parents and/or their siblings. To investigate this experimentally, Cooke and McNally (1975) tested three captive flocks for color preferences considering (1) bird-to-bird approach responses, (2) association preferences, and (3) mate selection. Their findings can be summarized as follow:

- Approach response. Young birds placed in a choice situation had a significant preference for birds of the color of their parents. Sibling color when different from parental color appeared to modify choices. If parents were removed during adolescence, early color preferences could be altered. The most recent association determined these preferences and no differences were detectable in the responses of goslings to maternal as opposed to paternal colors.

- Association preferences. In an open-field situation, birds usually associated with their peer group (siblings and nonsiblings) at both 1 and 2 years of age. The degree of this association faded from 1 to 2 years of age. When birds associated with non-peer-group birds, they showed a distinct tendency to associate with birds the same color as their peer group.

- Mate selection. In a flock raised as a single large group with virtually no parental contact, pair formation did not depart from randomness in terms of color, suggesting that nonrandom mate selection in lesser snow geese is additionally a function of early experiences. In other than captive flocks where parents and offspring were the same color, mate selection reflected preferences for familial color; where parents and offspring were of opposite colors, either parental or offspring color was chosen. Similar results were found in marked wild birds under field conditions.

If the parent is removed (as happens in the wild), color preference may be altered but is more likely to be maintained through association with birds of the familial plumage color. Directly or indirectly, parental color influences mate selection. However, the relative strengths of color discrimination in the two sexes is currently unknown, as is the reason for the occasional renegade bird who prefers a mate of a color inappropriate to its pedigree.

Here then is a serviceable example of the role of genetically fostered early "learning" in self, mate, and species recognition. This was investigated with captive animals maintained under conditions mimicking wild ones as closely as possible, at the expenditure of great effort and resources. Note the cogent combination here of data harvested from wild and captive animals. (See Section 8.4 for an additional example provided by a short-lived species.) This represents the initial experimentally verified mechanism explaining assortative mating in a wild bird species under field conditions. But we still know neither the origins of this mating system nor its evolutionary sequelae. Cooke, Finney, and Rockwell (1976) ask about the relative Darwinian fitness of the various types of pair bonds; for example, are white × white and blue × blue parents more reproductively successful than white × blue ones?

10.13   Turkey-pheasant hybrids

Early posthatching mortality is high for ringneck pheasant female × bronze turkey male F_1 offspring brought into existence by means of artificial insemination (Asmundson and Lorenz, 1955). But mortality is low for hybrids from the reciprocal cross, with any survival differences being due to veterinary management. Hybrids were raised in mixed flocks with several species and hybrids. The smaller pheasant female × turkey male hybrids from pheasant eggs fared poorly at first, while the larger hybrids out of turkey eggs did better. Later mortality was due to severely pendulous crops, remediable by surgery. Neither eggs nor sperm could be matured by year-old hybrids, and these birds never engaged in any sort of mating behavior, however preliminary. There was some secondary sexual modification of the skin on the head of males, but this was all. Clearly, normal behavior characteristic of turkeys and of pheasants was totally disrupted by hybridization.

While the total sterility of these F_1 generations precludes further genetic analysis (by means of backcrosses, etc.), only if hybridization is resourcefully attempted and reattempted can this particular, sometimes fruitful avenue of behavior genetics analysis be exploited, as shown in Chapter 5 in some sibling species and races of *Drosophila* and in Section 10.14 on duck hybrids. A prime example in which the hybridization technique was particularly rewarding was the study of nest building behavior by lovebirds, discussed in Chapter 5.

10.14   Duck hybrids

Species-specific behavior has been used for taxonomic purposes; interspecific hybrids (if they can be brought into being in sufficiently viable numbers, see Section 10.13) allow the employment of this behavior for genetic purposes. Ten male pintail ducks, *Anas acuta*, were confined with the same number of mallard, *A. platyrhynchos*, females (Sharpe and Johnsgard, 1966). Most of the eggs deposited never hatched, but a total of three females and four males were reared. These F_1 interspecific hybrids were bred with one another, and 19 F_2 female plus 23 F_2 male birds were produced. This F_2 generation was analyzed for pintail- versus mallard-like plumage and behavior. We present this material for two reasons: (1) because it represents a rare behavior-genetics analysis of birds and of a bird other than the best-known domestic chicken, and (2) because Sharpe and Johnsgard (1966) utilize an interesting hybrid index (see Table 8) for the semiquantitative evaluation of these precious interspecific hybrids as regards not only their anatomical but their behavioral

TABLE 8. Plumage and Behavior Indices of Hybrid Mallard (A. acuta) × Pintail (A. platyrhynchos) Ducks

Scale	Plumage	Bill pointing	Turning	Nod swimming	Down-up drinking	Burp call
			Behavior			
0	Wholly mallard phenotype	Mallard-like with tail raised	Toward female, mallard-like, partial turn	Mallard-like fully expressed	Present, mallard-like	Absent, mallard-like
1	Appearance approaching mallard phenotype			Intermediate		
2	Appearance intermediate between two species			Rudimentary		
3	Appearance approaching pintail phenotype	Pintail-like with tail lowered	Toward female, pintail-like, full turn	Absent, pintail-like	Absent, pintail-like	Present, pintail-like
4	Wholly pintail phenotype					

From Sharpe and Johnsgard (1966).

phenotypes. Table 8 provides details about the most pertinent behaviors scored. Figure 17 graphs the relations between the two hybrid indices; no F_2 hybrid displays were atypical of one or the other parental species. (Note that the cumulative behavior indices for 11 F_2 males ranges from 3 to 15.) We may safely conclude that both sets of features, feathery appearance and behavior, clearly are under genetic control. (Grunt-whistling, a male vocal courtship display aimed at pair formation, is not included in either Table 8 or Figure 17 because it is very similar in both parental species and the hybrids grunt-whistled like their ancestors.) The $r = +0.756$ line in Figure 17 represents a significantly positive correlation coefficient between the inheritance of these behavioral and plumage characteristics in the hybrid F_2 generation. (See Section 6.8 on the calculation and interpretation of r, the widely employed correlation coefficient, and Section 9.3 on correlations of behavior and morphology.) How might such a close relationship be exploited by the behavior geneticist?

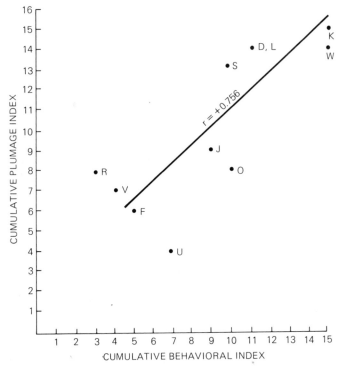

17. **Plumage and behavioral indices correlation** for 11 mallard (*A. acuta*) × pintail (*A. playtrhynchos*) F_2 duck hybrid males (letters). (From Sharpe and Johnsgard, 1966.)

10.15 Horses

In horses, the trotting gait is autosomally dominant to the pacing gait. In order of increasing speed, normal gaits are walk, singlefoot or rack or amble, canter, pace, trot, and gallop. Where Lf means left foreleg; Rf, right foreleg; Lh, left hindleg; Rh, right hindleg; and O, no feet on the ground:

pacing = Rh, Rh Rf, O, Lh, Lh Lf, O, Rh
trotting = Rf, Rf Lh, Lh, O, Lf, Lf Rh, Rh, O, Rf

Horses must, of course, be trained to race in harness, but there is an inherent tendency for them to move in one or the other distinct fashion. This seems to be simply genetically determined (Snyder and David 1957).

10.16 Cows

With regard to nonsexual behavior in cattle, Payne and Hancock (1957) investigated the effect of tropical climate on the performance of European type (Jersey and shorthorn) cattle utilizing six sets of identical twin heifer calves. The twins (one per set) observed in Fiji did not uniformly respond to stresses imposed upon them by a tropical climate: two performed (in milk production, butterfat production, nonfat solids production, feed intake, water intake, rectal temperature, and in respiratory rates) as did their cotwins in nontropical environments, while four did not perform like their own individual cotwins quartered elsewhere. The authors believe that "this suggests that there are profound differences in the reaction of individual temperate-type dairy cattle of the same breed to a tropical climate, and that these differences must be based at least in part on genetic differences between individuals."

Hancock (1954) has also intensively investigated grazing behavior in six sets of uniformly treated lactating monozygotic cattle twins. The animals were observed in a 1-acre paddock on 8 days at approximately monthly intervals, each occasion covering a 24-hour period (Fig. 18). The following data were collected on each day for each cow: (1) time spent grazing, (2) time spent standing or walking and loafing, (3) time spent lying down, (4) distance walked, (5) number of defecations, (6) number of urinations, (7) number of drinks, plus some observations of (8) number of bites of grass, (9) time spent ruminating, (10) number of boluses ruminated, and (11) number of bites to each bolus.

Exploiting the benefit monozygotic twins employed in such experiments confer by reducing error variance (Sections 6.3 and 6.4), striking

18. **Studying monozygotic cattle twins.** Observation methods used by Hancock. (Courtesy Ruakura Agricultural Research Centre, Hamilton, New Zealand.)

statistically significant correlations between cotwins were calculated for time spent grazing, a crucial activity of a dairy cow. Hancock comments on the habit of cotwins to graze simultaneously and to stand close together most of the time. "Twins provided a special case of gregariousness," he notes, in that they seek each others' company while grazing and when congregating to lie down. Such inclinations could lead to greater apparent similarities, perhaps imitative ones — a situation faced again, when human cotwins are raised together. (See also pertinent comments by Kilgour, 1975.)

In summary, the mean differences with regard to time spent grazing (4 minutes), loafing (7 minutes), and lying down (8 minutes) are negligible within sets of identical twins, while the between-set differences for time spent grazing (138 minutes), loafing (114 minutes), and lying down (60 minutes) are very great.

10.17 Bulls

Fortunately, one set of monozygous triplet bulls has been examined for behavioral likenesses, among other traits (Olsen and Peterson, 1951), albeit briefly. All three milking shorthorn males were reported as being alike in stubbornness and lack of interest in serving cows. They were, however, brought into service after some effort, at 13.5 months of age. At that time they were capable of producing but one ejaculate a week; 4 months later this unimpressively increased to two — low for bulls in general. With reference to these few ejaculates, volume, density, total number of sperm per ejaculate, motility, and presence of abnormal sperm were statistically not consistently different among these brothers. Olsen has studied three sets of monozygous triplet bulls (one milking shorthorn and two Guernseys), plus two sets of monozygous twin (Guernsey) bulls (Olsen and Peterson, 1952). Twinning occurs more frequently in cattle than in man, so there is more material of this sort available for use in behavioral and genetic studies. (Of all births in dairy cattle, 0.11 percent, and of all like-sexed twins, 10.6 percent, are estimated to be monozygotic twins; Hancock, 1954. In man, 0.0035 percent is the frequency of monozygotic twinning considering all populations; Levitan and Montagu, 1971.) Only cattle produce an unlike-sexed dizygotic twin called a *freemartin*. This occurs when the female cotwin is masculinized by anastomoses of blood vessels delivering male hormones into her fetal circulatory system.

10.18 Conclusions

With bovine clones, so to speak, we end this literature survey. We are concerned lest we have omitted worthy efforts, but the literature of behavior genetics is disparate and requires cataloguing. (For one example, there is the isolated report of a homogamic mating preference in one merino ram; Hayman, 1964.) This is in itself a good sign; it is indicative of a certain, albeit still not impressive, bulk to that literature. Our hope is that we participate in the stimulation of an increase. In particular, the question of the behavior genetics of domesticated species is one that may be of much greater importance to the animal breeder than previously realized. For an early appreciation of this, see Hafez (1968) on interspecific hybridizations, normal and aberrant mating behaviors, and selection for twinning, among other behaviors in domesticated animals. (In 1935, as an even earlier example, Hodgson briefly considered the effects of inbreeding on swine behavior.) We also recommend Hafez's (1969) anthology, *The Behavior of Domesticated Animals* — specifically Chapter 3, "The Genetics of Behavior," by J. Fuller; Chapter 12, "The Behavior of Horses," by E. Hafez and J. Signoret; and especially Chapter 13, "The Behavior of Rabbits," by V. Denenberg, M. Zarrow, and S. Ross. This last chapter contains a Section III B that summarizes all that is now known about the behavior genetics of rabbits. These include a variety of maternal behaviors: nest building and nest lining, cannibalism, aggressive protection of the young, changes in housing, retrieving, and suckling.

Domesticated species, willfully regulated in their reproductivity by man for his own purposes, provide rewarding material for further study. A final tangential query: is man himself a domesticated, wild, or weed (uncultivated) species? See Section 13.6 for a discussion of the evolution of our own species.

Note the phylogenetic series incorporated into the different sections composing this chapter. Apparently, organisms "lower" on this scale are currently offering more and most information concerning the chemical bases of behaviors. Too, we have had to rely to a considerable extent upon hybrid analyses; they are uniquely useful when genetic control is limited. And this is the sort of limitation removed only by further and extensive study.

In several sections of this composite chapter, we report *correlations between morphologies and behaviors* in organisms as diverse as wasps and fish and ducks and, in Chapters 9 and 12, even in mice and men.

And finally, for reasons itemized above, we predict that much of our future progress as behavior geneticists will be biometrical, with applications to this wide and perhaps wider variety of subject material.

Man: Certain Discontinuous Traits

In this and the next chapter we consider man, and consequently these two chapters include what are assuredly the most complex materials a student of behavior is obliged to cope with. This complexity stems from our inability to mold and to extend pedigrees to extract maximal genetic information. We simply cannot "make" the appropriate crosses and backcrosses with human subjects. Nor can we obtain information of the type that can be derived from inbred strains and their hybrids and from selection experiments.

We lose another advantage awarded by experimental animals — the capacity to control and to define the environments in which our experiments take place (Sections 6.1 and 6.2).

Chapters 11 and 12 encompass the *limitations of the methodologies* available to the behavior geneticist studying humans. The severity of these limitations can be assessed by comparison with the methods usable with experimental animals. Probably the major issue is the possibility of defining the environment for experimental animals and the impossibility of defining it for man. If our readers can *objectively* assess data concerning the possible genetic bases of human behavior, we have achieved one of our primary aims.

Our primer for analyzing behavioral traits in man is Chapter 7; there, distinction is drawn between threshold (present-absent) traits and continuously varying traits. While in some instances this distinction is cloudy, the present chapter is concerned with the former category and Chapter 12 with the latter. Complete coverage is not attempted; only representative examples are cited. Because man is anthropocentric, the literature is vast.

Here, we insert for didactic purposes *only*, the history of the muddled geneticomedical interpretations of kuru, a degenerative, always fatal neurological disease peculiar to the Fore tribe of New Guinea. Kuru means trembling. By now, this cerebellar malfunction has been properly identified as of viral etiology, transmitted when the warring Fore males cannibalize and share their dead enemy. Indeed, the earlier onset of the disease in males than in females had been recorded (see Leader, 1967; Leader and Leader, 1971, for additional information). The infectious microorganism is a slow virus, one whose striking effect is remote in time from entry into its host. Chimpanzees inoculated with extracts from the brains of kuru victims show equivalent symptoms — paralysis — only 18 to 30 months after injection.

Kuru was initially assigned a wholly genetic base:

- *KuKu* females and males suffered early onsets of the disease
- *Kuku* females had late onsets
- *Kuku* males were healthy
- *kuku* females and males were healthy

The high incidence of the symptoms in Fore natives was ascribed to genetic drift (cf. Stern, 1960 and 1973). This disease illustrates some of the problems encountered in the study of the genetic basis of human behavior. The difficulties are compounded in the study of primitive tribal groups whose cultures differ markedly from our own and are obscure to us for that reason. In the case of kuru, it was only after the anthropologists provided appropriate information that the initial simple genetic interpretation was found to be incorrect.

11.1 Stuttering

Stuttering, also called stammering, is a particular type of speech disorder beginning during childhood and persisting into adulthood in about 20 percent of cases (Van Riper, 1971). Sex has a marked effect, with males affected almost four times as often as females. Stuttering is also strongly familial; it "runs in families." Though little is known of its cause, environmental (nongenetic) factors play a demonstrable role. Even monozygotic twins are not always concordant for stuttering. In the presence of such environmental influences, the phenotype shows no Mendelian patterns of genetic transmission, and the observed familial pattern can often be "explained" by either a multifactorial or a single-major-locus model. This is especially so if only presence or absence of the trait is considered (Kidd, Reich, and Kessler, 1973, and references therein). How then can we even tentatively distinguish between these alternatives?

The possibility of wholly cultural inheritance must be considered. Cultural heredity is, after all, much more flexible than the biological sort; it is capable of rapid and distinct alteration from generation to generation. Children may mimic an older or colateral relative with whom they have sufficient contact. There is also great variation in degree and frequency of stuttering, both among individuals and temporally for one individual, probably in response to environmental input, particularly its emotional aspects. This view of the "inheritance" of stuttering has generally been ascendant in recent decades, but Van Riper (1971) has concluded that it does not adequately explain all aspects of the familial and developmental patterns of stuttering. Kidd, Reich, and Kessler (1973) have located a firm basis for assuming some genetic variation contributing to stuttering.

Garside and Kay (1964) noted that stuttering females have a higher frequency of stuttering relatives than do stuttering males. Though they could not exclude single-locus inheritance, they preferred a polygenic model with two thresholds — a higher one for females. With a higher threshold, females would be less frequently affected, but those affected would be more "genetically loaded" than affected males and, hence, have more affected relatives. Similar observations and explanations exist for cleft lip and palate (Woolf, 1971a) and for other congenital anomalies: males are more commonly affected, the trait clusters familially, but there are no overt Mendelian patterns of transmission. (These traits are assuredly not X-linked because affected father plus son is a common family situation.)

More recent and more sensitive genetic analyses utilizing the sex effect in searches for differences in dominance variances (Kidd, Reich, and Kessler, 1973, and Chapter 7 of this text) have tried to discriminate between the alternatives of multifactorial (or polygenic) and single-major-locus transmission. Though Kidd and coworkers could not exclude either mode of inheritance, the single-major-locus model gave a better fit to family data. Using the same concept of two thresholds but only one locus with two alleles and the requisite environmental factors, their analysis suggests that the gene for stuttering occurs at a frequency of approximately 8 percent (0.08) with the normal (nonstuttering) allele at 92 percent. The environmental factors involved interact in ways such that approximately 25 percent of male heterozygotes, but only 5 percent of female heterozygotes, stutter at some time during their lives. All homozygotes for the stuttering gene stutter at some time, irrespective of sex. Thus environmental factors affect only the heterozygotes, and differently according to sex. This model not only accounts for the greater frequency of affected relatives of female stutterers (a fact used by Garside and Kay, 1964, to support polygenic inheritance), but also explains the observed high incidence of affected sisters of female probands. This latter observation cannot be explained by polygenic inheritance. (Because this observation may be due merely to sampling error, the polygenic model of stuttering still cannot be excluded.) Additional research, such as twin, adoption, and linkage studies (the last the most difficult), are needed to resolve these uncertainties about stuttering. Nevertheless, Kidd, Reich, and Kessler's (1973) inquiry, though it produced no definitive explanation of stuttering, has pertinence as a method of genetic analysis for other human traits which show a sex effect and familial concentration, but no Mendelian patterns — e.g., dyslexia, hyperkinesis, enuresis, and the congenital anomalies mentioned above.

11.2 Enzymes and behavior: the porphyrias

What is the relation between the genetics of behavior and that of molecules? Stein, Wise, and Berger (1972) have speculated on the possible chemical cause of schizophrenia, noting the (progressive) brain damage which might be caused by a genetically determined enzyme deficit. Eiduson et al. (1964) considered this point among others, specifically in their excellent chapter on biochemical genetics and behavior. They offered a list of 20 hereditary metabolic disorders culminating in neurological and/or behavioral disturbances:

- Tay-Sachs disease (amaurotic idiocy)
- Gargoylism
- Niemann-Pick disease
- Acanthocytosis
- Congenital agammaglobulinemia
- Wilson's disease
- Phenylketonuria
- Albinism
- Maple sugar urine disease
- H disease

- Galactosemia
- Cretinism
- Adrenal hyperplasia
- Porphyria
- Congenital nonhemolytic jaundice with kernicterus
- Gaucher's disease (infantile type)
- Fanconi's disease
- Idiopathic spontaneous hypoglycemia
- Argininosuccinicaciduria
- Cystathioninuria

We consider some of these conditions elsewhere; and porphyria (Greek for purple, referring to the color of urine containing excreted porphyrin, a hemoglobin derivative) is interesting in the current context. Here is an instance where the chemistry of the biosynthesis is well known: enzymatic defects are involved in all forms of porphyria, but there is insufficient information about the behavioral abnormalities associated with acute toxic intermittent porphyria and variegated chronic porphyria. A third type, porphyria erythropoietica, is not associated with behavioral disorders.

Abdominal pain is often the initial complaint in autosomally dominant acute porphyria; less frequently, the first symptom is partial paralysis, and in this case the neurological component is obvious from the clinical beginning. Repeated attacks leave residual damage including dementia (Stevenson, Davidson, and Oakes, 1970). Acute overt attacks may be foreshadowed by years of nervousness, hysteria, and psychoneurosis. During attacks, neurosis gives way to psychotic episodes and manic-depressive behavior; delirium with hallucinations occurs. Korsakoff's psychosis is now seen: disorientation, polyneuritis, muttering, insomnia, and pain over areas more extensive than the abdomen, i.e., into the extremities.

Abdominal discomfort plus neurological involvement also characterize mixed chronic porphyria, but photosensitivity is prominent in this form. The symptoms associated with both these porphyrias (also autosomal and dominant), may be aggravated by the ingestion of certain drugs such as barbiturates or sulfonamides.

Porphyrins are fundamental to various kinds of cellular metabolism in that these compounds participate as intermediates in the synthesis of hemoglobin, myoglobin, cytochromes, catalase, peroxidases, and even plant chlorophyll (Eiduson et al., 1964). There is no cure for porphyria except treatments of the sort that tranquilize nervousness. But "in the patient with known disease who has been warned about the precipitating factors, the prognosis is now much better than this . . . mortality rate of 24 percent over a five-year observation period" (Tshudy, 1973).

11.3 Tasting abilities and other sensory perceptions

With Kalmus (1967, and references therein) we agree that the *behavioral consequences of inherited differences in sensory perception* is a subject so vast that one can only endeavor to provide suitable references and partially survey appropriate materials. In Section 11.4, vision is considered briefly. The reader is referred to McKusick's (1971) excellent itemization of the known genetic bases of sensory and other defects, especially the section on hereditary deafness.

Phenylthiocarbamide (PTC) tasting is the best-known human polymorphism for tasting ability, though the behavioral consequences of the three possible PTC genotypes (*TT*, *Tt*, and *tt*) and the two possible PTC phenotypes (T, taster, or tt, nontaster) are actually unknown. Taste differences involved in food choices, for instance, are likely to involve little or no PTC discrimination, but nontasters are reputedly less discriminating. PTC tasting depends ultimately upon $=\text{N}-\underset{\underset{\text{S}}{\|}}{\text{C}}$ recognition.

(Snyder and Davidson, 1937, and Barrows, 1945, have investigated other genetic variables concerning diphenylguanidine- and brucine-tasting deficiencies, respectively.)

Cavalli-Sforza and Bodmer (1971) have commented on the difficulties of ascertaining taste thresholds in subhuman animals. Fisher, Ford, and Huxley (1939) apparently attempted this with primates in the London Zoo. There, a chimpanzee spat in Fisher's face after ingesting a bit of PTC, which is bitter to tasters but neutral to nontasters. So apparently this particular tasting polymorphism exists in other primates as well as humans (Fig. 1), with approximately 25 percent of humans having the *tt* (nontaster) geno- and phenotype. This is not to imply that the *t* allele is perfectly recessive; nontasting is almost always a recessive condition, but recent studies are based on the response to serial dilutions of PTC, whereas older ones employed a single dilution, usually in the form of PTC crystals or PTC-impregnated paper. Utilizing these older testing conditions, Rife (1938) reported an almost 4 percent discordance in the PTC-tasting ability of monozygous twins.

What, then, is correlated with the inability to taste PTC? Adenomatous nodular goiter occurs more often in nontasters than in tasters and there are rumors that nontasters are more common among people who dislike the flavor of alcoholic beverages. Other types of goiters, e.g., toxic diffuse goiter, occur more often among tasters. Rimoin and

1. **PTC tasting in primates.** Percentages of tasters and nontasters in different primate genera. Note that no tasters were found among spider monkeys or woolly monkeys. (From Chiarelli, 1963.)

Schimke (1971) have commented on Shepard's (1961) data, again documenting the increased incidence' of PTC nontasters in families with athyreosis (a form of thyroid gland disgenesis resulting in cretinism) and other thyroid conditions, but finding no link via tasting perception among these conditions.

Fischer et al. (1961) and Fischer and Griffin (1960) tested for relations between taste thresholds and food dislikes (with 118 different foods) or for what may be more formally called the genetic aspects of gustation. They discerned three loci for taste: one for quininelike compounds including taste competence for sucrose and sodium chloride, one for 6-n-propylthiouracil and related compounds, and one for hydrochloric acid and other substances. The lower the taste threshold for bitter substances (including quinine and 6-n-propylthiouracil), the more foods disliked. These and other hereditary differences in taste acuteness may reflect "generalized" drug responsiveness and also certain Wechsler Adult Intelligence Scale personality indices. We agree with Spuhler and Lindzey (1967) that the frequencies of the nontaster allele in different human populations (see Garen's, 1961, survey) are too high to be maintained by the traditionally implicated mutation-selection balance of forces. This is truly a behavior-genetics balanced polymorphism for sensory perception, though its intricacies are still obscure. We may then inquire: Why?

Families vary profoundly in their taste thresholds, with reports ranging from fivefold intrafamilial differences to identical taste thresholds in identical twins. Hirsch (1964) reviewed this and other familial data regarding, for example, auditory acuity and discrimination. Ehrman (1972) has considered phenotypic assortative mating on the basis of sensory perception, especially with regard to auditory normalcy. Such assortative mating for hereditary sensory defects, e.g., deafness, may have the most profound genetic consequences. See, for instance, the pedigree in Figure 2.

Vandenberg (1967, 1972) has offered suggestions for research work, commenting that practically nothing is known about the genetics of

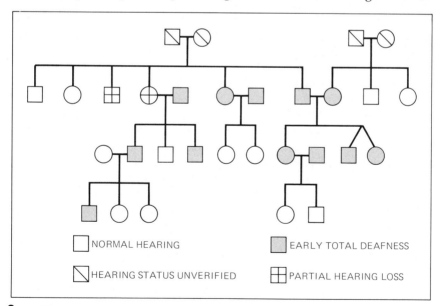

2. **Phenotypic assortative mating for deafness** demonstrated in a sample human pedigree. (From Ehrman, 1972.)

smell or of kinesthetic perception. His chapters and those of Hirsch and of Spuhler and Lindzey, cited above, are recommended as surveys of what is still to be done. What has been done is enticing, but we note the concentration of efforts upon PTC tasting because of its ostensible genetic simplicity. After all, man samples his environment via his senses — before he responds to it behaviorally. The genetic basis of his cues are important in three respects (Ginsburg, 1967): (1) as a clue to the evolutionary history of the species, (2) as a potential for further evolution, especially if conditions of life change, and (3) as a means of understanding individual variability and of dealing effectively with it in a given situation.

11.4 Color and other vision

Defective color vision of the kind generally known as red-green color blindness was recognized as early as the eighteenth century. (See Kalmus, 1965, and Cornsweet, 1971, on the physiology of normal and restricted color vision.) Normal people can match color by the mixing of colors from three spectral regions: red, green, and blue. Hence we can refer to normals as trichromates. In the most severe form of the red-green defect, the subjects are able to match color only when two hues are involved. These subjects are therefore known as dichromates. There are two types of dichromates, the protanopes (red-blind individuals) and the deuteranopes (green-blind individuals). The two corresponding anomalous (or weakened) trichromatic conditions are termed protanomaly and deuteranomaly: affected individuals are partially red-blind or green-blind, respectively.

These conditions are controlled by sex-linked recessive alleles at two closely linked loci, one for red-blind alleles and the other for green-blind alleles. Because the alleles are sex-linked, the frequency of the conditions is much higher in males than in females (Section 2.3). A characteristic incidence of color blindness in males is about 5 percent (0.05), so in females the expected incidence is $(0.05)^2 = 0.0025$ or 0.25 percent. In addition to the above conditions is a rare dichromatic defect, tritanopia and the corresponding tritanomaly, in which color discrimination in the blue-green region is affected. Autosomal incomplete dominance or autosomal recessive inheritance seems to be involved here.

The frequency of color-blind males varies from population to population and may be greater than 10 percent (Table 1). This proportion is far too high to be maintained by mutation alone and so suggests a genetic polymorphism, although as yet there is little idea of the selective factors involved. These factors surely concern, however, interaction between genetics first, and culture second. Color blindness is more common in societies that long ago abandoned hunting and gathering; its incidence seems to increase in industrialized societies, suggesting that selection for normal color vision must have been relaxed. Note, for instance, the three entries for Australia in Table 1. Could there have been, as Neel and Post (1963) suggested, a "transitory 'positive' selection for color-blindness"? Color-blind hunters (and soldiers) are said to possess a sharpened sense of form and awareness of edges; they "see through" camouflage. Judd (1943) proposed that normal-visioned observers be provided with filters to confer upon them, when needed, the enhanced color-blind capacities to distinguish elements of and in terrain. Would, therefore, a mixed scouting troop be advantageous — one including a small proportion of color-blind hunters (Pollitzer, 1972)?

The performance of color-blind drivers has been investigated by Cole (1970) and their traffic accident frequencies found to be insignificantly different from those of normal-visioned individuals (Gramberg-Danielson, 1962). This must in part be due to the more conspicuous use

TABLE 1.   Proportion of Males with Defective Color Vision in a Variety of Populations

Population	%*	Population	%*
Europe		*Africa*	
English	6.8– 9.5	Bechuana	3.4
Scottish	7.5– 7.7	Bugandan	1.9
French	6.6– 9.0	Bahutu	2.7
Belgian	7.5– 8.6	Batutsi	2.5
German	6.6– 7.8	Congolese	1.7
Swiss	8.0– 9.0		
Norwegian	8.0–10.1	*North America*	
Czechoslovakian	10.5	U.S. White	7.2–8.4
Russian	6.7– 9.6	U.S. Negro	2.8–3.9
Jewish (Russian)	7.6	Amerindian	1.1–5.2
Finnish (Leningrad)	5.7	Eskimo	2.5–6.8
Turkish (Istanbul)	5.1	Canadian White	11.2
		Mexican (urban)	4.7–7.7
Asia		Mexican (tribal)	0–2.3
Tartar	5.0– 7.2		
Chinese	5.0– 6.9	*South America*	
Japanese	3.5– 7.4	Brazilindian	0–7.0
Indian (caste Hindu)	0–10.0	"White" Brazilian	6.9–7.5
Indian (tribal)	0– 9.0	"Dark" Brazilian	8.8
Israeli	2.1– 6.2	Brazilian Japanese	12.9
Druse (Israel)	10.0	*Australia*	
Filipino	4.3	White	7.3
Fiji Islander	0– 0.8	Aborigine	2.0
Polynesian (Tonga)	7.5	Mixed	3.2

* Ranges of percentages represent different samples.
From Kalmus (1965).

of blue in green traffic lights. Operations involving color coding in industry require four classes of judgment (Cole, 1972): comparative judgment of colors, connotative recognition of colors (learning associations such as greenlike leaves), denotative recognition of colors (the correct naming of a simple color), and esthetic judgments.

Dunlop (1943) pointed out that mental and even nutritional maladjustments may affect color vision, so that a phenocopy of hereditary color blindness may be produced. *Phenocopies* are phenotypic modifications produced environmentally that mimic genotypic modifications (Section 2.5). Taylor (1971), while investigating the effects on employment of defects in color vision, found, to his surprise, that only 224 suitable career choices were voiced by 613 teenage boys with defective color vision: he "wondered then whether there was something about colour blindness which made the subject gravitate towards the wrong career." Color-defective art students have been studied by Pickford (1972):

> The influences of temperament and personality are important. Perhaps they may be summarised as follows: If a student is bold and ignorant of his defect or insensitive about its presence, he may be able to use colours in a striking way which may seem original, and he might suggest that he could start a new mode of colouring in painting. If he is sensitive about his defect he may become involved in attempts to learn selfconsciously to compensate for it or avoid its effects. This may lead to considerable anxiety about his work and examinations, especially if he feels that his defect will not be understood so that allowance can be made for it. . . . Consequently, some collaboration between schools of art and psychology departments will be called for if a proper handling of the whole subject is to be achieved.

The reader should also be aware of an array of genetic anomalies wherein albinism is combined with abnormal, often cross-eyed, vision. Similar abnormalities have been described in the "white tiger," Siamese cats, ferrets, hamsters, mink, and many other mammals (Guillery and Kass, 1973, and references therein).

11.5 Some mutations in *Homo sapiens*

Exploiting Victor McKusick's comprehensive and consistent efforts (1971), we itemize in Table 2 a few of those mutations, in the broadest sense, recorded in human subjects that are known to alter behavior. These alterations are most often *not* the primary effects of the mutant genes, but for us they assume not less than the greatest importance.

We adhere to McKusick's (1971) method of classification and omit all itemization of hereditary deafness, referring our readers to pages xxiii-

TABLE 2. Some Mutations Affecting Behavior in Man

Autosomal Dominant Phenotypes

10430 Alzheimer's disease of brain
Presenile dementia, sometimes with parkinsonism; like Pick's disease (lobar atrophy)
10850 Ataxia, periodic vestibulocerebellar
Vertigo, diplopia (double vision), and slowly progressive cerebellar ataxia in some
11340 Brachydactyly-nystagmus-cerebellar ataxia
Nystagmus, mental deficiency, and strabismus
11530 Carotinemia, familial
Nightblindness
12620 Disseminated sclerosis (multiple sclerosis)
Neurological disorder
12640 Double athetosis (status marmoratus or Little's disease with involuntary movements)
Infantile cerebral palsies
12770 Dyslexia, specific (congenital word blindness)
Speech defects associated in many instances
12820 Dystonia, familial paroxysmal
Paroxysmal dystonia; unilateral dystonic postures without clonic movements or change in consciousness
13040 Electroencephalographic peculiarity
Occipital slow beta waves (16 to 19 per second) replace alpha waves
13630 Flynn-Aird syndrome
Neuroectodermal syndrome with visual abnormalities including cataracts, atypical retinitis pigmentosa, and myopia; bilateral nerve deafness, peripheral neuritis, epilepsy, and dementia

Autosomal Recessive Phenotypes

20130 Acroosteolysis, neurogenic
Abnormality of peripheral sensory nerves, perhaps insensitivity to pain
20420 Amaurotic family idiocy, juvenile type (Batten's disease in England, Vogt-Spielmeyer's disease on European Continent)
Rapid deterioration of vision and slower but progressive deterioration of intellect; seizures and psychotic behavior
20700 Anosmia for isobutyric acid
Inability to smell isobutyric acid (sweaty odor)
20790 Argininosuccinicaciduria
Mental and physical retardation; convulsions and episodic unconsciousness
20910 Atonic-astatic syndrome of Foerster
Muscular hypotonia, static ataxia, monotonous speech
21450 Chediak-Higashi syndrome
Photophobia and nystagmus
21870 Cretinism, athyreotic
Endocrine disorder (thyroid malfunction with profound mental and physical consequences)
21890 Crome's syndrome
Congenital cataracts, epileptic fits, mental retardation, and small stature

Table 2. *(Continued)*

22180	Dermo-chondro-corneal dystrophy of Francois
	Skeletal deformity of hands and feet; corneal dystrophy; abnormal electroen-cephalograms with seizures
23070	Gangliosidosis, GM (2), type III, or juvenile type
	Ataxia between ages of 2 and 6 years followed by deterioration to decerebrate rigidity; sometimes blindness occurs later

X-Linked Phenotypes

30050	Albinism, ocular
	Fundus is depigmented and choroidal vessels stand out strikingly; nystagmus, head nodding, and impaired vision
30160	Angiomatosis, diffuse corticomeningeal, of Divry and Van Bogaert
	Demyelinization
30170	Anosmia
	Inability to smell
30370	Color blindness, blue-mono-cone-mono-chromatic type
30540	Faciogenital dysplasia
	Hypermobility in cervical spine with anomaly of the odontoid resulting in neurological deficits
30700	Hydrocephalus due to congenital stenosis of aqueduct of Sylvius
	Mental deficiency and spastic paraplegia
30990	Mucopolysaccharidosis type II (Hunter's syndrome)
	Mental retardation and deafness
31170	Periodic paralysis, familial
31300	Spatial visualization, aptitude for
31330	Spinal ataxia
	Incoordination of limb movements

From McKusick (1971).

xxv in McKusick (1971) and to Section 11.3 in this book (and see Jay, 1974, on ophthalmic genetics as correlated with the McKusick catalog). We have arbitrarily selected 10 examples in each genetic category.

11.6 Epilepsy

The Greek stem from which the noun epilepsy is derived means "to seize upon," for epilepsy is the falling sickness. This refers to the collapse and prone position assumed at the onset and during seizures. In the United States alone 0.5 percent of the population are epileptic. They experience sudden, excessive, and unruly discharges of neuronal cells (Lennox, 1959).

One type of epilepsy is progressive and myoclonic (twitching, from the Greek words for muscle tumult). The full medical title of myoclonic epilepsy is Lafora-body-type myoclonic epilepsy. (Lafora bodies are intracellular bodies within the brain, diagnosed at autopsy). This feared epilepsy is progressive and is associated with dementia. We shall use data concerning it to demonstrate the oft-employed a priori method of determining modes of inheritance, according to the recommendations of Stern (1973, p. 206). The method is referred to as a priori because it assumes (1) a mode of inheritance and (2) that all families have an equal probability of being included in the data. The families itemized in Table 3 have produced one or more diseased offspring, but the parents are normal: if D = normal and d = epileptic, the crosses are assumed to be and are tested as being

$$Dd \times Dd \rightarrow \underbrace{1\ DD : 2\ Dd}_{3\ D};\ 1\ dd$$

with epilepsy due to autosomal recessive inheritance. Therefore, the probability of the production of a myoclonic epileptic is $\frac{1}{4}$. The first two columns of Table 3 list affected children from nine families. The third column lists the average number of affected children per sibship *expected* according to the a priori $\frac{1}{4}$ ratio. However, note that the numbers in this column exceed $\frac{1}{4}$ of the sibship. The reason is that those families $Dd \times Dd$ not producing epileptic offspring are excluded automatically from the data. As might be expected, this bias decreases as the sibship size increases. Entries in the fourth column are the numbers obtained when the number of affected children is multiplied by the number of sibships, and it should be compared with the observed data presented in the fifth column. Since 17 does not disagree with 16.666 (one cannot have two thirds of a child), we can conclude with assurance that this particular epilepsy fulfills the conditions imposed by autosomal recessive Mendelian inheritance.

Combining the assorted types of epilepsy (see Pratt, 1967, for a book-

TABLE 3. Autosomal Recessive Inheritance of Myoclonic Epilepsy, Tested via the *a priori* Method

Size of sibship	No. of sibship	Expected no. affected in each sibship	Total affected	
			Expected	Observed
1	1	1	1	1
4	1	1.463	1.463	2
5	1	1.640	1.640	2
6	3	1.825	5.475	7
8	1	2.222	2.222	1
9	2	2.433	4.866	4
	9		16.666	17

From *Principles of Human Genetics*, 3rd ed., by Curt Stern. W. H. Freeman and Company. Copyright © 1973.

long discussion of the genetics of these and other neurological disorders; and Alter and Hauser, 1972) and combining the data on like-sexed and unlike-sexed dizygotic twins (because they exhibit no significant differences in concordance, though age of initial seizure is lower in females), a table such as Table 3 in Chapter 7 can be constructed. From this, an H statistic (Equation 7.4) of 0.30 was calculated; it demonstrates a reasonable degree of genetic determination, though not as great as for the above data on myoclonic epilepsy. Remember though, that the data analyzed in Chapter 7, Table 3, dealt with a more heterogeneous group of epilepsies.

Lennox (1959) warned of the physician's responsibility to survey for and to counsel first-degree relatives of their patients about the three primary causes of epileptic seizures — heredity, infection, and trauma. Of the three, they singled out the first as the most important. Apparently, common childhood febrile seizures should also be carefully attended to, since they are indicative of slightly enhanced likelihood of the later development of epileptiform seizures both in the child experiencing even one convulsion and in his or her siblings. (The chances that any given child will have one or more childhood convulsions are approximately 1 in 70 or only three times the incidence of epileptics in the general population.)

The horns of the "trilemma" — genetics, environment, and genetic-environmental interactions — are nowhere sharper than in consideration of the mental illnesses that afflict the human species. This is shown in our discussions of manic depression and of schizophrenia in the following two sections. *The Epidemiology of Epilepsy: A Workshop* (edited by Alter and Hauser, 1972) demonstrates this nicely for epilepsy. We recommend especially Chapter 17, "Genetic Factors in the Epilepsies" (pp. 97-102) by Metrakos and Metrakos; Chapter 18, "Maternal Factors Influencing Risk of Epilepsy" (pp. 103-107) by Myrianthropoulos; and Chapter 19, "Inheritance of Focal and Petit Mal Seizures" (pp. 109-112) by Bray.

11.7 Manic-depressive mental illness

Kraepelin's (1896) description is still valid:

> Manic-depressive insanity includes on the one hand the entire province of so-called periodic and circular insanity, and on the other hand, the simple manias, the largest part of the clinical pictures designated as melancholias, and also a not inconsiderable number of cases with amentia. Finally, we count as well certain mild and very mild, partly periodic, partly enduring morbid pictures with similar coloring, which may start out as grave disturbances, but which alternatively may pass over without clearly defined boundaries into the realm of deviant personality organization.

Unipolar affective illness is depression alone, without the alternate manic phase characterized by inordinate exaltation, elation, and excitement; bipolar illness is manic depression. Both represent harmful extremes of emotion due to profound mood fluctuations.

Manic-depressive mental illness occurs in 0.5 percent of the general population as defined by Lynch (1969) and Stern (1973). It is rare (1.6 per 1000) in one northern Swedish isolate consisting of a few hundred people, where schizophrenia (discussed later in this chapter) occurs in relatively high frequency, approximately 9 per 1000 (Böök, 1953). *Isolates* are small discrete populations whose members are much more likely to breed with one another than they are with nonisolate individuals because of religious, ethnic, or other limiting reasons. Manic depression is most prevalent among another isolate, the Hutterites of the western United States and western Canada, in whom schizophrenia is rare (Eaton and Weil, 1955). Manic depression has an incidence of 4.6 per 1000 Hutterites, or 9.3 in persons aged 15 or older; schizophrenia has an incidence of 1.1 per 1000 Hutterites, or 2.1 in those aged 15 or older. The Hutterites represent a sect founded in the 1500s by Jakob Hutter; persecution caused their migration from Europe (Moravia, and then Hungary and Russia) to South Dakota in 1874.

There is no doubt at all that heredity plays a part in the etiology of manic depression. However, there is no agreement on the answers to such questions as: Are manic depression and schizophrenia distinct entities or does one blend into the other? Are unipolar depression and bipolar manic-depressive manifestations separate genetic entities themselves? Is the mode of genetic transmission polygenic or monogenic, or are these disorders instances of genetic heterogeneity (multiple genetic etiologies)?

This textbook cannot include a thorough discussion of the first question. Suffice it to quote Lynch (1969): "The potentialities for a cyclic psychosis may be considered to be genetically specific and unrelated to the potentialities for schizophrenic psychosis." By approving this quotation, we stand in agreement with the majority of students of this important problem (see Rosenthal, 1970, especially the sections titled "Biological Unity of Manic-Depressive Psychosis," p. 211, and "Specificity of the Manic-Depressive Genotype," p. 215).

The question of hereditary "loading" as it fosters unipolar or bipolar illnesses is easier to review. Is the genetic loading, i.e., the predisposing hereditary endowment (loading sometimes means the risk to other relatives of an affected proband), heavier when bipolar episodes are observed? We might also ask if it is heavier in instances of earlier onset so that a gradient can be envisioned, in order of increasing psychiatric severity and genetic predisposition: (1) unipolar and of late onset, (2) unipolar and of early onset, (3) bipolar and of late onset, and (4) bipolar and of early onset. Such a gradient is most easily interpreted in terms of a polygenic genetic etiology, but what if bipolar and monopolar illnesses are not related genetically? No monozygotic twin pair has ever been diagnosed as one unipolar manic plus one unipolar depressive, while there are many recorded monozygotic twins with one bipolar and one unipolar member (Zerbin-Rudin, 1969). Bipolar index cases often have unipolar relatives, and there is a 23.4 percent repeat risk for the parents of a bipolar patient. Twin studies show an impressive overall 95.7 percent concordance rate if the twins are monozygotic, and 26.3 percent if they are dizygotic.

For didactic purposes only we consider unipolar and bipolar affective illness to be expressions of the same genome *within a family unit*, as did Mendlewicz, Fleiss, and Fieve (1972). Eisenberg (1973) considered separate genetic bases in lucid comments on psychiatric intervention, (also see Winokur, 1973).

Consider the seven pedigrees constituting Figure 3. Deutan color blindness involves red-green color blindness and is a well-known example of a sex-linked recessive trait. Protan color blindness is also inherited as a recessive X chromosomal gene: the female homozygote is color-blind for red only, as is the male hemizygote. (See Section 11.4 for more detail concerning color vision; these conditions may represent two very closely linked genes at adjacent loci.)

Let us analyze the linkage or the independent assortment (the opposite of linkage) in these families, unit by unit. They not only afford us information about the possibility of X-chromosome transmission of manic-depressive illness, but also furnish examples of recombination in coupling and in repulsion of possibly linked genes (located on the same chromosome).

  • Family 1. II-2 has two sons, and how fortunate we are that these are male offspring, hemizygous (half a zygote) for hereditary factors on their single X chromosome so that sex-linked genes are expressed (for further explanation, see Section 2.1). II-2 refers to the generation in this particular pedigree, II, and to the individual, number 2. The number 59 above II-2s blackened circle is her age. II-2 then is a 59-year-old female, who, like her mother I-1, has bipolar manic depression and is color-blind. III-1 has been hospitalized for depression already and he is protanopic, like his uncle, II-1. At his age, he is still young for a definitive manic-depressive diagnosis. III-2, the brother, is neither mentally ill nor color-blind. If III-1 develops into a mentally ill adult, he represents a nonrecombinant X chromosome. The X received from his mother will have the genes at two loci, respectively, for red color blindness and for manic-depressive illness. Every normal human male receives his single X chromosome and hence all sex-linked material from his mother. He receives his relatively genetically inert Y chromosome

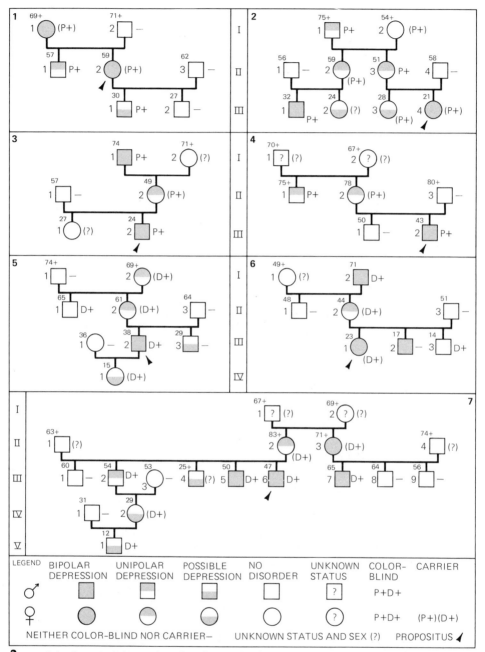

3. **Inheritance of depression disorders and color vision** in seven families. (From Mendlewicz, Fleiss, and Fieve, 1972.)

from his father. If III-1 is not manic-depressive, he is a recombinant, produced as a result of the normal crossing-over process taking place during the first meiotic division between homologous chromosomes, in this case in his mother's ovaries. He will not have received his mother's gene for mental illness. Is younger brother III-2 a nonrecombinant, as far as can be judged at age 27?

• Family 2. The proband (always indicated by an arrow, with one arrow per pedigree) has a maternal grandfather, I-1, and a first cousin, III-1, who are both color-blind and mentally afflicted. (Note that one is unipolar, while the other is bipolar.) All three individuals may be judged nonrecombinant, and the two genes are being inherited together on the same X chromosome in this pedigree.

• Family 3. III-2, our proband here, is not a recombinant. His mother, II-2, and his maternal grandfather, were ill and bore genes for color blindness. Once again, we have a familial mixture of the poles of manic depression, as we do in every pedigree in Figure 3.

• Family 4. The proband's older brother, III-1, is wholly normal; his maternal uncle, II-1, was color-blind and had been treated as a depressed outpatient, not hospitalized. The proband is a nonrecombinant having apparently inherited "intact" his mother's X chromosome.

• Family 5. The proband, III-2, is manic-depressive and color-blind. His younger brother, III-3, has normal vision but has been hospitalized for depression; he is still too young for a definitive diagnosis of monopolar depression. Their maternal uncle, II-1, is red-green color-blind and mentally normal. So the brothers represent the opposite of what uncle II-1 was. If II-1 was a nonrecombinant, then they are recombinant. This is our first pedigree with deuteranopia plus manic-depressive illness.

• Family 6. Proband III-1 has "mildly impaired color vision, having exhibited some signs of deutan color blindness of both eyes." She appears to be a nonrecombinant, and her brother, III-2, a recombinant, because their maternal grandfather, I-2, is color-blind and manic-depressive. Youngest brother, III-3, is entirely too young for definitive judgments about affective illness.

• Family 7. III-1, III-2, III-5, III-6, and cousins III-7, III-8, III-9 are all nonrecombinants. The mother of the first family was treated for depression several times, as an outpatient, but the prime reason for classifying these four as nonrecombinants is a living brother, III-1, who is neither depressed nor color-blind. The mother of the second family (left side of the pedigree) had bipolar depression, had been hospitalized for mania, and was heterozygous for color blindness (as indicated by her son, III-7).

What do all seven pedigrees indicate? That a dominant sex-linked single gene should be considered as the route of and the means of transmission of hereditary factors fostering manic-depressive illness. Women with but one dose of the putative allele exhibit affective mental illness. This is not to say that all cases of mania or depression or both are due to dominant factors borne on the X chromosome; *genetic heterogeneity* (see discussion of schizophrenia, Section 11.8) — different genetic bases for the same syndromes — may be operative here and may explain contradictory reports (Slater, Maxwell, and Price, 1971; Perris, 1971). The possibility of genetic heterogeneity should be kept in mind here and elsewhere where comprehensive genetic decisions become tortuous.

As an addendum, we can again use the data in these seven pedigrees to estimate the nearness on the X chromosome of the genes for manic depression, for protan color blindness, and for deutan color blindness. Edwards (1971) has offered tables and schemata for genetic mapping; the interested reader is referred to this work and to Cavalli-Sforza and Bodmer (1971). Suffice it to state here that the closer the loci, the more unlikely it is that recombination between them will occur.

11.8 Schizophrenia

Schizophrenia, which appears to be an array of profound behavioral disorders, presents the behavior geneticist with an urgent and exceptionally difficult problem, perhaps the most difficult presented in this text. The history of the classification and treatment of schizophrenia is itself rather schizoid. Estimates of occurrence rate range from less than 1.0 percent (Dunham, 1965) to 2.9 percent with 1.5 percent hospitalized (McNeil, 1970). Annual expenditures for care of these patients in the United States are currently in excess of $800 million, and the affliction is by no means restricted to any one country, race, or socioeconomic group.

Wienckowski (1972) commented:

> More than 2 million Americans at one time or another have suffered from the tragic mental disorder of schizophrenia. Half of the Nation's beds in mental hospitals are now occupied by schizophrenic patients. It is estimated that 2 percent of the population will have an episode of schizophrenia some time during their lives; in certain social settings — the urban slum, for example — the prediction rises to 6 percent, or more than 1 in 20 persons.

Rosenthal (1970) remarked:

> I was asked to write about the promise that genetics holds for the understanding, prevention and treatment of mental illness in general and schizophrenia in particular. That is better than being asked to write about man in relationship to his universe, but not much better.

What advice can the geneticist offer physicians and therapists? Schizophrenia is characterized by disturbances of thinking (delusions, bizarre or illogical responses), perceptions (hallucinations), and affective responses (loss of interest, volition, and the capacity to experience ordinary pleasures). Kraepelin (1896) described the four classic subtypes:

- Simple. Gradual loss of interest and personal attachments leading to an indifferent, apathetic person who is almost totally isolated from human society

- Hebephrenic. Disturbed thinking, notably shallow affective responses, childlike behaviors and mannerisms

- Catatonic. Stuporous, often mute, remaining in one position for hours to weeks

- Paranoic. Characteristic persecutory or grandiose delusions, often auditory hallucinations

These subtypes may be compounded in one individual at various times. The schizophrenic tends to withdraw from reality to the degree that he is unable to distinguish clearly between his inner fantasies and the realities of his physical environment.

The theories that attempt to explain the genetic component(s) underlying the manifestation of schizophrenia constitute two broad schools: monogenic (hypothesizing a single crucial genetic locus) and polygenic. Numerous studies documenting families with histories of schizophrenia demonstrate a high rate of concordance (occurrence of the disorder in two or more individuals of a family or group) for siblings (10 percent, Ödegaard, 1963) and even higher concordance rates for monozygotic twins (50 percent, Gottesman and Shields, 1966). Evidence such as this indicates that a functional genetic component of some sort is present in schizophrenia, though such a component need not be omnipresent. (One wonders if environmental conditions alone can induce schizophrenic episodes.)

We consider here only one of the monogenic theories of the inherent basis for schizophrenia. Older monogenic proposals have been thoroughly amended so that they incorporate minor modifying genes, thereby contradicting any monogenic hypothesis. Heston, a psychiatrist whose publications (1966 on adopted children of schizophrenic mothers, 1970, 1972) on these matters are highly recommended, has offered data (Fig. 4) that show that the recorded cases of schizophrenia *plus* schizoidia approximate the expectations for control by a single dominant gene. *Schizoidia* may be defined as a pre- or potentially schizophrenic mental state characterized as "very withdrawn," "always frightened," "defense system never developed," "can't stand to have demands made upon oneself," "feels worthless," "quite regressed — wants to be treated like a baby, and despises oneself for these inclinations" (Landis, personal communication; see also Landis and Tauber, 1972). The schizoid patient is in touch with reality and realizes that there is trouble.

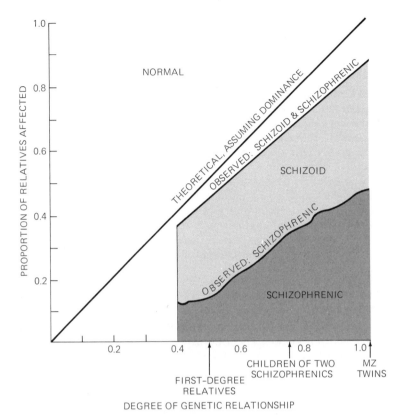

4. **Basic for monogenic theory of cause of schizophrenia.** Observed cases of schizoidia plus schizophrenia approximate proportions expected if major influence of a single autosomal dominant gene is assumed. See text for details. (From Heston, 1970. Copyright 1970 by the American Association for the Advancement of Science.)

For contrast, consider the withdrawal from reality permeating the following case history (Beckett and Bleakley, 1968) of a paranoid schizophrenic. Note the *autism* (condition of being wholly self-centered in thought and behavior, dominated by the subjective) that is displayed and realize the poor prognosis:

The patient was a twenty-three-year-old, single male.

Present Illness: The patient had just completed his M.A. degree in electronic engineering. During his final examinations he had the flu and began to worry excessively about his heart. He complained of general fatigue and of his brain being numb. He became very confused, felt that his food was being poisoned, and refused to wash. When his parents insisted upon his washing, he became impulsive and aggressive. He repeatedly referred to his mother as a murderess and a witch and said that she was hypnotizing or poisoning him. He was [hospitalized] after unsuccessfully attempting to stab her.

Past History: The patient was the elder of two siblings. He was a healthy infant and was easily toilet-trained. Although he was a brilliant scholar, he was inclined to be shy and sensitive with other children. At seventeen he suffered from pains in the legs and throat and was out of school for a year. During this time he was irritable and wanted to be left alone. Since he had entered college his only known social contacts had been with his family at mealtimes. The remainder of his time had been spent in his room in study or walking about the streets by himself.

Interview: On the ward he had been observed to have sudden fits of rage, while at other times he would sit in a corner convulsed with laughter. In appearance he was haggard and untidy, his movements seemed uncoordinated, and he talked to himself in a rapid and incomprehensible rush of words. He accused the interviewer of interfering with his thoughts by constantly sending coded electronic messages by microwave which he could not decode, and as a result he would not talk during the interview.

The diagonal in Figure 4 represents the theoretical expectation for control by a single dominant gene. The small unshaded area in the lower left corner is blank for degrees of genetic relations (as defined in Section 6.9) lower than 0.4. Note that the expected coefficient for first-degree relations (between parent and offspring, sibs, and dizygotic twins) is 0.5. Relations between two parents and their offspring have an expected coefficient of $\sqrt{2} \times 0.5 = 0.71$ (Equation 6.18). In the case of monozygotic twins the expected coefficient of relationship is 1.0.

The polygenic models of the etiology of schizophrenia generally concur with the diathesis-stress theory proposed by Rosenthal (1971) and by Gottesman and Shields (1971, 1972, 1973). According to this theory, the schizophrenic does not inherit schizophrenia itself but is genetically predisposed to develop the condition. The environment of the individual determines the likelihood of the affliction's being manifested. Environmental stress induces schizophrenia in the predisposed individual. Essentially, the diathesis-stress theory is based on a threshold polygenic model. The genetic basis for such a model is the additive effect of a number of relatively minor genes which exhibit little or no dominance or recessiveness.

A further elaboration on the polygenic theme has been described by Erlenmeyer-Kimling and Paradowski (1966) as:

a heterogenous [different genetic bases for phenotypically similar conditions] collection of entities stemming from a number of different, independently acting genic errors. . . . According to this model, a number of different primary enzymatic defects could feed into a final common pathway or intermediary mechanism. . . . The extent to which the final pathway is disrupted might be different, however, depending upon the route taken in reaching it, so that variations in predisposition could exist between the different genotypes. Multiple allele series at a given loci are also possible — with different alleles producing different degrees of effect. The activity of the various genes will, furthermore, be modulated by the total genotypic background against which the alleles are placed. . . . Finally, both the degree of predisposition and the influences of environmental factors will cooperate to determine whether schizophrenia, psychological disturbances of lesser sorts, or perhaps no symptomatology becomes manifest at the behavioral level.

A recent, definitive survey by Gottesman and Shields (1972) provides an excellent commentary on this entire matter.

The biochemistry of schizophrenia has been the subject of extensive research in recent years. The resulting literature details the intricate methodology applied to the investigation of the various theories currently proposed. We do not attempt to condense this literature, but refer the reader to more comprehensive sources (Kety, 1967; Omenn and Motulsky, 1972; McGaugh, 1972).

Table 4 summarizes the two major theories about the genetic basis of schizophrenia. Note the overlap, and hence confusion, manifest in at least the first two rows.

We, in turn, can now summarize in Table 5. Some calculated degrees of kinship and inbreeding are offered for explanation and for more general use. Coefficients of relationship (Section 6.8) are included for comparison. The morbidity risk for schizophrenia does not uniformly reflect the coefficient of relationship; cousins and half sibs share a 0.25 coefficient of relationship, yet concordance is more than three times greater among half sibs. This variance indicates the influence of environmental factors on the manifestation of schizophrenia. Though the genetic predisposition of cousins is identical to that of half sibs, the environment of cousins is likely to be different because they generally do not live in the same household or share the same living conditions or environmental stresses. Differences in environment may account for a large part of the variance in concordance evident among individuals with the same coefficient of relationship.

With regard to screening, it must be admitted that the current outlook is not sanguine. Mednick et al. (1971) have carefully assessed perinatal conditions and infant development (for 1 year) in children born to schizophrenic parents. They found that:

- Such children have a higher incidence of a somewhat lower birth weight. This may be associated with developmental abnormalities (e.g., rapid breathing, neurological and reflex abnormalities, inability to stand alone at 1 year of age). Offspring of nonschizophrenic parents do not evince these developmental abnormalities.

TABLE 4. Comparisons of Two Major Theories About the Genetic Basis for Schizophrenia

Aspects of the illness	Monogenic-biochemical theory	Diathesis-stress theory
Biological unity	Homogeneity: one gene, dominant, recessive, or intermediate. Trait is qualitative, discontinuous.	Homogeneity or heterogeneity. Trait may be qualitative or quantitative.
"What" is inherited?	A specific but as yet unknown error of metabolism due to a mutant gene.	(1) A single gene. (2) Several major genes. (3) Polygenes. In any case, a "constitutional predisposition."
Manifestation	Very high: almost everyone (67%–86%) with the genotype, but some have constitutional resistance to expression.	Considerably lower than monogenic-biochemical. Depends on whether predisposed schizophrenic encounters sufficient stress and how predisposed he is.
Role of environment	No special environments needed to precipitate illness. Incidental of minimal. Proponents like to cite a constant rate of schizophrenia in all cultures.	Necessary to precipitate the illness. The stressors are seldom defined: head trauma, disease, alcohol, parturition, exhaustion, etc., but usually psychological.
Clinical subtypes	Of secondary interest, usually thought to reflect other inherited or constitutional factors which influence the form in which the illness is expressed.	Usually holds that they represent different predispositions interacting with different kinds of stressors.
Severity of illness	Reflects the degree of metabolic disturbance.	Reflects the amount of inherited predisposition and the intensity of the stressor.
Remission	For some reason, the effect of biochemical disturbance clear, but personality defect remains.	Either the physiological aspects of the disease process are reduced or the stressors are reduced.
Premorbid personality	Varies in usual ways. When aberrant, the deviations are thought to be early signs of the metabolic disturbance.	Can provide clues about the nature of the predisposition inherited, as introversive personality or high anxiety.
Research strategy	(1) Search for the biochemical aberrancy and its corrective. (2) Estimate gene frequency in population, mutation rate, mode of inheritance, etc.	Learn about the nature of the predispositions, the stressors, and the nature of the interaction.
Example of problems posed by previous findings	Why does the distribution of illness in kindreds vary so markedly, showing dominant, recessive, or intermediate patterns?	Why does the illness continue when the ostensible stressor has been removed?

From Rosenthal (1970).

- Female infants of afflicted parents suffer more from complications in the pregnancies that produce them than do male infants. (This was particularly the case in pregnancies initiated by schizophrenic fathers.) This is precisely the reverse of what is reported in "ordinary" pregnancies. Could a significant portion of male embryos and fetuses of schizophrenic parents be aborted?

- Children of schizophrenic fathers consistently score high for perinatal and developmental problems. Do such fathers foment a stressful, deleterious environment in addition to their genetic influence?

Heston (1971) commented cogently on the Mednick survey and, in a recent personal communication, reported verification of its findings, noting that he does not "think early diagnosis will be helpful in view of our ignorance of anything to try to do to prevent the development of the disease. Of course, we should be doing longitudinal studies of *all* children of schizophrenics, but until we understand much more about the disease, I hope we do not find a way of certainly identifying preschizophrenics." Nonetheless, it appears that about half the children of schizophrenic parents suffer from an assortment of perinatal problems.

To conclude, albeit unsatisfactorily, this section, we refer the reader to the symposium of the Behavior Genetics Association (see *Social Biology*, September 1973). In summarizing their paper on the genetics of schizophrenia, Kidd and Cavalli-Sforza (1973) provide the following analysis:

TABLE 5. **Risk of Schizophrenia and Similar Conditions Among Relatives of Schizophrenic Individuals**

Relation to propositus	Coefficient of relationship	Morbidity risk (%)*	
		Schizophrenia	Psychopathic conditions similar to schizophrenia
Unrelated	0.00	0.85	2.9
Stepsibs†	0.00	1.80	
Half sibs	0.25	7.10	
Sibs	0.50	7.0–15.0	9.7
Parents	0.50	5.0–10.3	
Children	0.50	7.0–16.4	32.6
Grandchildren	0.25	3.0– 4.3	13.8
Cousins	0.25	1.8– 2.0	10.2
Nephews, nieces	0.25	1.8– 3.9	5.1

* Based upon no specific genetic theory about the cause of schizophrenia, but rather on direct observations. Ranges of risks represent summaries of many studies, where morbidity risk data sources are cited, with the coefficients of relationship added.

† Step sibs are genetically unrelated; they are the offspring of two spouses from their marriages to other partners.

Modified from Stern (1973).

Data on the inheritance of schizophrenia indicate that both genetics and environment contribute in an important way to the manifestation of the disease. However, heterogeneity of the data makes an accurate analysis very difficult. For this reason, the solution is unlikely to come from statistical techniques alone. Future advances in physiological, toxicological, and biochemical research are more likely to supply a firmer basis to the understanding of the genetic component of this disorder.

11.9 Sex

As a relief from the complexities of schizophrenia, we offer a stellar lone example of the intimate relation between that which is passed from generation to generation via gametes and that which we know as behavior — the disorder called testicular feminization. It is male (affecting only that sex) pseudo (false) hermaphroditism (one individual in whom the characteristics of both sexes are combined). Afflicted individuals perceive themselves, and are accepted by their families from birth on, as females. They have a normal life span. When they are wed, it is to men, but their unions are wholly sterile. They possess blind-ended, small vaginas accompanied by the full feminine development of the mammary glands. In addition, however, testes are present and their herniation is often the initial symptom invoking the diagnostic genital-pelvic examination; amenorrhea is the next symptom. Bergsma (1973b) cautioned that "the diagnosis should be suspected in all girls presenting with 'inguinal hernia(s)': before exploration and herniorrhaphy [surgery] a buccal smear [to count the number of X chromosomes] should be performed." (Also see Rimoin and Schimke, 1971.)

The pedigrees constituting Figure 5 are consistent with two types of inheritance and it is still impossible to distinguish clearly between them:

- An autosomal dominant such that a $Trtr$ female (tr from the word transform) is normal, though development of secondary sexual characteristics may be tardy and the individual often has sparse axillary and pubic hair. If she conceives XY offspring, half of them are intersexual because a $Trtr$ male is feminized

- An $X+X^{tr}$ female appears as above and is essentially normal and fertile. An $X^{tr}Y$ male is feminized.

No individual homozygous for this gene is known and linkage studies have not proved fruitful (Holmberg, 1972), so under what circumstances would it be possible to choose between these two alternatives?

Since Morris first described this syndrome in 1953, its cause has been postulated to be the insensitivity of peripheral target tissues to androgenizing secretions of the testes and adrenal glands; there is no

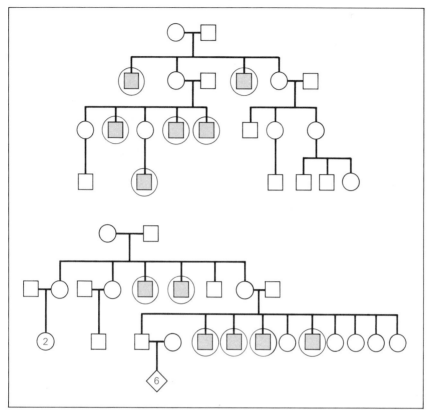

5. **Two types of inheritance of testicular feminization.** Afflicted individuals, depicted as squares (male) inside circles (female), are genetically male but phenotypically female. The gene appears to be dominant and autosomal or recessive and sex-linked; if autosomal, it is sex-limited (to genetic males) in expression. (From *Principles of Human Genetics*, 3rd ed., by Curt Stern. W. H. Freeman and Company. Copyright © 1973.)

response to circulating testosterone, even to injections of this hormone. There is a sharp deficiency of germ cell maturation in the small nodular testes, a lack of male secondary sexual characteristics including those of personality, total sterility, and the absence of body hair. (An incomplete form of testicular feminization presents similar symptoms coupled with postpubertal virilization.)

And what of the psychotherapeutic aspects of this gross syndrome? What sex does or should a feminized male relate to? Gayral et al. (1960) commented on 11 cases in five families (note the incidence of repeats here). In all instances they elected not to inform their patients of the true, currently incurable, nature of their affliction. Instead, the "women" were told of ovarian abnormalities resulting in lifelong sterility. Responses ranged from ostensible indifference to overcompensation to neurotic depression. In every case, the patient was feminine in appearance, bearing, and temperament. This, in the absence of any aspect of menstruation, even that of cyclic moodiness. The sexual drive and capacity for conjugality were normal, and the authors concluded that too much information, however accurate, would only result in at the least confusion and at the most severe depression. The parents of feminized males might therefore be counseled to the extent of citing the possibility of the production of additional similarly sterile offspring without being given more information. We wonder how our readers judge such profound decisions. Should all psychotherapeutic efforts be directed toward reassuring the individual afflicted with testicular feminization that she is a social, psychological, and emotional human female?

Man: Continuous Traits

12.1 Intelligence: genotype and environment

The attribute most important for achievement in school is what both psychologists and laymen call *intelligence*. Motivation, personality, and interpersonal relations at home and in the school are also significant factors, affecting both achievement in school and performance on tests designed to measure "intelligence." Empirical correlations have established that the IQ test as a measure of intelligence reflects ability to learn in school in many societies. It is clear too, from their academic performance, that schoolchildren vary in their capability for learning, especially abstract learning. The IQ score is an attempt at quantifying intelligent behavior (Section 7.4). Stern (1973) wrote: "Intelligent behavior is regarded as behavior which, on the basis of inherited capacity, makes good use of the social inheritance, such as language and numbers or scientific and moral concepts." Psychologists have divided mental abilities into distinct, so-called primary abilities, such as ability to visualize objects in space, to memorize, or to reason inductively. There may also be a general ability underlying intelligence, in addition to these primary abilities. Since primary abilities may vary somewhat independently of each other, individuals with the same overall IQ may differ in their mixtures of primary abilities. Although most work on intelligence is in terms of single scores obtained from intelligence tests, primary abilities are emphasized in some newer studies.

We first consider generalized intelligence tests. The designers of intelligence tests have attempted to make them independent of environmental influences within a given society. The interpretation of results of a test given to individuals in different societies is highly complex, since a different society implies a differing environment at least, and possibly a different average genotype. The reader is referred to Section 7.5 for discussion of large genotype-environment interactions obtained in rats, which could be quantified, since in experimental animals both genotype and environment can be defined. This is not possible in man. Even within a fairly homogeneous group, it seems impossible to achieve complete freedom from nongenetic influences. Human intelligence always operates in a cultural organization, and so it may not be possible to obtain tests that are entirely culture free. One aim of investigators has been to devise "culture-fair" tests in which the effects of cultural differences or groups within a given society are minimized. Even so, factors such as the greater eagerness of middle-class parents than of lower-socioeconomic-class parents for the intellectual success of their children seem difficult to eliminate. Indeed, in the search for a culture-free test, we may eliminate not only environmental sources of variation, but also genetic sources.

Despite these difficulties, some conclusions have been reached. Erlenmeyer-Kimling and Jarvik (1963) carried out a literature review of IQ and certain other general intelligence tests and calculated correlation coefficients between various groups of individuals reared together — unrelated individuals, foster parent and child, sibs, monozygotic (MZ) and dizygotic (DZ) twins. Another set of correlation coefficients was calculated between various groups of individuals reared apart — unrelated individuals, sibs, and MZ twins. The data were based on 52 separate studies, and the approximate median correlation coefficients obtained for each relation group are given in Table 1.

Recall from Section 6.8 that the coefficient of relationship is the coefficient of the additive genetic variance (V_A) in the covariance between relatives and reflects the share of genes due to common ancestry. An observed correlation coefficient close to the coefficient of relationship reflects a heritability close to unity if environmental complications can be eliminated and if the dominance variance is small (where relevant). Table 1 shows that the closer the degree of relationship, the higher the correlation coefficient, in both the reared-apart and reared-together categories; this implies a likely genetic component. However, comparing those reared apart with those reared together reveals an environmental component, since for the three cases where comparisons are made, the correlation coefficients for those reared together are greater than for those reared apart. The difference is particularly large for unrelated persons, for whom a correlation coefficient of zero would be expected; for unrelated persons reared together the coefficient is +0.23, a figure that shows the effect of environment quite clearly. Thus we can conclude that intelligence is controlled by heredity and environment but that heredity may be more important. This is the conclusion we arrived at in Section 7.5, where some twin studies are considered (Chapter 7, Table 5). The genotype-environment correlation of +0.25 obtained using Cattell's multiple abstract variance analysis (MAVA) method (Section 7.5) shows the relevance of environment; it is also close to correlations obtained between intelligence and social status.

A quite high degree of genetic determination for IQ is apparent from Section 7.6 using Burt and Howard's (1956) data. A large dominance component also is apparent in these data. Jinks and Fulker (1970) carried out a full biometrical analysis of a number of sets of IQ data and found dominance to be of some importance generally in the direction of high IQ. This implies that there has been considerable directional selection for increased intelligence during man's evolutionary history. An analysis of the IQ scores of 3558 individuals in a pedigree study carried out by Reed and Reed (1965) on mental retardation agrees with a dominance hypothesis (Eaves, 1973). In fact, assuming a correlation between mates (positive assortative mating) of 0.3 and assuming complete dominance, Eaves computed $V_A = 0.43$, $V_D = 0.215$, and $V_E = 0.18$. The importance of assortative mating is discussed in Section 7.6; it is necessary merely to reiterate here that recent biometrical-genetic analyses of intelligence invoke this as a significant factor.

Thus far we can conclude that IQ is under genetic control to quite a

TABLE 1. Approximate Median Correlation Coefficients Between Related and Unrelated Persons Reared Together and Apart for Intelligence Test Scores, and Coefficients of Relationship for Each Category

Persons	Reared apart	No. of studies	Reared together	No. of studies	Coefficient of relationship
Unrelated	−0.01	4	0.23	5	0
Foster parent and child			0.20	3	0
Parent and child			0.50	12	0.50
Sibs	0.40	2	0.49	35	0.50
DZ twins					
Like sex			0.53	9	0.50
Unlike sex			0.53	11	0.50
MZ twins	0.75	4	0.87	14	1.00

After Erlenmeyer-Kimling and Jarvik (1963).

large extent. Not only is additive gene action relevant but also directional dominance for high IQ. Furthermore, the value of the additive genetic variance is inflated unless a component for assortative mating is estimated separately. The distribution of IQ in populations is continuous, so it is likely that many genes are involved. Under various simplifying assumptions, the probable number of genes can be estimated. The estimates vary from 22 genes, with an average of 100 genes (Jinks and Fulker, 1970). There is therefore no doubt that IQ is under polygenic control and should show all the features of a polygenic trait, although environmental effects are likely to be larger than for morphological traits. Even so, certain rare genes such as the one responsible for phenylketonuria (Section 2.1) can reduce IQ dramatically. The same may be true for certain chromosomal abnormalities such as Down's syndrome (Section 4.3).

Penrose (1963) discussed the effect of these more discrete abnormalities on IQ by using a regression approach (Section 6.9). Burt (1943) has supplied data showing that a child's IQ is on the average between the father's IQ and the mean IQ of the population, which is expected to be 100 (see Penrose, 1963, for discussion). For example, one group of fathers with an average IQ of 117.1 had children with an average IQ of 109.1. Conversely, another group of fathers with an average IQ of 86.8 had children with an average IQ of 92.0. The IQ of the children, therefore, on average regressed nearly one half the way toward the mean of the population. If IQ were determined entirely by additive genes, a regression of one half the way to the population mean would be expected (Equation 6.19). For sibs, ignoring the complication of dominance, a similar result is expected. For half sibs, nephews, and nieces, who are more remote relatives, a regression three quarters of the way toward the mean is expected. For persons having IQs of 50 and above, observed data are in good agreement (Table 2).

TABLE 2. Mean IQs of Defective Patients and Their Relatives

				Relatives' mean IQ	
	Type of relation to patient	No. of pairs	Patients' mean IQ	Observed	Expected on additive gene hypothesis
Patients with IQ ≥ 50	Sib	101	65.8	84.9	82.9
	Half sib, nephew, niece	143	63.2	89.5	91.8
Patients with IQ < 50	Sib	120	24.2	87.4	61.1
	Half sib, nephew, niece	90	33.3	95.1	83.3

After Penrose (1963).

However, propositi with IQs less than 50 have relatives with IQs considerably higher than expected on an additive gene hypothesis. The interpretation is that these defectives have low IQs because they are homozygous for deleterious recessive genes or are karyotypically abnormal. In other cases, new mutations (mainly dominant) or environmental accidents such as birth trauma may be responsible. In all cases, these rather discrete but rare events lead to a breakdown in additivity. Therefore, for very low IQs, genes of major effect, chromosomal aberrations, and environmental accidents may play a part, in contrast with IQs closer to the expected mean of 100, where polygenic inheritance is normal. This conclusion has been confirmed in other studies such as that of Roberts (1952) based on the sibs of defectives.

This survey of intelligence as assessed mainly by the IQ test shows clearly that genetic and environmental influences play a part. Since our conclusion so far is in favor of a considerable degree of genetic determination of IQ, it seems appropriate to conclude by seeking out various situations in which specific environmental influences might predominate. One category is adopted children. The review of Erlenmeyer-Kimling and Jarvik (1963) gave a median correlation between foster parent and child of 0.20. This can be argued as resulting from (1) the similar environment of the foster parent and child and possibly (2) selection on the part of adoption agencies for similarity of the foster and biological parents (for example, the chances are that the presumed better-endowed children would be placed in similarly better-endowed adoptive homes). The obvious "genetic" experiment is to study adopted children with no open or hidden selective placement.

Despite these difficulties, it is worth looking at studies on adopted children. An obvious hypothesis is that if heredity has something to do with IQ, adopted children should be less similar to their adoptive parents than the children of a control group are to their biological parents. For a group of adopted children in Minnesota homes (Table 3) there was

TABLE 3. Mean IQ of Adopted and Own Children According to Fathers' Occupation

Occupation of father	Adopted children		Own children	
	No.	IQ	No.	IQ
Professional	43	112.6	40	118.6
Business, management	38	111.6	42	117.6
Skilled trades and clerical	44	110.6	43	106.9
Semiskilled	45	109.4	46	101.1
Relatively unskilled	24	107.8	23	102.1

After Stern (1973).

a continuous decline in mean IQ with the decline in the occupational status of the father from the professional to relatively unskilled occupations. This decrease in IQ was from 113 to 108, which is a rather narrow range, even if we attribute it merely to an environmental effect on test performance. In contrast is the control group of children reared by their own parents: here there was also a steady decline in IQ according to the occuptional status of the father, but the range was almost three times as great, from 119 to 102. This latter range, therefore, shows a much more pronounced correlation with the occupation of the father than is the case for the adopted children. In conclusion, it seems reasonable to assume that the greater differences between the scores of own children compared with adopted children are due to the fact that the own children resemble their parents more than do the adopted children, because they have inherited parental genes. (Evidence for IQ differences under genetic control among occupational classes appears in Section 12.2.)

A particularly instructive analysis carried out by Skodak and Skeels (1949) comes from the study of parent-child resemblance in intelligence during the development of the child. The performances of a group of children adopted during the first months of life were measured and were correlated with educational level of biological mothers and fathers and of adoptive mothers and fathers (Fig. 1). For comparison, data are available from another study on the correlation of the IQs of children and their biological parents, in whose homes they had been reared (Honzik, 1957). Up to about 2 years of age, there was little correlation between the performance of the child and the education of either the biological or the adoptive parent, irrespective of whether the child was reared by his own or by adoptive parents. However, as age increased, a strong rise in the correlation between the IQ of the child and his biological parents occurred, irrespective of whether the child was reared by his own or by adoptive parents. This correlation reached 0.3 by about 4 years and increased little thereafter. In complete contrast, however, was the low and declining correlation between children and their unrelated adoptive parents. It is difficult to escape the conclusion that the intelligence of a child does indeed depend on the genes received from his biological parents. Studies of adopted children, while perhaps showing a small environmental component in the determination of intelligence, clearly show the much greater importance of genetic influences. This study confirms the conclusion of Cavalli-Sforza and Feldman (1973) that only through careful studies on both adopted and biological offspring can the relative roles of cultural and biological inheritance be assessed rigorously (see Section 7.5).

We conclude this section by considering the relation between IQ and family size. Several family studies have shown fairly consistent negative

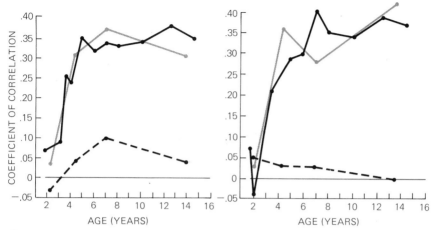

1. **Relation between parents' and childrens' IQs.** Correlation coefficients between education of biological and adoptive parents and IQs of children, according to ages of the children. **Left.** Correlation between child's IQ and mother's education. **Right.** Correlation between child's IQ and father's education. The three graphs in each chart are for (1) child reared by own parents and biological parent; (2) adopted child and nonrearing biological parent; (3) adopted child and adoptive parent. (Graph **1** data from Honzik, 1957; Graph **2** and **3** data from Skodak and Skeels, 1949.)

correlations between intelligence and family size. Correlation coefficients in the region $r = -0.20$ and -0.30 have been found. Table 4 presents data based on a study carried out at the Minnesota State School and the Dight Institute of Human Genetics (see Maxwell, 1969, for a discussion of Scottish data showing similar results). If studies are confined to particular social classes, smaller but still negative correlations again occur. From this result, predictions have been made that the

TABLE 4. Relation Between Family Size and IQ of Children

Family size (children per family)	IQ of children	Total no. of children studied
1	106.37 ± 1.39	141
2	109.56 ± 0.53	583
3	106.75 ± 0.58	606
4	108.95 ± 0.73	320
5	105.72 ± 1.15	191
6	99.16 ± 2.17	82
7	93.00 ± 3.34	39
8	83.80 ± 4.13	25
9	89.89 ± 2.94	37
10	62.00 ± 7.55	15

Data from Higgins, Reed, and Reed (1962).

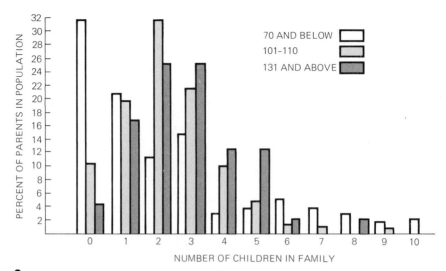

2. **Family size and IQ.** Distribution of size of family according to IQ of parents for three IQ categories: 70 and below, 101 to 110, and 131 and above, measured in percentages. (After Higgins, Reed, and Reed, 1962.)

average IQ of the population should drop from two to four points per generation. However, this has not happened; indeed, there has been a slight tendency for the reverse to occur. The explanation lies in the omission in much of the literature of families with no children. Thus many studies are biased, since both infertility and the probability of nonmating are ignored. Higgins, Reed, and Reed (1962) showed that more than 30 percent of persons whose IQs are 70 or less have no children, compared with 10 percent of those with IQs in the range 101 to 110 and only 3 to 4 percent of those whose IQs exceed 131 (Fig. 2). This contrasts sharply with the data in Table 4. Presented another way (Table 5), the reproductive rate of all siblings in families including unmarried

TABLE 5. Relation Between IQ and Reproductive Rate of All Siblings Including Unmarried Brothers and Sisters

IQ range	Average no. of children	No. of children
0–55	1.38 ± 0.54	29
56–70	2.46 ± 0.31	74
71–85	2.39 ± 0.13	208
86–100	2.16 ± 0.06	583
101–115	2.26 ± 0.05	778
116–130	2.45 ± 0.09	269
131 and above	2.96 ± 0.34	25

Data from Higgins, Reed, and Reed (1962).

sibs is low for IQ values less than 55 and steadily increases to nearly three children for IQ values of 131 and above. Based on a consideration of figures of this nature, Higgins, Reed, and Reed argue that the IQ level of the whole population should remain relatively static from one generation to the next, and certainly should not drop to any degree. The negative correlations disappear when a more complete sample is considered. This result points out a continuing difficulty in the handling of human data — the problem of the exact constitution of the sample used.

12.2   IQ, occupational class, and social mobility

One of the most comprehensive studies on IQ differences among social classes was published by Burt (1961). The data were for schoolchildren and their parents in a typical London borough. The classification was based on the head of the household's occupation; six classes were considered, ranging from higher professional to unskilled. Table 6 shows the means and standard deviations of the IQ distributions of parents and their offspring, according to the occupational class of the parents. The data show a very large and significant direct association of mean IQ with occupational class. The mean difference between the highest and lowest classes is more than 50. In spite of the significant variation between the observed means, the residual variation in IQ of parents within each class is still large. Thus the mean standard deviation of the parental IQs for the different classes is 8.6 or almost two thirds the standard deviation for the whole group.

The offsprings' means are between the parental and overall population means, regardless of whether the class IQ is above or below 100. This regression toward the mean is expected, as discussed earlier (Table

TABLE 6.   Distribution of Mean IQ with Standard Deviations According to Six Occupational Classes*

	Proportion in each class (%)	IQs	
		Parents	Offspring
Higher professional	0.3	139.7 ± 4.7	120.8 ± 12.5
Lower professional	3.1	130.6 ± 6.7	114.7 ± 11.2
Clerical	12.2	115.9 ± 9.3	107.8 ± 13.6
Skilled	25.8	108.2 ± 9.9	104.6 ± 14.3
Semiskilled	32.5	97.8 ± 9.9	98.9 ± 13.8
Unskilled	26.1	84.9 ± 10.9	92.6 ± 13.8

* For precise definitions of occupational classes, see Burt (1961).
From *The Genetics of Human Populations* by L. L. Cavalli-Sforza and W. F. Bodmer, W. H. Freeman and Company. Copyright © 1971.

2). Finally, the standard deviations of the offsprings' IQ distributions are almost the same as those of the general population, averaging 13.2. Assuming segregation of the many genes controlling IQ, this is expected.

On the basis of the data in Section 12.1, the differentials between occupational class may be assigned to causes that are mainly genetic. However, because of the fairly wide range of IQ values within each class, Burt (1961) argued for appreciable mobility among classes to maintain IQ differences. He estimated that to maintain a stable distribution of IQ among classes, a minimum of 22 percent of offspring would have to change class with respect to their parents each generation. This is below the 30 percent intergenerational social mobility in Britain. From interviews, he concluded that intelligence and motivation are the most important factors influencing social advancement, while home background and education have a lesser but still substantial influence, as might be expected.

A few recent papers (Waller, 1971; Gibson and Mascie-Taylor, 1973) have looked at social mobility in relation to the discrepancies between the general intelligence of sons and the social class into which they were born. Waller's (1971) study involved 173 males and their 131 fathers who were representative of the nonfarm white population of Minnesota; he subdivided the population into five social classes. Waller's findings supported the hypothesis that social mobility is correlated with the discrepancies between the general intelligence of sons as measured by IQ and the social class into which they were born. Figure 3 shows the relation between the percentage of sons moving up or down from the father's social class and the differences in father-son IQ. As the difference in IQ score (written as son minus father) increases, so does social mobility. Indeed for IQ score differences of 30 or more the mobility is in the 80 percent region. Therefore, differences in ability, which from the evidence cited must have a considerable genetic component, provide a situation leading to considerable mobility among classes. In this way social classes in an open society are prevented from congealing into castes.

Gibson and Mascie-Taylor (1973) considered university scientists and their fathers. Again, the differences in IQ between fathers and sons are correlated with the son's mobility on the socioeconomic scale relative to the father's occupation. It is argued that if IQ is correlated with social mobility and has significant heritability, then social mobility will lead to nonrandom transfer of genes from class to class. Hence, it is expected in theory that social classes will come to differ genetically to some extent (Thoday and Gibson, 1970). This, of course, does not exclude the addition of differences due to environmental, including cultural, reasons.

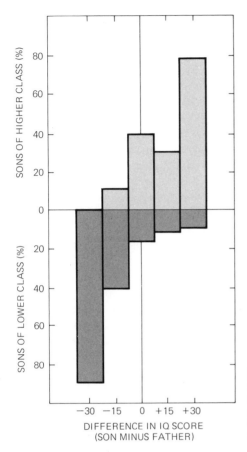

3. Relation between IQ and social mobility. Percentage of sons moving up or down from father's social class, by differences in IQ score. Note that percentages increase as IQ differences increase. (From Waller, 1971.)

The conclusion depends on the assumption that the relation between IQ phenotypes and social mobility implies some significant relation between IQ genotype and social mobility.

12.3 IQ differences between blacks and whites

One of the features of the range of variation in IQs just discussed is that they are maintained partly by mobility among occupational classes based to some extent on the selection for higher IQ in the higher occupational classes. In the United States blacks and whites enjoy no comparable mobility; skin color is an effective bar to mobility among races. Let us then look at the IQs of blacks and whites. In North American samples the average IQ of blacks is about 85, while the average IQ of whites is about 100. Many studies have shown results of this nature. Data obtained in one such study, based on IQ tests given to

1800 black elementary schoolchildren (Kennedy, Van De Riet, and White, 1963) in the southern United States, are shown in Figure 4. The distribution is compared with a 1960 sample of the United States white population. In this case the mean IQ difference is 21.1, which is fairly extreme, most differences being in the range 10 to 20. Even though there is a considerable degree of overlap between the two distributions, about 95.5 percent of blacks have IQs lower than the mean white IQ of 101.8, and 18 percent of blacks, compared with 2 percent of whites, have IQs less than 70. The IQ difference is usually less for the northern than the southern states, and clearly many intangible factors are involved. However, the qualitative point that blacks have lower IQs than whites is generally a reproducible result (for further details, see Jensen 1972).

Figure 4 shows clearly that individual differences in IQ *within* any one race greatly exceed differences between races. The white distribution is, however, more spread out, since the standard deviation for blacks is 25 percent less (12.4 compared with 16.4 for whites). This is an observation characteristic of a number of IQ studies. Even so, there is no denying that the mean IQs of blacks and whites differ; but can this be interpreted genetically? Some writers have asserted that the answer is yes. Jensen (1972, p. 163) and certain others argue that the various lines of evidence produced "make it a not unreasonable hypothesis that genetic factors are strongly implicated in the average Negro-white intelligence difference." Quantitatively, Jensen (1973) believes that between one half and three quarters of the average IQ difference between United States blacks and whites is attributable to genetic factors. The remainder is due to environmental factors and their interaction with the genetic factors. Jensen has strong support for his views, especially from the physicist Shockley, who received a Nobel prize for his contribution in the invention of the transistor. Shockley has coupled his views with a strong advocacy of more research effort devoted to the black-white IQ difference. It comes as no surprise that others believe the differences are almost exclusively environmental (Pettigrew, 1971; also Bodmer and Cavalli-Sforza, 1970).

As we have seen, stratification of occupations in whites is maintained by substantial social mobility across classes. There are, however, no direct comparisons for the mean IQ differences *between* blacks and whites because of the effectiveness of skin color as a bar to mobility across classes. What then of the environments of blacks and whites? The predominantly black schools of the United States are less adequate than the white schools, so that an equal number of years of schooling does not mean equal educational attainment. A number of students of child development have noted the developmental precocity of black infants, particularly for motor behavior, which as Jensen (1972) argues could

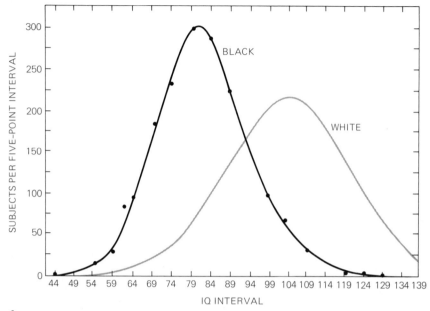

4. **IQ in United States blacks and whites.** Values for blacks were obtained from 1800 southern school children. Values for whites reflect "normative" sample of white population. (From Kennedy, Van De Riet, and White, 1963.)

hardly be environmental, since it is present in 9-hour-old infants. This behavioral precocity is paralleled by certain physiological indices of development such as bone development and brain wave development. Yet after a few years the blacks drop behind (see Coleman et al., 1966). Just as average school environment may differ for blacks and whites, so too may the home environment, in that blacks frequently live in economically deprived areas. The very early home environment may be of substantial importance for intellectual development, and some data clearly show the detrimental effects of severe sensory deprivation very early in life (see Pettigrew, 1971, for references).

Coleman et al. (1966), in a survey of the entire United States, assessed several environmental variables and socioeconomic indices generally thought to be major sources of environmental influence in determining individual and group differences in scholastic performance. Included were factors such as reading material in the home, cultural amenities in the home, structural integrity of the home, foreign language in the home, preschool attendance, parents' education, parents' educational desires for the child, parents' interests in child's schoolwork, time spent on homework, and child's self-concept (self-esteem). Coleman et al. found all these factors to be correlated in the expected direction with scholastic performance *within* each of the racial groups studied. But comparisons *among* groups show that the most environmentally disadvantaged group is the American Indian, on every environmental index lower than the blacks. The whites are, as expected, highest. However, the American Indian achievement and ability scores exceeded those for blacks on nonverbal intelligence, verbal intelligence, reading comprehension, and mathematics achievement. It is difficult to interpret this result either genetically or environmentally. The only valid way of interpreting results of this nature is to test different races under *identical* environments, and such data are unavailable. While identical environments might be created by having black children adopted into white homes and vice versa, even in this case (because of possible prejudice), the environments could not be regarded as identical; thus the required data are apparently unobtainable.

Pettigrew (1971), discussing the role of the black, has commented that he is not expected to be bright and therefore expects to fail; this leads to a lack of self-confidence, a lack of interest in learning, and a lack of progress. Furthermore, Pettigrew (1971) has found that blacks give more correct answers when tested by blacks than when tested by whites for certain tests incorporating intelligence. Another likely factor is nutrition. Pettigrew quotes a study by Harrell, Woodyard and Gates (1956) in which dietary supplementation during the last half of pregnancy had directly beneficial effects on the IQ scores of the children later. When a group moves from a restrictive environment to a stimulating environment average IQ is expected to improve (as predicted from Cooper and Zubek's (1958) experiments with rats; Section 7.5). Perhaps the most dramatic evidence comes from the Osage Indians. This group occupied land on which oil was discovered, a circumstance that afforded them a living standard vastly superior to that of other Indians. On performance and language tests they were found to be superior to the level of comparable whites of the area. Similar increases in IQ have been recorded among whites in mountain areas of eastern Tennessee between 1930 and 1940. This was a period during which broad economic, social, and educational improvements occurred, and the average IQ increased from 82 to 93. Other studies are cited in Pettigrew (1971).

For all these reasons, we find it difficult to agree with Jensen's conclusions; we do not regard it possible to prove his hypothesis that much of the difference between the IQs of blacks and whites is due to genetic causes. On the other hand, we do not regard it possible to disprove his hypothesis. The experimental situation needed for proof is not available — a problem inherent in work on our own species. If a controversy of this type occurred in laboratory rodents, for example, it would have been resolved long ago because genotypes can be replicated and environments controlled.

12.4 Primary mental abilities

A more recent approach to the issue of mental abilities as measured by IQ comes from the construction of tests designed to measure a number of *separate* abilities (Vandenberg, 1967). One such battery is the Chicago Primary Mental Abilities Tests constructed by Thurstone and Thurstone (1941), which has been used in several surveys. Table 7 presents a compilation of the results of four such surveys for verbal, space, number, reasoning, word fluency, and memory scores. The four studies agree on hereditary components as assessed by *H* statistics (Equation 7.10) for verbal score and word fluency score. There is also good agreement for a hereditary component for spatial score (an ability to deal mentally with two- and three-dimensional patterns). A rather lower significance for hereditary factors is apparent for memory score. With regard to the remaining two scores, the British study (Blewett, 1954) disagrees with the American studies. The number score is based on very simple arithmetic tests, and no evidence for an hereditary factor was found for the British study, while the other three (American) studies suggested hereditary factors. For the reasoning score the reverse occurred, as Blewett's subjects gave evidence for a hereditary component while none of the three American studies did. Vandenberg (1967) commented on the need for caution in the interpretation of these results because of variations between groups in socioeconomic experience or educational practices and because of simpler interpretations such as variations in sample sizes or different instructions. Even so, on the basis of all the data together, significant hereditary variation is suggested for number, verbal, space, and word fluency scores. Vandenberg went on to show that these four components are at least somewhat independent of each other from the genetic point of view. Possibly educational

TABLE 7. *H* **Statistics Calculated from Scores of DZ and MZ Twins on Six Items of Chicago Primary Mental Abilities Test**

	Study			
Test item	Blewett (1954)	Thurstone et al. (1955)	Vandenberg (1962)	Vandenberg (1964)
Verbal	0.68**	0.64**	0.62**	0.43**
Space	0.51*	0.76**	0.44*	0.72**
Number	0.07	0.34	0.61**	0.56**
Reasoning	0.64**	0.26	0.29	0.09
Word fluency	0.64**	0.60**	0.61**	0.55**
Memory		0.38*	0.21	

* $P < 0.05$
** $P < 0.01$ as measured by an F test consisting of V_{DZ}/V_{MZ}. This test of significance is frequently used by Vandenberg. It is related to the H statistic by $V_{DZ}/V_{MZ} = 1/1-H$ (see Chapter 7).
After Vandenberg (1967), where the original sources are specified.

practices and/or socioeconomic experiences are more important for reasoning and memory than for these four components, leading to more ambiguous results.

The finding of four genetic components corresponding to four of the scores is, if substantiated in the future, a result of some considerable significance. It indicates the likelihood that intelligence is made up of several separate contributions and that the IQ test evaluates an aggregate of these and no doubt other contributions. The trend toward such refined analyses of complex traits should lead to a better understanding of the composition of the evolutionary units underlying human intelligence.

There are many other tests for mental abilities to which subjects have been exposed, and not much can be said about the degree to which the traits examined are under genetic control. Considerable research is being undertaken in this area, especially in view of the ongoing controversy about the interpretation of IQ differences between blacks and whites.

12.5 Personality

Despite a predominant interest in intelligence, there has recently been a growing emphasis on personality. Multifactorial techniques have led to a tendency to concentrate on specific aspects of personality, rather than personality as a whole. Griffiths (1970) defined personality as "the more or less stable organization of a person's emotional, cognitive, intellectual and conceptual, and physiological behaviour which determines to a large extent his adjustments to environmental situations." Defined in this way, intelligence is just one aspect of personality. Advances in the assessment and production of reasonably reliable and valid tests of personality have made the assessment of genetic differences more reliable than previously.

The multifactorial tests include scales purporting to measure specific personality traits. Two examples of such tests are the Minnesota Multiphasic Personality Inventory (MMPI) and the California Personality Inventory (CPI). Gottesman (1965) used the MMPI in a study of 34 MZ and 34 DZ adolescent twin pairs in Minnesota (where a high proportion of subjects were of Scandinavian origin) and in another study of 82 MZ and 86 DZ pairs in Boston. The MMPI consists of 550 questions yielding scores on 10 aspects of personality (Table 8). There is reasonable agreement in the rank ordering of the H statistics with the exception of paranoia. Reasonably high H statistics were recorded for social introversion and psychopathy, and also, as might be expected, for the two psychotic scales (depression and schizophrenia). Even so, results varied according to age and sex, and the lack of complete agreement in the ordering of the H statistics argues for an effect due to geographical area or to the origins of the populations studied.

Jinks and Fulker (1970) analyzed neuroticism in the data of Shields (1962) as assessed by a questionnaire designed to give a measure of both neuroticism and extroversion. The subjects were MZ twins brought up together and apart and DZ twins. Estimates of H and E (Equations 7.10 and 7.11) statistics are given in Chapter 7, Table 5. The general conclusion of Jinks and Fulker was that the data for neuroticism could be explained by a model of additive gene action, with no dominance. This means that intermediate expression for neuroticism is favored; i.e., stabilizing selection for an intermediate optimum is likely, extremes being at a reproductive disadvantage (see Section 9.3 for the rationale behind this argument). Gottesman (1965) speculated similarly for a number of such traits, since he considered that extremes would be disadvantageous, but previously little evidence was available. Jinks and Fulker found an indication of positive assortative mating in the data, which was not, however, significant. They also considered that cultural and class differences had little or no effect on this major personality dimension.

Extroversion was analyzed by the same questionnaire by which neuroticism was assessed. This trait, together with neuroticism, completes a broad two-dimensional view of major personality tendencies as described by Eysenck (1967). The environment was found to be more relevant for extroversion than for neuroticism, in that the introvert genotype was found to be more modifiable than the extrovert genotype by the within-family environment. Even so the degree of genetic determination was high. An interesting point comes from Shields (1962), who

TABLE 8. H Statistics Derived from Scores of MZ and DZ Twins on Minnesota Multiphasic Personality Inventory

Personality trait	Minneapolis study		Boston study	
	H	Rank	H	Rank
Hypochondriasis	0.16	7	0.01	10
Depression	0.45	3	0.45	1
Hysteria	0.00	10	0.30	7
Psychopathy	0.50	2	0.39	2
Masculinity/femininity	0.15	8	0.29	8
Paranoia	0.05	9	0.38	3
Psychasthenia	0.37	5	0.31	6
Schizophrenia	0.42	4	0.33	4
Hypomania	0.24	6	0.13	9
Social introversion	0.71	1	0.33	4

After Gottesman (1965).

discussed how one of a pair of monozygotic twins brought up together assumes the dominant role and thereafter assumes the role of leader that develops throughout the lives of the twins.

The first study to use factor scores was that of Eysenck and Prell (1951). It marked the beginning of the trend away from single measures to combined measures. Data presented in Table 9 for neuroticism, extroversion, autonomic activity, and intelligence, show high H statistics especially for neuroticism. Many other studies have been devoted to nature-nurture analyses of personality, especially twin studies (see Mittler, 1971), but no simple summary of the literature is possible. Eysenck's more recent work is of interest in that some attempt is being made to relate personality to constitutional variables such as conditionality, speed of cognitive functioning, and body type. It is an area of extreme complexity and challenge for both psychologists and geneticists. Future work must involve the study of interactions between genetic and environmental factors, in the hope of gaining some understanding of their relative contributions and the actual processes and measures involved.

12.6 Sensory, perceptual, and motor tasks

Electroencephalographic (EEG) records are much more similar in MZ twins than in DZ twins. Most early work depended on the visual inspection of EEG records rather than the more precise analyses made possible by the use of computers (Juel-Nielsen and Harvald, 1958). Computer analysis has opened up new possibilities for the study of genetic aspects of the central nervous system. Generally, MZ twins are much more similar than DZ twins, as expected. Mittler (1971) commented that genetic factors could play a more important role in the development of visual-spatial abilities than in the various traits discussed so far, which mainly involve components of intelligence and personality. Precise methods of EEG quantification can be expected from work on evoked cortical potentials. Specific signals, such as flashes of light or pure tones, are used, and the exact constituents of the cortical

TABLE 9. Intraclass Correlations and H Statistics for Factor Scores of Various Personality Traits

Trait	r_{MZ}	r_{DZ}	H	Source
Neuroticism	0.85	0.22	0.81	Eysenck and Prell (1951)
Extroversion	0.50	−0.33	0.62	Eysenck (1956)
Autonomic activity	0.93	0.72	0.75	Eysenck (1956)
Intelligence	0.82	0.38	0.71	Eysenck (1956)

response to these signals are analyzed. For example, Dustman and Beck (1965) reported on a comparison of visually evoked potential to 100 light flashes in 12 pairs of MZ twins, 11 pairs of DZ twins, and a control group of 12 pairs of unrelated twins matched for age. They analyzed the wave components for the first 250 milliseconds and the first 400 milliseconds and compared central with occipital readings. Generally, MZ twins showed higher intraclass correlations than DZ twins; for the occipital reading for 250 milliseconds an H statistic of 0.57 was obtained.

Variants of sensory and perceptual functions for which a relatively simple genetic mechanism can be found are considered in Chapter 11. A number of detailed tasks involving visual perception have been carried out. As Fuller and Thompson (1960) commented, much of this work was motivated by the idea that afterimages, flicker fusion, and susceptibility to illusions are valid indices of personality. A summary is given in Table 10 for comparisons of MZ and DZ twins. A deficiency of much data is the failure to determine possible effects of prior experience upon simple perception.

An *afterimage* is assessed by fixating on, for example, a square on a neutral background for a fixed period of time, then judging the size of the afterimage by "projecting" the afterimage on screens at greater or lesser distances than the fixation distance. The data in Table 10 are for afterimages projected on screens at 50 and 200 cm after fixation at 100 cm. The H statistics obtained are high to very high.

TABLE 10.   Intraclass Correlations and H Statistics Derived from Scores of MZ and DZ Twins on Perceptual Tasks

Task		r_{MZ}	r_{DZ}	H
Size of afterimage	(1)	0.71	0.08	0.68
	(2)	0.68	0.00	0.68
	(3)	0.98	0.22	0.97
	(4)	0.75	0.23	0.67
Eidetic imagery	(1)	0.50	0.10	0.44
	(2)	0.66	0.15	0.60
	(3)	0.67	0.05	0.65
Critical flicker fusion		0.71	0.21	0.63
Muller-Lyer illusion	(1)	0.53	0.39	0.22
	(2)	0.55	0.05	0.52
	(3)	0.51	0.37	0.22
	(4)	0.57	0.28	0.40
Autokinetic phenomenon		0.72	0.21	0.64

From Mittler (1971).

Eidetic imagery indices are obtained by using a variety of complex visual stimuli, such as pictures with large colored areas, and recording the degree of image persistence reported by the subject. From the sum of several tests a composite eidetic score can be obtained. Each stimulus is presented with and without a slight flicker. H statistics are high, but not on the whole as high as for the size of afterimage. *Critical flicker fusion* also gives a high H statistic.

The *Muller-Lyer illusion* is produced by a pair of arrows whose shafts are of equal length, but whose arrowheads point outward or inward (Fig. 5). The shaft of the arrow with the outgoing heads looks longer, although both shafts are the same length (see Gregory, 1966, for a discussion of these and other illusions). The subject is asked to judge which shaft seems longer. The H statistics for this task are lower than for the others.

In the test for the *autokinetic phenomenon*, the subject is instructed to fixate on a stationary light and describe what he sees. If movement is reported, the subject traces the path and is scored by the length of the line drawn. For this test the H statistic is high.

Generally, therefore, all these visual perceptual tasks show some genetic basis. The H statistics are similar in magnitude to those obtained for mental abilities and personality traits. It is surprising, however, how little work has been done on the possible genetic basis of such traits. Comprehensive studies such as have been done for mental abilities and personality on a variety of relatives have not been reported for visual perception. However, Cummins (personal communication) has looked at parent-offspring relations for two perceptual tasks: the kinesthetic-spatial aftereffect and the visual-spatial aftereffect. A spatial aftereffect

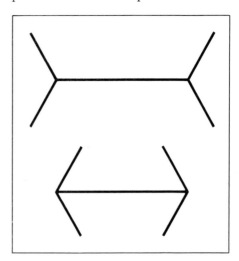

5.   **The Muller-Lyer illusion.** The shafts of the two arrows are the same length.

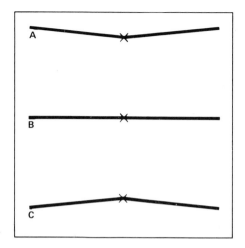

6. **Pattern demonstrating a spatial aftereffect.** After prolonged fixation on the apex of *A*, subject sees *B* as bent in opposite direction *C*.

occurs when a subject fixates at the apex of a bent line (Fig. 6A) for 2 or 3 minutes and then shifts his gaze to the midpoint of a straight line (Fig. 6B). The straight line is judged to be bent slightly in the opposite direction (Fig. 6C). There are a variety of aftereffects, and Cummins' study involved the two aftereffects of visual tilt and kinesthetic tilt using stimulus objects slanted at 15° (see Day, 1969, for a detailed description of the traits). The regression coefficient of offspring on midparent for the kinesthetic aftereffect was 0.42 ± 0.27, giving a heritability in the narrow sense of 0.84 (Equation 6.20), but for visual aftereffect the obtained value of 0.04 ± 0.15 indicates little genetic determination. However, the correlation between parents for the visual aftereffect came to 0.58, which was significant at the 5 percent level, indicating likely positive assortative mating. The corresponding value for the kinesthetic aftereffect was 0.06, which did not deviate significantly from zero. While these results are preliminary and are based on a small number of subjects, they indicate the suitability of these and related traits for detailed biometrical genetic analyses such as have been carried out for mental abilities and personality traits.

For motor skills, twin studies have provided evidence suggesting genetic control — e.g., for pursuit rotor, spool-packing, and card-sorting tests. As already stressed, psychologists have paid far less attention to the biological variables discussed in this section than to cognitive variables. It is, however, difficult to avoid the conclusion that there must be a substantial genetic component in sensory, perceptual, and motor behavior; whether the genetic component there is more important than for mental abilities and personality traits is difficult to assess from the evidence reviewed.

Spuhler and Lindzey (1967) discussed variations in sensory, perceptual, and motor processes among races. Although the earliest psychological comparisons of races dealt with simple sensory processes and modes of response, there has, until recently, been little systematic work in this area. Late in the last century it was found that for reaction time to visual, auditory, and tactile stimuli, American Indian subjects had the lowest average latency. They were followed by an African-Caucasian hybrid group, with a Caucasian group the slowest of all to react. Early in this century, the visual acuity of Torres Strait Islanders was found to be superior to that of European groups. Other variations among races for sensory motor processes discussed by Spuhler and Lindzey include weight discrimination, pain threshold, and olfactory acuity. These early studies suggest the possible existence of appreciable racial differences in behavior; little can be said about the degree to which such differences may be genetic. After this series of early studies, the investigation of the more complex processes discussed in previous sections of this chapter predominated in psychology, and only recently has there been a resurgence of interest in sensory-perceptual-motor studies. For example, recent work shows the Muller-Lyer illusion to be almost four times as common among American subjects as among Bushmen subjects. In the face of limited evidence, however, Spuhler and Lindzey (1967) wrote: "If we except PTC tasting, color vision, and certain perceptual illusions, there appears to be little compelling evidence at present either for racial differences or racial equality in simple sensory or motor processes." If the effort spent on analyzing personality and mental abilities is applied to sensory and motor processes, significant advances would surely occur.

12.7 Behavior and morphological variation

Sheldon (1940, 1942) looked into possible relations between human structure (i.e., somatotype) and behavior. He found quite high correlations, but in spite of this, few further studies have been carried out. Lindzey (1967) stressed the reluctance of psychologists generally to give serious consideration to the study of morphology and behavior. In Chapter 9, an association between morphology and behavior in mice is postulated, and found to exist for many traits, but not for traits incorporating a high component of learning and reasoning.

Sheldon (1940, 1942) put forward a classification of morphology based on three extreme physical types, with ratings for each dimension derived from a standardized set of photographs. Three extreme phenotypes (somatotypes) are envisaged:

- Endomorphy. The individual high for this component is characterized by softness and spherical appearance associated with an underdevelopment of bone and muscle, a relatively low surface/mass ratio, and highly developed digestive viscera. Since the functional elements of these structures derive primarily from the endodermal layer, the term endomorphy is used.

- Mesomorphy. The individual high for this component is hard and rectangular, with a predominance of bone and muscle, and so is equipped for strenuous and exacting physical demands. The term derives from the dominance of the mesodermal layer in this somatotype.

- Ectomorphy. The individual extreme for this component is linear, fragile, and characterized by a flatness of the chest and delicacy of the body. He is usually thin and light-chested. Therefore, he is made up more than the other somatotypes of ectodermal tissue, whence the term ectomorphy is derived. Ectomorphy is a physique poorly equipped for persistent physical action.

Sheldon's classification is not the only one that has been devised; others are discussed in Hall and Lindzey (1957), Lindzey (1967), and Stern (1973).

Some fairly clear associations exist between behavior and morphology in man. For example, the frail ectomorph cannot employ physical or aggressive responses with the same effect as the robust mesomorph. Height, weight, and strength put limits upon the adaptive responses an individual can make in a given environment. Lindzey (1967) has quoted evidence for clear and consistent behavioral differences among individuals who vary in morphological development. In general, an individual who is physically extreme in some sense, such as being excessively fat or thin, is exposed to a somewhat different set of learning experiences than someone who is more average physically. Note that "average" varies among ethnic groups.

A striking set of examples comes from the somatotypes of athletes (Carter, 1970). Almost all groups of championship athletes are highly mesomorphic. The most mesomorphic are the weight lifters followed closely by Olympic track and field throwers, football players, and wrestlers. The least mesomorphic men are the distance runners. Women athletes range from the track and field jumpers and runners, who are the least mesomorphic, to the gymnasts, who are the most. It is not surprising that champion performers at various levels of a particular sport exhibit similar patterns of body size and somatotype, but the patterns tend to become more extreme as the level of performance increases. Extremes at the behavioral level correspond to extremes at the morphological level. Conversely, certain somatotypes found in nonathletes are not found at all in groups of championship athletes.

Extreme somatotypes in athletes can be made more extreme by training, but training is unlikely to convert a nonathletic somatotype to an extreme athletic mesomorph.

Sheldon (1942) claimed striking associations between morphology and temperament. He selected three components of temperament:

- Visceratonia. Individuals high in this component are characterized by general love of comfort, sociability, gluttony for food, enjoyment of people, and affection. Such a person is relaxed in posture, reacts slowly, and generally is an easy person with whom to interact.

- Somatotonia. A high score is ordinarily accompanied by a love of physical adventure, risk taking, and a strong need for muscular and vigorous physical adventure. Such individuals are aggressive, with tendencies toward action, power, and domination.

- Cerebrotonia. A high score implies restraint, inhibition, and the desire for concealment. Such an individual is secretive, self-conscious, and afraid of people.

Therefore, we have Sheldon's classification for both physique (structure) and temperament (function). A priori, we can pair them as follows: endomorphy-visceratonia, mesomorphy-somatotonia, ectomorphy-cerebrotonia. Based on 200 subjects in each of the temperament classes, and using scoring systems for physique and temperament, Sheldon obtained the correlations presented in Table 11. The results clearly show large positive correlation coefficients for the above pairs and negative correlations elsewhere: there is a marked association between the structure or physical traits of the individual and his functional or behavioral qualities. The positive correlations are, however, extremely high and have been criticized by psychologists on the grounds that Sheldon himself executed both sets of ratings. Later studies (for a review see Lindzey, 1967) also show positive associations between morphology and temperament, but at rather a lower level (Child, 1950; Parnell, 1958; Walker, 1962).

For individuals showing criminal behavior, Sheldon found an excess of mesomorphs, in particular endomorphic mesomorphs, among delin-

TABLE 11. Correlation Coefficients Between Physique Components and Temperament Components

	Visceratonia	Somatotonia	Cerebrotonia
Endomorphy	0.79	−0.29	−0.32
Mesomorphy	−0.23	0.82	−0.58
Ectomorphy	−0.40	−0.53	0.83

From Sheldon (1942).

quent youths. A number of other surveys (Eysenck, 1964; Lindzey, 1967) confirm this, including one involving female delinquents (Epps and Parnell, 1952). Several investigators have found an association of somatotype with schizophrenia (Heston, 1970; Parnell, 1958), since mesomorphs are underrepresented among schizophrenics and ectomorphs are overrepresented. Paranoids on the other hand are often mesomorphic (Parnell, 1958).

In man, therefore, behavioral traits not ostensibly having a learning component may have an association with morphology. It is a neglected field compared with the studies on mental abilities discussed earlier in this chapter. Hall and Lindzey (1957) wrote: "It seems safe to say that Sheldon's work is of assured interest to posterity, for whatever its shortcomings may be it has led to findings that the future will have to reckon with."

12.8 Criminality

In Dostoevski's *The Brothers Karamazov*, the Russian monk advises:

> Remember particularly that you cannot be a judge of anyone. For no one can judge a criminal until he recognizes that he is just such a criminal as the man standing before him and that he perhaps is more than all men to blame for that crime. When he understands that, he will be able to judge. . . . If I had been righteous myself, perhaps there would have been no criminal standing before me.

Unsatisfactory home life, poor upbringing, poverty, ignorance, mental deficiency, absence of parents, cultural conflicts, and other environmental inputs have been repeatedly implicated in the cause of criminality and antisocial behavior. What of the genetics?

Stern (1973) asked, in an end-of-chapter exercise:

> Among 278 sibs of criminals, Stumpfl found 103 who had a criminal record. This corresponds to 1 criminal out of 2.7 sibs of criminals. Among 62 nonidentical twin partners of criminals, Stumpfl and Kranz found 30 offenders. This corresponds to 1 criminal out of 2.1 nonidentical twin partners of criminal twins. It has been suggested that the last-named higher frequency of criminals (1 in 2.1) as compared to the first-named frequency (1 in 2.7) is due to the greater environmental similarity for twins than for ordinary sibs.
> (a) What is the statistical significance of the data?
> (b) What bearing has the answer to the preceding question on the suggested explanation for the different frequencies?

For adult crime, a concordance rate of 71 percent ($N = 107$) has been recently reported for MZ and 34 percent ($N = 118$) for DZ twins. Equivalent figures for juvenile delinquency, however, are 85 percent ($N = 42$) and 75 percent ($N = 25$). These data, therefore, must not be judged conclusive or even indicative, since it is not possible to disentangle heredity and environment (Eysenck, 1964). The reasons should, at this point in our text, be obvious.

The factors discussed below, having some hereditary bases, may be important determinants of the commission or noncommission of crime (Rosenthal, 1971):

- A large number of criminals have low IQs.

- Juvenile delinquents and criminals have a higher incidence of abnormal EEGs than the general population. Rosenthal (1971) cited some prison samples with a 75 percent rate of EEG abnormality. The relation between heredity and both normal and abnormal EEG has been summarized by Omenn and Motulsky (1972; also see Section 12.6). Table 12 gives the genetic basis of a number of EEG patterns.

  On the basis of limited EEG observations, individuals producing either a monotonous tall alpha or a beta wave pattern tend to marry equivalent producers assortatively. For example, 17 out of 56 beta-wave producers married beta-wave producers, while only 5 out of 54 non-beta-wave producers married beta-wave producers. Those who form monotonous tall alpha waves (unusually regular alpha waves of high amplitude) similarly show an unexpectedly high proportion of marriages with people of the same EEG type.

  Occipital slow rhythms may be associated with psychopathy; persons with such an EEG type apparently exhibit an accumulation of psychological peculiarities. If this relation is established, it could be the very first physiologically characterized normal variant in man showing a qualitative influence on personality without impairing intelligence. Mittler (1971) may be consulted on the use of twins to elucidate the genetic components of electrocortical brain activity. Monozygotic twins normally have high concordance for EEG types.

TABLE 12. Genetic Basis of Variants of Human Electroencephalograms

Rhythm	Genetic basis	Population frequency (%)
Normal alpha (8–13 cps)*	Polygenic	
Low-voltage alpha	Autosomal dominant	7.0
Quick alpha (16–19 cps)	Autosomal dominant	0.5
Occipital slow (4–5 cps)	?	0.1
Monotonous tall alpha	Autosomal dominant	4.0
Beta waves	Multifactorial	5.0–10.0
Frontal beta groups (25–30 cps)	Autosomal dominant	0.4
Frontoprecentral beta (20–25 cps)	Autosomal dominant	1.4

* cps, cycles per second.
After Omenn and Motulsky (1972).

- Glueck and Glueck (1956) stated that some 60 percent of delinquents are mesomorphs of athletic build, with reference to Sheldon's (1942) somatotypes as discussed in Section 12.7. Being more athletic, are such individuals more prone physically to express their dissatisfactions and/or to attempt to eliminate them?
- Some male individuals have an extra Y chromosome and so are XYY (see Section 4.3 on the *behavior* of males with more than one Y chromosome). Apparently, even an oversized Y chromosome may have deviant behavioral implications. Nielson and Henriksen (1972), studying imprisoned Danish youths, found long Y chromosomes four times more frequently than in control males. Criminal records were also more frequent among the fathers and brothers of these imprisoned subjects than among the fathers and brothers of the controls. However, only this one population has been surveyed.

We envision and welcome future efforts to elucidate the roles and interplay of matters genetic and of matters environmental in the production of a criminal, much as has been attempted for IQ. Surely in many instances it appears that certain physiques and perhaps EEG patterns are associated with criminality. So far, we know more about possible physiques and specific nervous system changes that are associated with criminality than we know about such physical parameters that are associated with intelligence. In other words, we may be closer to the actual genes related to criminality than to the genes related to intelligence. Future research on criminality should be rewarding.

General readings

Cancro, R. (ed.). 1971. *Intelligence: Genetic and Environmental Influences*. New York: Grune & Stratton. The product of a wide-ranging conference on intelligence held at the University of Illinois.

Mittler, P. 1971. *The Study of Twins*. London: Penguin. A very readable account of twins in behavior-genetic research covering most of the traits listed in this chapter.

Osborne, R. H. 1971. *The Biological and Social Meaning of Race*. San Francisco: Freeman. A volume including discussions on the heredity-environment issue with respect to variations in intelligence.

Penrose, L. S. 1963. *The Biology of Mental Defect*, 3rd. ed. London: Sidgwick & Jackson. A classic considering mental defect in the broadest way.

Behavior and Evolution

13.1 Behavior as a component of fitness

If we define the fitness of a genotype as its relative ability to contribute to future generations, what is the role of behavior in fitness? More precisely, fitness can be regarded as the average number of progeny left by the carriers of a given genotype relative to the number of progeny left by other genotypes. To this we must add the complication that the fitness of a given genotype depends on the environment(s) to which it is exposed. Consider, for example, *Drosophila pseudoobscura*. Population cages containing various chromosomal karyotypes in pairs usually gave stable equilibria at 25°C, since often the heterokaryotypes were found to be fitter than the corresponding homokaryotypes (Wright and Dobzhansky, 1946). This is a situation for which a stable equilibrium is expected, as shown in Section 4.2. Also as expected from theoretical considerations, the stable equilibria occur irrespective of the initial frequencies of the karyotypes. However, at 16.5°C little change occurred in the frequencies in the population cages, and at 22°C an intermediate situation arose whereby some but not all populations showed stable equilibria (Van Valen, Levine, and Beardmore, 1962). These findings demonstrate the dependence of the equilibria, and hence the relative fitness of genotypes, on the environment, in this case temperature variations. Furthermore, fitness estimates are applicable only to genotypes in a given population, since genetic backgrounds vary and influence fitnesses, as is shown by the frequent breakdown of heterokaryotype advantage in between-population crosses in *D. pseudoobscura* (Dobzhansky, 1950). The gene complexes within populations are coadapted within and between chromosomes but not between populations. Therefore, we cannot speak of fitness as a property applying to a specific gene or karyotype without qualification. We may conclude that the dependence of fitness on environments and on the whole genome makes it impossible to define fitness as an invariant parameter associated with a particular genotype or karyotype.

It is not hard to see that most if not all of the behavioral parameters discussed in this book contribute in some way to the overall fitness of an organism; indeed, no behavioral trait can be regarded as neutral so far as fitness is concerned. Even a trait's lack of *obvious* connection with fitness does not mean that there is no effect: the lack of an obvious effect may just reflect our ignorance, hopefully temporary. Further, to consider behavior properly as a component of fitness, it is necessary to go outside artificial laboratory situations to the real world — a problem that presents peculiar difficulties associated with the species selected. It is also necessary to consider the contribution of behavior to alterations in the gene pool as well as the action of genes controlling or adjusting behavior, which has been the main theme of the book so far. This is because fitness is defined in terms of contributions of genotypes to future generations, which must mean that the effect of behavior on evolutionary processes is an issue of central importance. Finally, it will be seen in this chapter that when the researcher moves out of the laboratory into the wild, he often finds it impossible to dissociate behavioral from ecological factors.

Unfortunately, in any one experiment only a few (or only one) fitness factors are normally measured. A question of evolutionary significance concerns relations between fitness factors. There is evidence in *Drosophila melanogaster* that males which mate quickest also copulate more often more successfully and leave more progeny (Fulker, 1966). For polymorphic inversions in *D. pseudoobscura* at 25°C, heterokaryotypes are superior in innate capacity for increase in numbers. This is defined by Andrewartha and Birch (1954) as the maximum rate of increase attained by a population under a specific environment. The heterokaryotypes are also superior to the homokaryotypes as regards population size, productivity, egg-to-adult viability, and mating frequency. For mating behavior, the karyotype of the male is important for mating frequency in *D. pseudoobscura* as in *D. melanogaster* (Spiess, Langer, and Spiess, 1966). The results for these various traits in *D. pseudoobscura* were mainly derived by different experimenters in experiments carried out at different times (for references, see Parsons, 1973).

The relative associations among these components in a given population have been inadequately explored, even though they are of considerable importance in the study of the overall fitness of organisms.

Prout (1971a,b) described an experimental system for estimating certain components of fitness simultaneously in *D. melanogaster*. Fourth-chromosome recessive mutants, eyeless (ey^2) and shaven (sv^n), were used. This chromosome is very short (Chapter 2, Fig. 3), and recombination is not relevant as a source of complication. The estimated components of fitness were larval viability in each sex, and from adults two components were obtained, one representing female fecundity and the other male mating ability (virility). The adult components were the more important, so that ey^2ey^2 and ey^2/sv^n females were superior to sv^nsv^n, and the heterozygous males were superior to both homozygotes. In males, the depressed values of the two homozygotes varied with the female genotype to which the males were mated, indicating mating interactions. Larval fitness components were small compared with the adult fitness components. Prout, therefore, stressed the need to define a small number of components of fitness that encompass the entire life cycle and that are accessible for experimental evaluation. He tested his fitness estimates by attempting to predict the performance of experimental populations segregating for these same mutants. The results were in reasonable agreement with prediction. Therefore, the fitness estimates can account for most of the performance of the experimental populations. Further work is needed integrating fitness parameter estimates with population performance using an approach of this nature; in particular, generalizations over a series of environments seem essential.

The above evidence, although indirect, indicates that in each instance male mating behavior is an important component of fitness. This agrees with the earlier experiments of Merrell (1953), who found gene frequency changes in experimental populations of *D. melanogaster* to be predictable from male mating behavior variations. In the Neotropical South American species *D. pavani*, males heterozygous for various gene arrangements were superior in mating activity to the corresponding homokaryotypes in the same population (Brncic and Koref-Santibañez, 1964). On the other hand, in *D. persimilis*, both sexes appeared to be important in the mating frequency of various karyotypes (Spiess and Langer, 1964b). Even so, it may be concluded that, at least in the laboratory, male mating behavior differences between genotypes are of importance in modifying gene pools in subsequent generations. Male mating behavior, therefore, forms an important component of fitness. The evolutionary significance of this result is, however, difficult to assess without extrapolation to nature.

13.2 Habitat selection: *Drosophila*

Since this is a text on behavior genetics, our discussion of habit preference is restricted to comparisons within species and between closely
related species. Consider Drosophila first. A general account of the
behavioral and ecological genetics of this genus appears in Parsons
(1973). For example, there are food preference differences among species
associated with seasonal and geographical variations in distribution
(Dobzhansky and Pavan, 1950). One factor seems to be differential
attraction to different yeast species (Dobzhansky et al., 1956). In the
extraordinarily diverse Hawaiian *Drosophila* fauna, the distribution of a
number of plant species and a number of other ecological factors such as
wind intensity, humidity, temperature, and light intensity are clearly
important (Carson et al., 1970). Moderate wind currents and light intensities, humidities below 90 percent, and temperatures above 21°C seem
to be avoided by many species. Therefore, it is not surprising that during
overcast weather, when the humidity approaches 100 percent and especially if misty rain is falling, the flies of these species tend to move
upward into the available vegetation and can be found on the undersur-
faces of leaves and plant limbs up to about 10 ft from the ground.
However, on cloudless sunny days when humidity falls, the flies rapidly
disappear, presumably seeking out small poorly lighted areas where
humidity is high and light intensity is low. Here then we see intimate
behavioral adaptations to the prevailing environment.

Drosophila shows a range from species dependent on a particular plant
species (monophagous species) to those using a variety of host plants
(polyphagous species). A number of the polyphagous species of
Drosophila can be cultivated on laboratory media; such cultivation is a
much more difficult proposition for monophagous species. Presumably,
monophagous species are adapted to their own specialized niches,
whereas polyphagous species have less specialized requirements. The
species of *Drosophila* on which most behavior-genetics work has been
done generally fall into the less specialized class as regards nutritional
requirements. Even so, subtle behavioral and associated ecological differences are found among certain closely related species.

Some species that are difficult to cultivate on laboratory media have
quite elaborate behavioral patterns. We refer to the Hawaiian species,
many of which are geographically very restricted. In some instances
specialized forms of behavior not found elsewhere in the world occur
(Spieth, 1958; Carson et al., 1970). Field and laboratory studies indicate
that the males of many species patrol and defend a small but definite *lek*
(courting and mating territory). The territories are not randomly deter-

mined; each species has preferences apparently controlled by ecological factors such as light, humidity, temperature, and spatial conditions. The territories are close to, but separate from, their feeding sites. Associated with this is the development of sexual dimorphism. These species still show the basic drosophiloid behavior pattern, but superimposed on this are territoriality, aggression, and advertising in males, associated with the spatial separation of feeding and courtship sites. Males are not defensive at feeding sites, where they may be judged as gregarious, but once at their specific leks they become pugilistic. Unfortunately, genetic studies on these species are fewer than those on cosmopolitan species, but future behavior-genetics investigations of the complex behavioral patterns of the Hawaiian species and their associated morphologies should contribute greatly to our understanding of the evolutionary biology of the genus. In fact, the Hawaiian flies form a group of species in which the integration of genetics with behavioral and ecological studies is essential.

The Hawaiian species show such diversity that of the world fauna of 1000 to 2000 species, up to 500 identified species occur in this and in closely related genera (drosophiloids) in Hawaii, and there are probably 200 or more species belonging to the closely related genus *Scaptomyza* and other related genera (scaptomyzoids). Presumably this burst of diversity in the Hawaiian islands represents an adaptive radiation analogous to that of Darwin's finches on the Galápagos islands (Dobzhansky, 1968). Probably the adaptive radiation arose from the chance arrival of one or two species (Carson et al., 1970). Future work on the fascinating diversity of species now developed will clearly be of importance for the behavior geneticist as well as for the evolutionary biologist — two categories that are not mutually exclusive.

We conclude this section by comparing some closely related species. The species *D. melanogaster* and *D. simulans* are morphologically almost identical, and so are sibling species (see Section 4.2 for definition). Though frequently collected in the same locality, they are completely distinct species. One reason is that hybrids between them are sterile. It is instructive to review (Parsons, 1975) some of the rather subtle behavioral and ecological differences found within and between the two species, since in the laboratory they survive in almost identical culture systems, implying that their needs are at least similar. We discuss below some of the relevant studies.

- Sexual behavior. Ethological isolating mechanisms almost entirely prevent cross mating. The sexual behavior of the two types of males can be readily dissected into the same basic elements of courtship — orientation, vibration, licking, and copulation — as described in Section 3.2 where differences between mutants of *D. melanogaster* are discussed. However, males of

D. simulans take longer to begin courtship and consequently have longer bouts of simple orientation; in other words, the courtship behavior of *D. melanogaster* is more active than that of *D. simulans* (Manning, 1959). There is, therefore, no difference in the basic organization of the sexual behavior of the two types of males, but *D. simulans* males are slower to rise to sexual excitation than *D. melanogaster* males. Females of *D. simulans* are more responsive to the visual aspects of the male's courtship, and less responsive to those stimuli perceived by their antennae, than are *D. melanogaster* females. In fact, *Drosophila* species can be grouped in three classes on the basis of components of mating behavior in relation to light-dependence (Grossfield, 1971): (1) species unaffected by darkness, which include a number of cosmopolitan and broad-niched species such as *D. melanogaster;* (2) species inhibited by darkness, so that facultative dark mating occurs, which include *D. simulans;* (3) species in which mating is completely inhibited by darkness. The only cosmopolitan species showing light-inhibition is *D. simulans,* and it is suggested that its unique situation may be accounted for by its close relation to *D. melanogaster* and may reflect behavioral divergence of the two. The group in which darkness inhibits mating includes a number of specialized narrow-niche species, including the Hawaiian species, in which, as we have seen, visual cues are of critical importance.

Finally, even though isolation is nearly complete, a few hybrids can be brought into being under laboratory conditions. The degree of isolation can be shown to vary among different strains (Parsons, 1972b), but it is consistently strong. Environmental factors known to affect levels of isolation in the laboratory include age, whether single or mass matings are used, and, in the latter case, the proportion of males (for references, see Parsons, 1975).

- Dispersal activities. McDonald and Parsons (1973) found that the dispersal activity of *D. melanogaster* exceeds that of *D. simulans.* Comparison of the two species dispersing toward a light source, and without a light source, showed that *D. simulans* is more dependent on the presence of light than *D. melanogaster,* as was found for mating behavior. Similarly, for phototactic responses along a gradient of light intensities, *D. simulans* shows greater phototaxis than *D. melanogaster* (see Parsons, 1975). *D. melanogaster* shows a more even distribution over the various light intensities than *D. simulans.* Therefore, in both cases the behavior of *D. melanogaster* is less light-dependent than that of *D. simulans,* arguing that *D. melanogaster* may be regarded as the broader-niched species.

- Oviposition. In competition experiments, *D. simulans* tends to oviposit in the center of food cups and on food having a surface crust; *D. melanogaster* does not. In other words, desiccation makes the medium less favorable for *D. melanogaster* (Barker, 1971).

- Larval dispersal. Larvae of both species are found equally in the upper section of the medium, but lower down the proportion of *D. simulans* larvae exceeds that of *D. melanogaster* (Barker, 1971). This may provide an explanation for Sturtevant's (1929) observation that *D. melanogaster* may be affected more rapidly in old dry cultures than *D. simulans.*

• Ethanol in the medium. *D. melanogaster* is more tolerant of 9 percent ethanol than *D. simulans* both as larva and adult. Furthermore, as regards behavior, *D. simulans* adults show an aversion for ovipositing on sites containing 9 percent ethanol, whereas *D. melanogaster* shows a slight preference for doing so (McKenzie and Parsons, 1972). This explains the almost exclusive presence of *D. melanogaster* inside a winery near Melbourne, Australia, while immediately outside the winery both species occur, with *D. simulans* usually in excess. Furthermore, release-recapture experiments during vintage suggest that *D. melanogaster* moves toward the cellar in a regular fashion, while *D. simulans* moves away from it (McKenzie, 1974). Thus the distribution of the two species at vintage may be a function of their dispersal activities. This indicates, in agreement with the laboratory experiments, that the presence of alcohol in the environment alters the behavioral pattern of the two species in the wild.

• Temperature and desiccation. These factors are more ecological than behavioral, but in the avoidance of extremes of high temperature and low humidity, it is clear that behavior must play a part, and its importance may vary between species. There are known variations among strains within the two species for tolerance to these two stresses. It seems likely that *D. melanogaster* can tolerate a greater range of temperatures than *D. simulans* (for references, see Parsons, 1975), indicating that *D. melanogaster* may have a broader niche similar to that for light-dependence in mating behavior, dispersal activity, and phototaxis. Levins (1969) concluded that acclimation to dry heat by *D. melanogaster* depends more on developmental flexibility and physiological acclimation than on genetic differentiation of populations for its adaptation, whereas *D. simulans* shows lower developmental flexibility and depends more on genetic differences. While behavioral factors are clearly relevant, their relative importance in these instances in the two species is unknown.

• Some general ecological factors. Finally, a number of factors are known which have only minor behavioral components but which differentiate the two species. El-Helw and Ali (1970) found that *D. simulans* is more tolerant to natural yeasts in the medium than *D. melanogaster*, a fact which may be associated with the observation above that *D. simulans* burrows deeper into the medium. Minor differences in development rate, survival, fecundity, fertility, hatchability, and adult viability have been found showing a general superiority for *D. melanogaster*. Many of these experiments were carried out at 25°C, a temperature that is often lethal to *D. simulans* in the laboratory (Parsons, 1975). Indeed, in population cages *D. melanogaster* usually displaces *D. simulans* at 25°C, but at 15°C the reverse may occur (Moore, 1952).

No doubt all these effects found within and between these two sibling species are relevant in determining their distributions in the wild. The two species coexist to a fairly high degree in many regions, but the above considerations suggest the existence of subtle behavioral and ecological differences, so that the niches they occupy must differ. The one general conclusion is that *D. melanogaster* is more broad-niched than *D. simulans*. Genetically this is supported by the observation that gene-enzyme variability may be higher in *D. melanogaster* than in *D. simulans* (for references, see Parsons, 1975). This agrees with arguments and data presented by Beardmore (1970) that within a species an association is expected between the ecological heterogeneity to which a population is exposed and its genetic variability. It seems a reasonable argument for comparisons between very closely related species, but other factors must be taken into account for more distantly related species (Selander and Kaufman, 1973). The comparison between these two sibling species is discussed in some detail to show the subtle interplay of behavioral and ecological factors in determining the habitats and isolation among species.

Another pair of sibling species about which we have much information is *D. pseudoobscura* and *D. persimilis*, which are sympatric in some regions. Isolation is maintained by the following factors:

• The two species have somewhat different habitat preferences. *D. persimilis* occurs in cooler niches and *D. pseudoobscura* in warmer niches.

• The two have different food preferences, including differential attraction to different yeasts.

• Many species of *Drosophila* show high activities in the early morning and evening. As shown in Table 1 for flies collected in the Yosemite region of California, among flies attracted to yeasted baits in the morning, the proportion of *D. pseudoobscura* was lower and that of *D. persimilis* higher than among flies attracted during the evening period of activity (Dobzhansky et al., 1956).

• For a population where the two species were sympatric, the mean photo-response (affinity for light) was greater for *D. persimilis* than for *D. pseudoobscura* (Rockwell, Cooke and Harmsen, 1975).

• Sexual isolation is associated with differing male courtship songs for the two species (Ewing, 1969). Males of *D. pseudoobscura* produce two songs controlled by the wings, a song with a low repetition rate consisting of trains of 525-Hz pulses at 6 per second, and a song with a high repetition

TABLE 1. **Number of *D. pseudoobscura* and *D. persimilis* Collected Morning and Evening in Yosemite Region of California**

Month	Morning		Evening	
	D. pseudoobscura	*D. persimilis*	*D. pseudoobscura*	*D. persimilis*
June	68	111	682	432
July	210	297	694	446
August	65	75	681	443

After Dobzhansky et al. (1956).

rate in which 250-Hz pulses are repeated 24 times per second. In *D. persimilis* the low-repetition-rate song is absent or occurs in a very abbreviated form, and the high-repetition-rate song consists of 525-Hz pulses repeated 15 times per second.

The first four of these factors are not entirely effective because both species have been found feeding side by side in the same slime flux on the black oak *Quercus kellogii* (Carson, 1951). This suggests that the absence of interspecific matings in natural habitats is largely due to ethological isolation; in any case, when interspecific matings take place in the laboratory, fewer sperm are transferred than in intraspecific crosses, the F_1 males are sterile, and the F_1 females have reduced vigor.

In the laboratory hybridization occurs relatively readily. Virgins were about 4 days old in many experiments (see Section 8.4 about this specific age). However, when flies of both sexes were placed together a few hours after emergence, the proportion of hybrids was lower. Spieth (1958) has suggested that this higher level of isolation in older flies may be due to individuals of both species maturing together and then acquiring the ability to discriminate between the species before sexual maturity. Furthermore, a *D. persimilis* female once having mated with a *D. persimilis* male will not accept a *D. pseudoobscura* male subsequently. This indicates that the high level of isolation in the wild may be not merely innate but partly learned. More evidence of the effect of experience on mating behavior is presented in Section 8.4, where it is noted that Drosophila females prefer to mate with the type of male previously accepted by them.

Laboratory experiments indicate other relevant variables. The degree of isolation has been found to be temperature-dependent (Mayr and Dobzhansky, 1945), being relatively low for flies raised at 16.5°C. However, the level of sexual isolation can be increased or decreased by selection (Koopman, 1950; Kessler, 1966), showing that the degree of sexual isolation itself is under genetic control (Section 5.3). Further discussions of the genetics of sexual isolation are part of Section 5.3.

13.3 Habitat selection: Rodents

Selection of an optimal environment is important not only in Drosophila, but in any organism that exists in nature over a wide spectrum of habitats. Temperature is one of the prime factors involved in many adaptations, and at first it seems that those organisms possessing mechanisms of thermal adaptation have a reproductive advantage over those that do not. *Poikilotherms* (animals not having an internal mechanism for regulating body heat) adapt by immobility and reduced metabolic rate during periods of cold or by physiological and behavioral adaptations allowing maximum utilization of heat and protection from the cold. *Homeotherms*, on the other hand, have an internal mechanism for regulating body heat and can function efficiently within a wider range of temperatures. Even so, they still possess various special ways of combating extreme variations, and many of these ways are behavioral — burrowing, seeking shade, basking, shivering, migrating, and various locomotor activities.

The laboratory experiments described in Section 9.3 show that mice, when faced with a thermal gradient, select a preferred temperature that presumably denotes a physiological adjustment of importance for optimal functioning. The data discussed in Section 9.3 show that the temperature preference of mice has a fairly close association with various physiological and morphological traits, so that it can be regarded as an innate trait permitting the selection of a most favored habitat.

Studies in the deer mouse, *Peromyscus*, have shown behaviors predictable from the habitats occupied by them in nature. The prairie deer mouse of the Midwestern and Plains states of the United States (*P. maniculatus bairdii*) is a strictly field-dwelling subspecies that avoids all forested areas, unlike the closely related woodland form, *P. maniculatus gracilis*. Some work has been done on trying to identify the environmental cues that cause the mice to choose a place to live. Harris (1952) presented individual prairie and woodland deer mice with a choice between a laboratory "field" environment and a laboratory "wood" environment. Each type of mouse exhibited a clear preference for the artificial habitat most closely resembling its natural environment. Furthermore, laboratory-bred *Peromyscus* which had had no previous experience with either of the natural habitats exhibited a preference for the type of habitat it normally inhabited in nature. Habitat selection can be regarded as basically genetic in nature, under the control of natural selection for the field or wood environment.

Ogilvie and Stinson (1966) found the thermotactic optima of *P. maniculatus bairdii* and *P. maniculatus gracilis* to be 25.8°C and 29.1°C, respectively, in agreement with the warmer wooded environment of *P. maniculatus gracilis* and the cooler prairie environment of *P. maniculatus bairdii*. Another species, *P. leucopus*, which comes from an area where the ground temperature is 3° to 4°C higher than the woodland regions of *P. maniculatus gracilis*, preferred an even higher temperature, 32.4°C. Therefore, *Peromyscus* species show considerable differences in temperature preferences which are associated with habitat. It can be concluded that the animals tend to select those habitats that most closely resemble their natal homes and that this preference is influenced by genotypes. Wecker (1964), in experiments comparable to those of Harris (1952), argued for a behavioral feedback to the genotype which occurs simply by keeping a given population restricted to its habitat.

In the two *P. maniculatus* subspecies, genetic differences are apparent in their reaction to sand (King, 1967). Over a 24-hour period, the median amount of sand removed from a tunnel by *P. maniculatus gracilis* was 0.1 lb, while *P. maniculatus bairdii* removed 5.9 lb. The differences are compatible with our knowledge of the life history of the mice, since *P. maniculatus gracilis* is semiarboreal and *P. maniculatus bairdii* terrestrial. Other differences are such that *P. maniculatus bairdii* matures more rapidly in locomotor responses than does *P. maniculatus gracilis*, and conversely *P. maniculatus gracilis*, the semiarboreal species, develops a greater clinging response. These results are compatible with the life histories of the two subspecies. Various differences in developmental rates as assessed by morphological and physiological traits are found between the two subspecies, suggesting the likelihood of developmental differences in the morphology and biochemistry of their central nervous systems. Here then we have good evidence for an association of morphological, physiological, and behavioral traits which are assuredly related to habitat selection. This type of relation, postulated in Section 9.3 for the mouse, a mammal in which genetic studies have been well developed, is now supported by the work on *Peromyscus*, an organism frequently observed in field studies.

Evidence for the genetic control of habitat preference also emerges from an examination of the north-south distribution of *Peromyscus* in Canada and in the United States (King, Maas, and Weisman, 1964). From north to south the following taxa occur: *P. maniculatus gracilis, P. maniculatus bairdii, P. polionotus,* and *P. floridanus* (Fig. 1). This distribution approximates the rank-order difference in the net amount of material used by the four species and subspecies to construct nests in the laboratory. *P. maniculatus gracilis* uses more nest material than *P. maniculatus bairdii*, and both use more material than either *P. polionotus* or *P. floridanus*. Only *P. polionotus* and *P. floridanus* build nests of about equal size. In the north, large nests are presumably built to provide insulation from colder weather, while in warmer climates nests of this size are unnecessary. The strains tested were descendants of mice collected in the wild and then bred in the laboratory. Therefore, the differences are at least partly genetic in origin and so contribute to the confirmation of behavioral adaptations as having evolved through the influence of natural selection.

The work on *Peromyscus* stresses the need to study wild house mouse populations in more detail, since genetic studies can be readily done in house mice. Laboratory work (Section 9.3) certainly indicates the likelihood of variations between wild strains for traits similar to those described in *Peromyscus*. Recent studies (Lynch and Hegmann, 1972) have revealed differences in nesting behavior in house mice, assessed by the

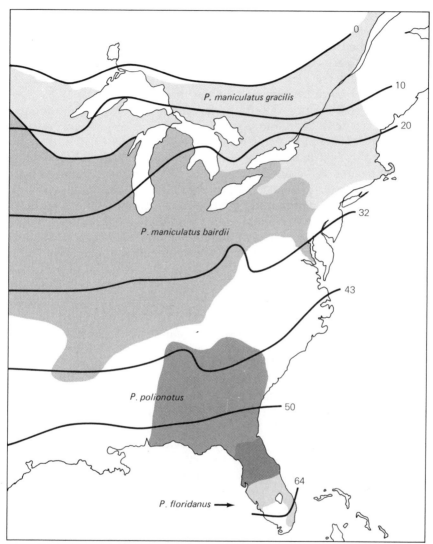

1. **Geographical range of Peromyscus.** *Dark lines* mean January isotherms in degrees Farenheit. (From King, Mass, and Weisman, 1964.)

utilization of cotton in nest building, among five inbred strains. Furthermore, Lynch and Hegmann (1973) found that differences between two strains, BALB/cJ and C57BL/6J, were greater for animals tested at 5°C than for those tested at 26°C. Data ought to be collected over a series of environments, since this result implies the existence of genotype-environment interactions.

It is clear that the behavioral differences between *P. maniculatus bairdii* and *P. maniculatus gracilis* are maintained by selection. Stocks maintained in the laboratory for 12 to 20 generations made no selection when offered a choice of field or forest habitat. However, if the laboratory-bred prairie mice were raised in the field, they selected the field habitat significantly more frequently than the forest (Wecker, 1964). Recently captured prairie mice raised in the laboratory, however, definitely selected the field. Therefore, a genetic change occurred among the mice raised in the laboratory for several generations. The inherent tendency to select the field habitat was diminished, but it could be regained by early exposure to the environment that had previously imposed selection for this trait. This effect shows that both heredity and experience can play a role in determining the preference of the prairie deer mouse for the field habitat. Presumably in the wild there is behavioral evolution from the learned behavior to an innate response. The learned behavior that initially arose becomes innate and hence under genetic control by natural selection (Wecker, 1964). Evolutionary changes that increase hereditary control are advantageous because they tend to limit the number of possible ways an organism can respond to a particular environmental stimulus (Waddington, 1957). This is beneficial in that natural selection favors those responses conducive to survival. So long as the environment remains stable, the population as a whole eventually becomes adjusted to the ecological situation it is best able to exploit.

It is not surprising that changes occur in a laboratory regimen, since natural selection for the habitat preference is relaxed as the laboratory habitat differs from the natural. For traits important in habitat selection, as pointed out in Section 9.3, it is expected under natural conditions that stabilizing selection occurs to keep them within relatively narrow limits. Animals showing behavior away from the norm in a certain population are unlikely to breed with other members. Furthermore, animals occupying the most suitable habitat in a heterogeneous environment have less need to utilize the physiological and behavioral adaptations an animal possesses to combat imperfect environments. Another advantage of occupying a suitable habitat is that the chances of interbreeding by individuals with similar genotypes are greatly enhanced, thereby ensuring their continuity. Ultimately, such a process could lead to reproductive isolation among populations, a thing that has occurred frequently in evolution.

13.4   Population dynamics: Rodents

Clearly, behavioral mechanisms are important as evolutionary forces leading to changes in the gene pool of a species, as illustrated by the discussions on habitat selection. Concerning the dynamics of populations in general, our knowledge is much more restricted. For rodents, definitive measures are required of genetic changes resulting from population parameters such as migration, aggression, mating systems, differential fertility, and differential mortality. In recent years the importance of behavior as a force in evolution has become particularly evident from studies on the population structure of the house mouse and the vole.

Earlier ecological work showed that the home range of house mice is relatively small. Southern and Laurie (1946) demonstrated that the home range of house mice occupying corn ricks must be about 50 ft², with vertical movement being less than lateral movement, and similar evidence has been found in wild house mice in Canada and the United States. Petras (1967) studied house mice in the buildings of six neighboring farms in southeastern Michigan over a 4-year period. Small breeding units seem usual with low population sizes. In fact, estimates of "effective" population size (defined in Li, 1955, and based on the number of breeding individuals) ranges from 6 to 80. Such estimates were obtained from both genetic and ecological data. The genetic data were based on the frequency of two loci controlling biochemical polymorphisms — the esterase-2 (*Es-2*) locus and the hemoglobin (*Hb*) locus. These loci were polymorphic but showed a deficiency of heterozygotes, which can be explained by a subdivision of the population into a number of separate small breeding units in which there is an excess of homozygotes compared with heterozygotes, based on an expectation of random mating (Li, 1955). In other words, this subdivision into small breeding units leads to a situation equivalent to inbreeding. The theory is too complex for a general text of this nature. The first detailed population study which could be explained only by assuming small breeding units was described by Lewontin and Dunn (1960) with reference to data for a polymorphism at the *T* (for tail) locus in mice, which controls certain aspects of development of axial structures in the caudal region of the spine.

Ecological data cited by Petras (1967) were in agreement, showing strong territoritality, leading to the subdivision of populations into small breeding units (*demes*), several of which existed within single farm buildings. Migration rates were low and Petras quoted work showing that not more than 5 percent of all mice in a farm building move either in or out of isolated buildings. Between farms, migration is probably negligible, since populations separated by land lacking groundcover have an extremely restricted exchange of genes. Crowcroft (1966 and earlier papers) reported experiments with wild mice placed into a large enclosure (250 ft²). They dispersed into spatially distinct breeding areas or territories. Aggressive behavior was rare within a family group, where social hierarchies were presumably established, but when unrelated mice were introduced into the family group, markedly aggressive behavior followed, often resulting in the destruction of the intruder.

Reimer and Petras (1967) used both wild and laboratory stocks in a study of the breeding structure of the house mouse in a population cage. Mice were released into a cage consisting of a series of nest boxes connected by runways. The mice formed small breeding colonies, each consisting of one dominant male, several females, and several subordinate males, due primarily to male territoriality. Between demes migration was rare and occurred mainly through migrating females. Such breeding colonies were found to be stable over several generations. Therefore, it seems that mice arrange themselves into small breeding units due to territoriality on the part of the males. Recent work described by DeFries and McClearn (1972) and performed in the laboratory has shown evidence for an association between the social dominance of males of different genotypes as assessed by fighting and Darwinian fitness as assessed by the proportion of litters sired by dominant males. Thus, even though artificial laboratory conditions were used, the data fit with those obtained under more natural conditions and, furthermore, indicate a genotypic basis for social dominance.

Selander (1970) reported on the biochemical genetics of wild house mice by analyzing allele variations at various genetic loci for hemoglobin and esterase variants. Between regions in Texas, there were gradients of allele frequencies (clines). However, a marked heterogeneity of allele frequencies was found when samples of wild mice were collected from different barns within the same region. These differences were found even for barns a few yards apart, agreeing with the behavioral and ecological evidence already cited. Within the same barn a complex mosaic pattern was found for each locus, with small regions of high and low allele frequencies. The clustering of similar genotypes is regarded as a direct outcome of the demes found in wild mice. The population structure is therefore a mosaic of small breeding units (demes) where the effective population size may be extremely small. This means that chance plays an extremely important role in determining gene frequencies at the very local levels. Because the small effective population sizes are greatly dependent on social behavior, that behavior must have important effects on the genetic composition of mouse populations.

A feature of small rodents is the population cycle that occasionally leads to increases of colossal proportions followed by declines to very low numbers. The population cycle of small rodents has long been a classic problem in population ecology. There are two opposing schools of thought about what stops population increase in small rodents. One school considers that extrinsic agents such as food supply, predators, or disease stop populations from increasing. The other school, which is favored as the more important, looks to intrinsic factors — the effects of one individual upon another. Krebs et al. (1973) recently reviewed evidence in the field vole, *Microtus*.

First consider demographical changes during a population cycle. In *M. pennsylvanicus* (Fig. 2), a typical cycle may, once begun, continue through the winter. The peak phase begins with a spring decline in numbers, and then a summer or fall increase restores the population to its former level. The decline phase is most variable and may begin in the fall of the peak year or be delayed until the next spring. The decline may be very rapid, as in Figure 2, but often is gradual and prolonged for a year or more. Then follows a phase of low numbers about which little is known. This pattern is characteristic of several species of voles. The immediate cause of the fluctuations resides in changes in birth and death rates. The percentage of adult females captured that are visibly lactating is reduced in both the peak phase and the decline phase, and in these phases, the death rate of juvenile animals increases dramatically. By contrast, the death rate of subadult and adult animals is not increased in the peak phase but is increased in the decline phase, along with that of the juveniles. Thus, in a peak population, if an animal survives through the early juvenile phase, it has a good chance for adult survival. However, declining populations are characterized by a low birth rate and high mortality rate for both juveniles and adults.

A fencing experiment was carried out in *M. pennsylvanicus*. On two

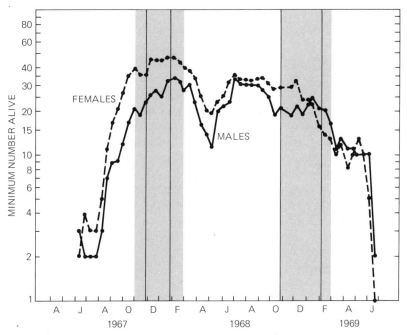

2. **Population density changes in the vole,** *M. pennsylvanicus,* on a grassland area in south Indiana. Shaded areas indicate winter months. (From Krebs *et al.,* 1973; copyright 1973 by the American Association for the Advancement of Science.)

adjacent 0.8-hectare fields, one fenced and one unfenced, vole population sizes increased rapidly, but in the early peak phase there was a sharp divergence such that the fenced population increased to 310 animals — about three times the unfenced population. The overpopulation of the fenced group led to habitat destruction and overgrazing, followed by a sharp decline with symptoms of starvation — a situation that did not occur in the unfenced population. The same result was found in *M. ochrogaster*. The conclusion is that fencing a *Microtus* population destroys the regulatory machinery that normally prevents overgrazing and starvation. Dispersal is the obvious process prevented by a fence, especially as there was no indication that predation pressure was changed by the fence.

Two ways are envisaged in which dispersal may operate in support of population regulation. First, dispersal may be related to population density so that more animals emigrate in the peak and decline phases. These animals are at a great disadvantage from environmental hazards such as other voles, predators, and weather. Second, the *quality* of the dispersers may be more important than the numbers; if animals of only a certain genotype can tolerate high densities, dispersal may be a mechanism for sorting out these individuals. An experiment was done in which two areas were kept free of *Microtus* by removing all animals by trapping for 2 days every 2 weeks. Between the episodes of trapping, voles were free to colonize the areas. Dispersal was most common in the increase phase of a population fluctuation and least common in the decline phase. In fact, Krebs et al. (1973) considered that much of the loss rate in increasing populations is due to emigration. Conversely, little of the heavy loss in declining populations is due to dispersal and so most losses must result from death in situ.

For the polymorphic serum protein loci *Tf* (transferrin) and *LAP* (leucine aminopeptidase), evidence has been found of large changes in gene and genotype frequencies in association with population changes. The LAP^S allele frequency (distinguished by slow electrophoretic mobility) dropped about 25 percent in *Microtus* males beginning at the time of high losses and 4 to 6 weeks later declined an equal amount in females. This type of observation strongly supports the hypothesis that demographical losses are genetically selective and that losses are not equally distributed over all genotypes.

The transferrin genotype frequencies of dispersing *Microtus* females are compared in Figure 3 with resident or nondispersing females. Clearly heterozygous females Tf^C/Tf^E (Tf^C and Tf^E are alleles of the transferrin gene) are more common in dispersing than in resident populations. In fact, 89 percent of the loss of heterozygous females from the resident populations during the population increase was due to dispersal. Certain genotypes show a tendency to disperse, a possibility suggested in the literature (e.g., Lidicker, 1962) but not previously demonstrated in animal populations. Therefore, intense selection pressure must be occurring. If the intrinsic factor of behavioral interactions among individual voles is the primary mechanism, then the behavioral characteristics of individuals should change during the cycle. This has been tested for males of *M. pennsylvanicus* and *M. ochrogaster*. Laboratory studies showed significant changes in aggressive behavior as assessed by paired round-robin encounters in the laboratory during the population cycle (Myers and Krebs, 1971): individuals in peak populations were the most aggressive. Furthermore, males of *M. pennsylvanicus* which dispersed during peak periods tended to be even more aggressive than the residents. This needs to be looked at further in the field, as Krebs et al. (1973) pointed out.

The results are consistent with the hypothesis of genetic and behavioral effects proposed by Chitty (1967) and displayed in Figure 4. Chitty's model of population regulation is based on a behavioral polymorphism in which there are socially tolerant and intolerant animals with respect to crowding, such that changing density acts as a selective force on the behavioral types. According to this hypothesis, as the population size increases so too does mutual interference and selection for aggressive behavior. Emigration is one of the ways through which selection takes place. As a selecting force this occurs to the greatest extent during the increase phase, and there is now clear evidence (see above) for genetic differences between the emigrants and resident animals. As Krebs et al. (1973) pointed out, the greatest gap in our knowledge is in the behavior-genetics area. For example, there is no knowledge about the heritability of aggression in voles. This example is documented, not because the story is complete, but because it shows the likely importance of genetic and behavioral changes associated with population cycles — a result that, if generalized, seems of far-reaching significance, perhaps even for our own species.

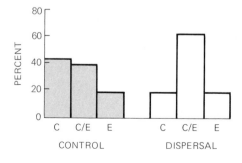

3. **Transferring genotypes during increase phase** of the vole, *M. pennsylvanicus*, in the fall of 1969, for dispersing females compared with control or resident females. *C* and *E* are alleles. (From Krebs *et al.*, 1973; copyright 1973 by the American Association for the Advancement of Science.)

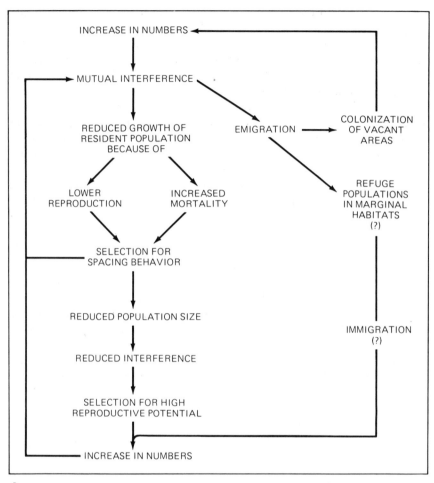

4. **Population regulation in small rodents.** Model modified from Chitty's original proposal. (From Krebs *et al.*, 1973; copyright 1973 by the American Association for the Advancement of Science.)

Vale, Vale, and Harley (1971) studied 45- to 55-day old male house mice from five inbred strains in population numbers of 2, 4, and 8 animals per cage. Agonistic behavior and social grooming were recorded over a period of 10 days, at the end of which adrenals, testes, and seminal vesicles were removed and weighed. There were differences among all strains for five behavioral measures and the three weight measurements (Table 2). Population number was associated with significant effects in two of the behavioral measures, number of fights and number of attacks, and in adrenal weight. There was a positive association between prolonged aggression and adrenal weight, as has been reported elsewhere (references in Vale, Vale, and Harley, 1971), and also a positive association between social grooming and gonad weight, both revealed by principal component analysis. For the two variables, number of attacks and adrenal weight, there were genotype-population number interactions indicating that not all strains act in an identical way with respect to aggression as population numbers are increased. That is, population number increase does not necessarily lead to increased aggression or adrenal weight in all genotypes. This agrees with Chitty's (1967) model of socially tolerant and socially intolerant animals with respect to crowding. It begins to close the behavior-genetics gap.

13.5   Genetic and linguistic differences in tribal man

Howells (1966) analyzed population structure in Bougainville in the Solomon islands, considering 18 ethnic groups representing most of the islands' territory (Fig. 5). Linguistic and cultural evidence makes it clear that the existing ethnic differentiation has not been purely a local process, since at least three distinct immigrant groups are involved. The physical environment varies from coastal beaches to hill slopes and ridges to mountainous areas along the central spine. Based on data collected on over 1300 males by Oliver, a number of "distances" between the 18 ethnic groups were computed, as well as correlation coefficients between the 153 possible pairings among the 18 groups. The distances relevant for our discussion are:

- Geographic distance (GEOG). Measured between centers of group areas.

- Linguistic distance (LING). A measure taking into account the number of shared words between languages.

TABLE 2.   Summary of Results of Analyses of Eight Variables in Males of Five Inbred Strains of House Mice

	Effects		
Variables	Strain	Population number	Interaction
No. of chases	$P < 0.01$	$P > 0.05$	NS
No. of attacks	$P < 0.0001$	$P < 0.025$	$P < 0.01$
No. of fights	$P < 0.01$	NS	NS
No. of social grooms	$P < 0.0001$	NS	NS
No. of tail pulls	$P < 0.0001$	NS	NS
Adrenal weight	$P < 0.0001$	$P < 0.005$	$P < 0.005$
Testis weight	$P < 0.0001$	NS	NS
Seminal vesicle weight	$P < 0.0001$	NS	NS

NS, not significant.
After Vale, Vale, and Harley (1971).

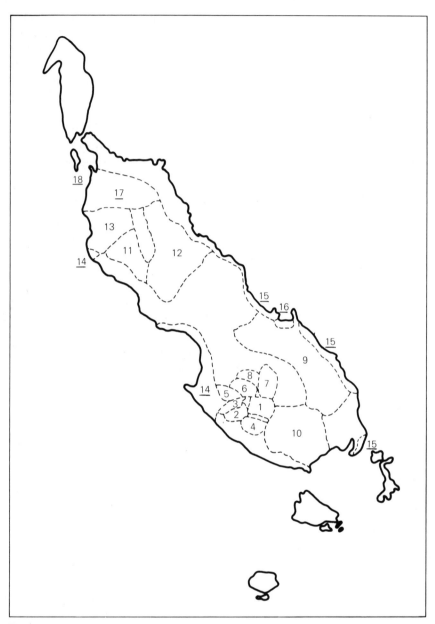

5.   **Map of Bougainville** showing location of populations studied. **Underlined groups,** 14 to 18, are Melanesian speaking. (From Howells, 1966.)

- Size distance (SIZE). Penrose's (1954) size distance, which measures general size independent of proportional differences. It was based on eight measurements including sitting height, arm length, chest breadth, head length, and four other head measurements.

- Shape distance (SHAPE). Penrose's (1954) shape distance, which measures differences in proportions keeping size constant. It was based on the same eight measurements as SIZE.

- Morphological observations (SCOPIC). Anthroposcopic or nonmetric observations of a number of traits including hair form, hair texture, head hair color, eye color, height of eye opening, forehead slope, nasal root breadth, nasal tip inclination, frontal nostril visibility, and integumental lip thickness. A generalized mean distance was obtained as described in Howells (1966).

The correlation coefficients among the measures are all positive (Table 3). Of particular interest to us is the association between the linguistic distance and the three biological distances (SIZE, SHAPE, SCOPIC), since the biological distances can be regarded as under strong genetic control. High and significant correlations were found between these three biological measures, especially SHAPE and SCOPIC, and LING. In other words, there is an association between biological (genetic) and linguistic divergence on Bougainville; or, there is an association between biological and cultural divergence using language as a measure of culture. Such linguistic divergence is presumably due to a combination of migration of differing groups to Bougainville and the occurrence of linguistic drift taking place in an initially genetically homogeneous population, which subsequently leads to the isolation and linguistic differentiation of the

TABLE 3. Coefficients of Correlation Among Various Distance Measures in Two Studies

Measure*	GEOG	LING	SIZE	SHAPE
GEOG				
LING	0.58			
SIZE	0.13	0.31		
SHAPE	0.24	0.43	0.36	
SCOPIC	0.22	0.42	0.45	0.28

Measure†	1	2	3
1 Geographical			
2 Linguistic	0.506		
3 Serological	0.406	0.565	
4 Anthropometric	0.170	0.547	0.416

* Modified from Howells (1966).
† Modified from Friedlaender *et al.* (1971).
In both examples $P = 0.01$ for a deviation from zero when the correlation coefficient is 0.22.

respective speakers. Especially if, for some reason such as terrain, populations speaking the same language were rather isolated, this can easily occur. Once having developed, a linguistic difference can discourage social contact and act as a barrier to gene flow so that genetic divergence evolves. Indeed, Friedlaender et al. (1971), in a consideration of 18 villages in Bougainville, found that intermarriage among villages was largely confined to the same linguistic group; there was little migration *among* linguistic groups. In their research they used blood group polymorphism and anthropometric data as measures of biological distance and came to conclusions qualitatively similar to Howell's (Table 3).

The other measure in Table 3 (GEOG) is of less relevance for the behavior geneticist. For Howells' data there is a high correlation with LING, but lower correlations with the three biological measures (SIZE, SHAPE, SCOPIC). An association of geographical distance with the other four variables seems, in any case, inevitable. However, the correlations of LING with the biological measures are higher in all cases than for GEOG. Friedlaender et al. (1971) reported results that are essentially similar (Table 3): a high correlation between geographical and linguistic distances, and the correlations of the linguistic distances with the two biological distances exceeding those with geographical distance. Taken together, the two surveys therefore agree in showing an association between biological and linguistic divergence, which means an association between genetic and linguistic divergence.

Studies made in other parts of the world are in general agreement: e.g., in Ruandi-Urundi and Kivu in Central Africa (Hiernaux, 1956) and in New Guinea (Livingstone, 1963). The same conclusion holds for Australian Aboriginal tribes in the Northern Territory of Australia (White and Parsons, 1973). Here sociocultural divergence also was found to be associated with linguistic and genetic divergence. A particularly interesting example based on genetic distances from blood groups comes from the relatively close relation between the Murngin tribe, who live in Northeast Arnhem Land (the northern section of the Northern Territory), and the Aranda tribe, who live in the arid country in the center of Australia (Fig. 6). Dermatoglyphic data are in agreement. Clearly, this result is at variance with expectation based on geographical distance. The probable explanation is that the Aranda represent a relatively recent migration of northern individuals southward (Birdsell, 1950). Linguistic data support this explanation, since the Murngin and Aranda languages are more similar than are the languages of the Aranda and the desert tribes adjacent to them in Central Australia.

In Australia as in Bougainville, gene flow between tribes is assumed to be low (Tindale, 1953). This argues for language and associated sociocultural factors as a barrier to gene flow by discouraging social contact. Be-

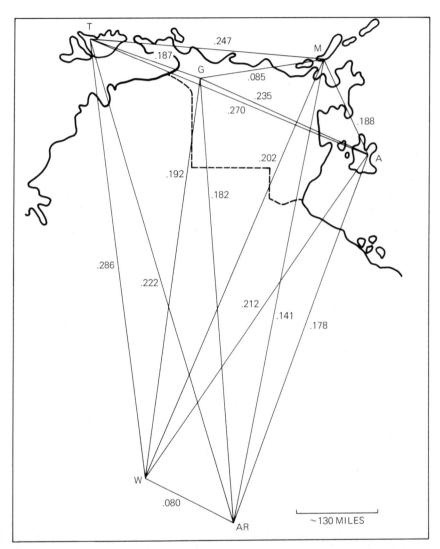

6. Genetic distances between Australian Aboriginal tribes based on blood groups and serum protein systems. Tribes are (A) Andilyaugwa, (AR) Aranda, (G) Gunwinggu, (M) Murngin (T) Tiwi, (W) Wailbri. Note the location of the two tribes discussed in the text — Murngin in Northeastern Arnhem Land and Aranda in Central Australia. **Dashed line** indicates division between Arnhem Land and Central Australia. (From White and Parsons, 1973.)

fore European settlement the Aborigines were subdivided into tribes that to some extent were genetically discrete, the isolation being maintained by low migration rates between tribes. While it is difficult to know what actually happened in the past, it is clear from such tribal groups that remain that linguistic and genetic heterogeneity occurs among tribes and that the two are associated. The resultant population structure is a mosaic of tribes with somewhat differing overall gene pools. The analogy with demes in mice is clear.

Birdsell (1973) discussed in some detail tribal size among the Australian Aborigines. Many tribes having a mean size of about 500 are dialectical tribes, i.e., those with no political organization or authority and so without leadership. An exception is the Aranda, whose total number was estimated as about 1500 at the time of historical contact. The Aranda tribe consists of three subgroups: the Northern Aranda, the Western Aranda, and the Southern Aranda. They recognize themselves as belonging to the same dialectical community, yet they are also aware of regional differences in speech. At the time of historical contact, then, it seems likely that this oversized tribe was in the process of differentiating into three new dialectical tribal units that would have averaged about 500. As Birdsell (1973) put it: "The tendency for oversized dialectical tribes to fail to maintain homogeneous speech throughout their bands is not unexpected, and may be explained in terms of the concept of density of communication." This division into new tribes must then be assumed to be associated with sociocultural, linguistic, and genetic differentiation; furthermore, the evidence obtained so far argues for positive correlations among all three factors.

The structure of the Australian Aborigines at historical contact consisted of a number of separate breeding units, so that across tribes a mosaic of allele frequencies is expected; even so, broad clines in allele frequencies are apparent on an Australia-wide basis (Kirk, 1966), as found by Selander (1970) in his study of mice over the state of Texas. The expected level of heterogeneity can be assumed to be lower than in wild mice, for example, where effective breeding units are normally less than 100. Presumably the relative differences in population size can be attributed at least partly to the development of language as a method of communication in the Australian Aborigines.

13.6 Evolution of behavior in man

The evolution of man has been described in many texts (Mayr, 1963; Dobzhansky, 1964; Harrison et al., 1964); here our aim is to mention some of those behavioral changes that are important in this evolutionary process. Contemporary man is the end point of a long evolutionary history, as was realized by Darwin (1871). From the fossil record it appears that the Hominidae (human) and Pongidae (ape) families separated during the Eocene geological period and the differentiation was well established by the Miocene and Pliocene epochs about 10 million to 15 million years ago. In the later part of the Pliocene, about 2 million years ago, the first definite hominid appeared, *Australopithecus*. This remarkable form was characterized by: (1) enlarged cranium and hence neural tissue, although his brain was only slightly larger than that of the modern chimpanzee and hardly more than one third the size of modern man's; (2) bipedal locomotion, freeing the hands for manipulation; (3) the use of tools, which follows from the development of bipedal locomotion; (4) the ability to communicate and to hunt in bands and the development of a carnivorous diet. This form existed until well into the Ice Age, perhaps up to 700,000 years ago.

The next major step was the appearance of *Homo erectus* about 600,000 years ago. This species was characterized by a brain size of about 1000 cc, approximately double that of *Australopithecus* and 75 percent of that of modern man. Fossils of *H. erectus* are often accompanied by stone tools, including axes, which he must have made and used. Since the fossil sites contain bones of large animals, which the hunters apparently killed, the existence of well-organized bands is implied. The existence of both stone tools and organized bands implies a form of speech between members of this species, a much more developed level of communication than that found between apes and other animals. Communication is quite general in the nonhuman primates, but not verbal communication, like man's. The genetically determined neural substrate is presumably not sufficient to support speech behavior (DeVore, 1965). Evidence from fossil sites indicates that the behavioral traits of *H. erectus* are closer to modern man's than are those of *Australopithecus*.

The first men, indistinguishable from ourselves, evolved about 35,000 to 40,000 years ago during the last advance of the glaciers. Their appearance was accompanied by a rapid expansion, diversification, and improvement of culture. They buried their dead together with flowers and implements carefully laid around the body, so it is not unreasonable to assume that they believed in an afterlife and had some form of religion. This was *Homo sapiens* — modern man.

The evolutionary trend, therefore, is toward the development of intellectual capacity — the feature that makes man's position unique. Morphological trends such as an increase in brain size from about 500 cc in *Australopithecus* to about 1400 cc in *H. sapiens* and the development of bipedalism, plus behavioral trends such as the development of the ability to communicate and make tools, all are in agreement. In modern man, in addition to the advances pioneered by *H. erectus*, we have: (1)

advanced toolmaking, (2) elaborate cultural organization, (3) additional increases in brain size, a (4) childhood and adolescence extended in time and providing a longer period in which cultural achievements can be assimilated, (5) a degree of control of the environment by means of advances in medicine and technology.

The increase of brain size is an example of directional selection that in terms of the paleontological time scale was very rapid. Since it was associated with the development of progressively more advanced intellectual capacity, there must have been a selective advantage in more efficient communication, perhaps related to tool development, the use of fire, and hunting in bands. Quite likely the period when man's brain was increasing most rapidly coincided with the evolution of his ability to invent and use language for communication. Speech is not only essential for these behavioral developments, it is also basic to the development of ideas and plans for the future. When we look at the recent evolution of man, the importance of behavioral changes in initiating new evolutionary phenomena is undeniable. It can reasonably be assumed that initially there were only minor modifications at the structural level, but that the evolution of a morphological change followed a permanent behavioral change. For example, the perfection of bipedalism may well have been speeded up by the preoccupation of the anterior extremities with manipulation, a behavioral characteristic. Similarly, the significant increases in brain size were associated with the development of an efficient system of communication, speech. It is remarkable that the development of language was accompanied by tremendous linguistic diversity and isolation. This was associated with genetic diversity as observed in those regions where tribes of *H. sapiens* can be still studied (see Section 13.5).

Quite likely the breeding structure of primitive man affected his evolutionary rate. If the leader of a group had several wives (*polygyny*), he contributed a greater than average share to the genetic composition of the next generation of his group. The tremendous reproductive advantage of a leader in a group or tribe would favor the characteristics of man, since for leadership certain physical and mental traits would inevitably be favored. These in turn would depend to a considerable extent on the genotype of the individual, so that the reproductive advantage would ultimately make a maximal contribution to the fitness of the entire group. The actual evidence for polygyny is difficult to obtain, but it may be an original condition in a few living tribes, and it is more or less developed in nearly all anthropoid apes (Bartholomew and Birdsell, 1953).

Modern man, appearing 35,000 to 40,000 years ago, was anatomically indistinguishable from ourselves: there has been little selective pressure since then for altered anatomical features. Has natural selection ceased? The answer must be negative. Man has changed from a species living in small hunting communities to a species of which many members live in large, highly organized communities. Up to this century the rate of population increase in man was quite slow because of an extrinsic factor — disease. There must have been a high premium placed upon genes for resistance to specific diseases. Some of these diseases must have become important because of man and his way of life. For example, Livingstone (1958) traced malaria to the slash-and-burn agriculture that opened the forest floor to stagnant pools and brought man into contact with insect vectors and hence malaria. One consequence of malaria was gene pool change. Because of their greater resistance to malaria, carriers of genes for sickle cell anemia, thalassemia, and glucose-6-phosphate dehydrogenase deficiency were favored, and this led to polymorphisms in regions where malaria was present.

Technological advances have brought man into contact with other diseases (Omenn and Motulsky, 1972). Rodents, attracted to settled populations, brought epidemic diseases. The practice of single-crop agriculture brought nutritional risks (since each cereal has its own limiting amino acids) and a propensity to protein undernutrition and endemic dysentery. Most bizarre is the culturally based occurrence of kuru (Chapter 11), a disease contracted by the cannibalistic practice of eating the brains of dead enemies. Today, contagious diseases have receded as a major problem, but a concomitant effect of the control of disease is a rate of population increase reminiscent of the voles during their increase phase (Fig. 2). In the vole situation, emigration of certain genotypes reduced the rate of increase. Until recently, emigration has been a factor in human populations, but for us this phase too is nearly over.

The extreme territoriality of small rodents has been noted. Beginning with population sizes of 32 and 56 Norway rats in different experiments in the presence of adequate food and water, Calhoun (1962) showed that rats confined in a 0.25-acre enclosure reached a population size of about 150 at 27 months. One would expect, from the very low adult mortality in uncrowded conditions, a population of 5000 at 27 months. However, at this stage infant mortality was high and pathological behavior was evident in both sexes, associated with social dominance of certain males. It seems that much of the abnormal behavior observed was a consequence of the disruption of territories and peck orders by the presence of excess numbers. As a consequence, normal behavior patterns broke down, and the predicted population growth curve did not occur because of insufficient space to develop normal social behavioral patterns. If we can argue from the rodents, it seems that space may well become a progressively more limiting factor for man (Hoagland, 1966); and as it becomes more limiting, various behavioral changes in populations may occur that may be to some extent under genetic control.

Is there evidence for stress syndromes in man leading to a reduction in the growth rate of human populations? The human pituitary responds under stress in a way similar to the response of other mammals, and there is indirect evidence that inmates of concentration camps experienced acute stress syndromes that accounted for their death. Concentration camps may be more analogous to highly congested animal populations than city slums, since even in very crowded cities the poor have some mobility, although the occurrence of street gangs and juvenile delinquency characteristic of crowded cities is a form of social pathology. The increased incidence of atherosclerosis and other cardiovascular pathology associated with urban living and its competitive stresses may be enhanced by crowding. In underdeveloped countries where high birth rates and recently lowered death rates produce a population growth of 2 percent or more per year, the use of health measures increasing life expectancies masks any growth-retarding effect of a stress syndrome. Therefore, direct comparisons with rodents are hardly possible. But although we are far from the rat colonies, space limitations are likely to become progressively more important; indeed, they are already so in the poorer sections of some of the world's largest cities. The tendency is likely to increase. More highly developed technology reduces the number of people needed to produce a given amount of food and encourages the drift of people from rural to urban areas, a tendency that was accelerated by the Industrial Revolution and that is continuing today.

Demographical, medical, and technological changes, associated with social change, are now occurring so rapidly that populations have difficulty adapting. Therefore, improved understanding of human behavior and its genetic basis is essential, and this is the task of the biologist and psychologist.

A final problem of a general kind is the effect of medicine on the gene pool of man. In any society where either preventive or curative medicine plays a significant role, a greater proportion of zygotes survive to reproduction than survived in previous generations. Genotypes normally eliminated by natural selection continue to survive and reproduce. For example, the fitness of diabetics was dramatically increased by the discovery of insulin. Thus medicine has enabled diabetics to survive, reproduce, and spread their deleterious genes into subsequent generations. The same applies to behavioral disorders. As a simple example, the IQ of phenylketonurics may be improved by a special diet, and this presumably results in an increase of their reproductive rates. Hence the average biological fitness of the population may be expected to fall because of the increased incidence of deleterious genes. In considering the possible future evolution of man, it is important to take account of the effect of medicine on the gene pool itself, as well as the more obvious demographical consequences. But we must be careful, for as we have already seen fitness depends on the environment. Under an optimal environment fitness differences between genotypes may be minimal. If the environment deteriorates in the future here on our earth, then the effects of natural selection will become more apparent, and fitness differences between genotypes may increase. By environment we mean both the physical and social environment and the rate of change of each. A major consideration is the effect of the various environmental pollutants to which we are exposed; although they may have few short-term effects, their longer-term effects may well be important (see Ehrlich and Ehrlich, 1970).

It is beyond the scope of this book to consider the future of man further, except to say that evolutionary changes will continue to occur. Evolutionary changes in behavioral traits may well assume progressively more importance, just as selection for disease resistance must have been strong in many communities before the advent of modern medicine. The dilemma is that man is the one species in which objective information on the genetic basis of behavior is difficult to obtain except for the simplest of traits. Furthermore, direct extrapolation from studies on experimental animals can and should be made only with the utmost caution.

General readings

Calhoun, J. 1962. Population density and social pathology. *Sci. Am.* 206(2):139–148. A very readable account of some classic experiments on the consequences of overcrowding in rats.

Parsons, P. A. 1973. *Behavioural and Ecological Genetics: A Study in Drosophila.* Oxford: Clarendon Press. A general account containing much source material from the genus *Drosophila*, concluding with a part on evolutionary implications.

Washburn, S. L., and P. C. Jay. 1968. *Perspectives on Human Evolution.* New York: Holt. One of the many volumes with discussions on human evolution.

Wecker, S. C. 1964. Habitat selection. *Sci. Am.* 211(4):109–116. A discussion of habitat selection in the deer mouse, *Peromyscus*.

Conclusions and Future Directions

Having completed our current account of behavior genetics, we are obliged to look back and inquire about the extent to which behavior genetics can legitimately be regarded as a distinct discipline capable of generating new questions and resolving important issues (e.g., Vale, 1973). There is no question in our minds that the special emphases of behavior genetics listed in Chapter 1 warrant a text on the subject. These emphases are (1) the difficulty of environmental control, (2) the difficulty of objective measurement, and (3) the study of learning and reasoning. They are discussed many times in this text. (Clearly, it would be difficult to do justice to these emphases in a general genetics text.) Basically, the behavior geneticist applies the principles of genetics directly to behavioral traits, and no new genetic principles have emerged from these efforts. But because of the special emphases of behavior genetics, certain aspects of genetics must be rigorously understood. The major example is quantitative genetics, which allows the disentangling of genetic and environmental influences, particularly in infrahuman animals.

When we proceed to evolutionary considerations though, our view is that the study of behavior and its genetic architecture is of critical significance. As stressed by Mayr (1963), "the shift into a new niche or adaptive zone is almost without exception, initiated by a change in behavior." Appropriate alterations in behavior may accompany, precede, or even follow morphological changes. Studies with the deer mouse, *Peromyscus*, reported in Section 13.3, show substantial behavioral variations among the various taxa that can be related to habitats. While there are some associated morphological and physiological changes, they are less obvious. A dramatic example of associated behavioral and morphological changes occurred during the evolution of man, discussed at the end of Chapter 13. An understanding of the genetic control of behavior allows a greater understanding of these evolutionary processes.

Caspari (1967) remarked that "all biological phenomena can be considered from two points of view: mechanism and evolution." This text considers behavior genetics from precisely these points of view. The two approaches are not always separated; however, the direction of our book's coverage is from the mechanism of the inheritance of behavioral traits to the evolutionary aspects of behavior. The means by which evolution proceeds are now reasonably well established through the sophisticated developments of population genetics. Less well known is the role of behavioral mechanisms in evolution. This becomes clear when the variety of mating systems in animals leading to nonrandom mating is considered — a complication that has not been sufficiently explored by population genetics. Only in those species groups whose evolutionary basis is known in great detail is the role of behavioral mechanisms and their contribution to evolutionary processes beginning to be understood.

It is clearly not possible to consider behavioral factors in evolution in isolation, since many of the factors determining the distribution of organisms are wholly or partly ecological. Ecological genetics is well established through the efforts of Ford (1971) and others. It deals with genetic changes in natural populations resulting from ecological variability in space and/or time. The experimental material constituting ecological genetics is scattered throughout the discrete disciplines of genetics and ecology.

Behavior genetics has a more diffuse spread and encompasses, in addition to genetics and ecology, the disciplines of ethology, psychology, and (in its human context) sociology. This diffuse spread has meant that the literature is widely scattered, a complete overview of behavior genetics has rarely been attained. It is not surprising, therefore, that recognition of behavior genetics as an entity has come more recently than recognition of other subdivisions of genetics, such as ecological and developmental genetics. This recognition of behavior genetics as a viable section of genetics must be appreciated if the discipline is to develop further and contribute to science as a whole and to evolutionary biology and psychology in particular.

Even so, some achievements creditable to behavior genetics can be listed. First, it has enabled the asking of meaningful questions concerning the heredity-environment issue in man with regard to complex behavioral traits. While definite answers cannot necessarily be given, behavior genetics shows why. As Vale (1973) commented: "Psychology has everything to gain and genetics has little to lose if behavior geneticists turn their attention away from formal genetics and towards the best use of genotype in the analysis of behavior." Behavior genetics can be credited with the burial of the nature-nurture controversy in psychology. This profitless argument has raged for many years during this century (and outposts supporting it still exist). Behavior genetics has shown that full understanding of the inheritance of behavior in man depends not only on human studies, but also on an understanding of studies with experimental animals in which superior control of genes and of environments is possible.

Second, in some cases the behavior geneticist has shown how the mechanisms underlying behavior can be explained. For example, the embryonic fate maps which can be constructed for Drosophila permit the correlation of precise anatomical sites with abnormalities affecting behavior; more generally some of the behavioral mutants in Drosophila can be associated with abnormalities of the central nervous system. In the house mouse, a given genotype affecting behavior can often be regarded as controlling a series of interrelated traits, thus producing an overall complex phenotype that includes morphological and physiological components (Chapter 9 and Section 13.3). In man too, in a few cases involving traits controlled by major genes, such as phenylketonuria, there may be (as discussed in Chapter 1) a relatively direct correlation leading from the gene to the altered behavior.

Third, as shown in Chapter 13, the study of behavior at the intraspecific level, and at the interspecific level for closely related species, is producing results of significance for evolutionary biology, especially if ecological components are included. For this, further specialized studies considering both behavioral and ecological variables, such as those done in Drosophila (see Parsons, 1973), are necessary.

Predicting future trends is always risky. Obviously, further refinement and elaboration of the three points above will occur. Some possible future trends that come to mind are listed below:

- As we have seen, many studies are based on a restricted series of genotypes in a restricted set of environments. Calculated heritability values are usually based upon studies of one population in a single environment. When the number of genotypes and environments affecting trait expression is considered, we have seen the enormous problem of generalization. For two genotypes and two environments, six genotype-environment interactions are possible (Section 6.2), and the number increases rapidly as the number of genotypes and/or environments increases. The problem is incredibly complex if the environment is defined in the broadest sense, to include not only the physical environment, but also effects of previous experience. Future endeavors will have to come to grips with this problem.

• Related to the above point is the study of the behavioral responses of different genotypes to stress. The study of the great variety of neural, endocrine, and cardiovascular structures and functions relevant to stress in a number of animals will be of importance (see Emlen, 1973, for references). Therefore, a comprehensive understanding of genetic factors relevant to stress responses may be aided by behavior-genetics studies.

• The effect of drugs on behavior, in particular on learning, is well known. The use of differing genotypes should help provide more information of use in drug therapy and psychotherapy. Ultimately, progress with regard to learning and its genetic and biochemical bases may be predicted. Recent studies on learning in a variety of organisms including Drosophila appear to be probing in this direction.

• For complex behaviors, the trend of looking at components of a behavioral trait and studying their genetic bases is likely to develop further. Such an approach is described in Chapter 12 for primary mental abilities and may be expected to accelerate, especially now that vast amounts of data can be processed by computers. This approach ought to be particularly valuable in man. In some cases, complex traits with ambiguous hereditary mechanisms may be broken down into subunits, each showing distinct and simple hereditary mechanisms. Psychoses seem to be amenable to this approach, as is complex social behavior in animals. However, there are special problems even here, since extreme care in sampling techniques and experimental design is essential for meaningful results.

• The tendency of behavioral traits to change over time may merit more emphasis (Section 6.10). The behavior geneticist necessarily needs to know about cyclic changes that may be diurnal and/or seasonal. Furthermore, changes over a life span need further study, a point that undoubtedly applies to those traits that are modified by learning. At present, the competent experimenter controls age or time of day at measurement in carrying out a behavior-genetics analysis. However, the lability of behavioral traits over time has hardly been explored. This is certainly not surprising in man, who has a long life span, and it now is known to occur even in short-lived Drosophila. Traits particularly amenable to this form of analysis may be the sensory-perceptual ones; they can be studied quite easily utilizing behavior-genetics techniques.

• More generally, it seems that behavior genetics will play a major part in the slowly growing rapprochement between sociology and the biological sciences (in particular, in the study of human variation). For example, studies of early experience and prenatal environment should lead to an increased participation by social scientists in behavior-genetics research. Behavior genetics should play a central role in future integrated research projects involving social, behavioral, and biological scientists. For example, Eckland (1972) wrote: "At least in sociology, we never seem to run out of social and psychological hypotheses for almost any phenomenon. Yet, although our theories have much to recommend them, our tools and our findings leave much to be desired. In fact, when weighed in balance, the evidence on a number of points tends to push the search for explanation back to biology."

• Increased study of social structures and mating patterns in man (and other animals) is likely, with the aim of assessing their effects on the gene pool of populations. Indeed, random mating, so commonly assumed, probably rarely occurs. Therefore, the augmented study of mating patterns in many species can be expected, coupled with consideration of their evolutionary consequences. Such studies will be both theoretical and observational. This will lead into detailed considerations of those secondary sexual characteristics conferring some social benefits upon their bearers. Two main types of benefits can be envisaged. Their bearers either have superior competitive ability against others of the same sex (e.g., antlers in deer) or they possess increased sexual appeal in competition for the opposite sex. In the first case, selection is within one sex (*intrasexual* selection), and in the second case, selection is performed by the opposite sex (*intersexual* selection). Included in intrasexual selection is selection for social rank, especially if dominant individuals contribute more of certain genes — e.g., for size and aggressiveness — to future generations than those that are not socially dominant. Detailed studies of sexual selection in populations is of prime evolutionary significance and will need a combined laboratory-field approach based on a knowledge of ethology, ecology, and genetics. Clearly, this is a topic of prime importance in the evolutionary biology of species.

• Our book considers natural selection mainly in terms of individuals in a population — its action on genotypes through the differential reproductive success of individuals. In some species there are traits benefiting the population that are disadvantageous to those individuals possessing them. These are referred to as *altruistic* traits and must be considered at the level of the population as well as the individual. Examples include the sharing of food, warning signals, and the cooperative efforts as found in social insect colonies. Hamilton (1964) has considered the evolution of altruistic behavior as it occurs due to parental care. More study of species exhibiting altruistic behavior, and of the genes involved, is a topic for the future behavior geneticist.

Another example at the population level comes from a number of species, especially small mammals, whose population structure consists of small, semiisolated breeding units or demes (Section 13.4). If a gene decreases mortality or increases reproductive success in a given deme, then that deme may divide into daughter groups more rapidly than other demes. The result is that the frequency of the gene increases by *interdeme* selection after *intrademe* selection. Generally, this is a component of *group selection*, which has a number of advocates. In particular, Wynne-Edwards (1962) has invoked group selection to explain a large number of phenomena. He has argued that survival in a population which has run out of food or some other resource depends on certain individuals acquiring, through group selection, behavior patterns serving to pass information to others so that population sizes are regulated by reductions in reproduction-associated behaviors. Territoriality, as described in Section 13.4, is an example of population control. Whether the Wynne-Edwards model is correct or otherwise, as Wilson (1973) pointed out, his views have opened up a vast unexplored area of importance to both sociobiology and ecology. The behavior geneticist seems destined to play a part in looking at behavior-genetics variability in populations in which group selection either does occur or is postulated to occur.

• The genetic basis of behavioral factors involved in determining the niche of organisms has hardly been touched. As Emlen (1973) pointed out, there are numerous definitions of *niche*, but in brief the niche is the set of conditions, physical and biotic, in which an organism can live. In a broad niche, a greater range of behavioral responses is more likely than in a narrow niche. Conversely in a stable habitat, behavior is likely to be more stereotyped and responses rigid. Closely related species of *Drosophila* and rodents provide excellent experimental material to add to those studies already published. The assessment of behavioral factors in determining niche will involve the extremely difficult extrapolation of laboratory studies of behavioral traits to natural populations. In other words, it will involve the transition from a relatively determinate environment to one that, except for rare special circumstances, is essentially indeterminate.

This brief overview of likely developments leads us to the prediction that the future behavior geneticist will need training not only in basic genetics, animal behavior, and psychology, but also in population biology, ecology, quantitative genetics, and the principles of experimental design and statistics. An understanding of developmental biology is also important. The behavior geneticist must be an individual — quite rare in science today — who is both integrative in thought over a broad field, and yet able to do specific and meaningful research in a narrow field. This person must have the outlook of a broad evolutionary biologist and must also appreciate the efforts of the psychologist (and sociologist). Only then will he or she be able to put behavior-genetics studies in their context and see new challenges as they arise. It is this type of person who will aspire to see behavior genetics as a distinct discipline and to whom this book is directed.

Bibliography

Adler, J. 1969. Chemoreceptors in bacteria. *Science 166*:1588–1597.

Adler, J., G. C. Hazelbauer, and M. M. Dahl. 1973. Chemotaxis toward sugars in *Escherichia coli. J. Bacteriol. 115*:824–847.

Alawi, A., V. Jennings, J. Grossfield, and W. L. Pak. 1972. Phototransduction Mutants of *Drosophila melanogaster.* In *The Visual System.* G. B. Arden (ed.). New York: Plenum Press, pp. 1–21.

Alawi, A., and W. L. Pak. 1971. On-transient of insect electroretinogram: Its cellular origin. *Science 172*:1055–1057.

Alexander, R. D. 1962. The role of behavioral study in cricket classification. *Syst. Zool. 11*(2):53–72.

Alexander, R. D. 1968. Arthropods. In *Animal Communication.* T. Seboek (ed.). Bloomington: Indiana University Press, pp. 167–216.

Allard, R. W., S. K. Jain, and P. L. Workman. 1968. The genetics of inbreeding populations. *Adv. Genet. 14*:55–131.

Allee, W., A. Emerson, O. Park, T. Park, and K. Schmidt. 1949. *Principles of Ecology.* Philadelphia: Saunders.

Allison, A. C. 1955. Aspects of polymorphism in man. *Cold Spring Harbor Symp. Quant. Biol. 20*:239–255.

Allison, A. C. 1964. Polymorphism and natural selection in human populations. *Cold Spring Harbor Symp. Quant. Biol. 29*:137–150.

Alter, M., and A. W. Hauser (eds.). 1972. *The Epidemiology of Epilepsy: A Workshop.* Pub. No. (NIH) 73-390. Washington, D.C.: U.S. Department of Health, Education and Welfare.

Anderson, W. W., and L. Ehrman. 1969. Mating choice in crosses between geographic populations of *Drosophila pseudoobscura. Am. Midland Nat. 81*:47–53.

Andrewartha, H. G., and L. C. Birch. 1954. *The Distribution and Abundance of Animals.* Chicago: University of Chicago Press.

Archer, J. 1973. Tests for emotionality in rats and mice: A review. *Anim. Behav.* 21:205-235.

Armstrong, J., and J. Adler. 1969. Location of genes for motility and chemotaxis on the *Escherichia coli* genetic map. *J. Bacteriol.* 97:156-161.

Ashburner, M., and E. Novitski. 1975. *Genetics and Biology of Drosophila.* New York: Academic Press.

Asmundson, V. S., and F. W. Lorenz. 1955. Pheasant-turkey hybrids. *Science* 121:307-308.

Bamber, R., and E. Herdman. 1932. A report on the progeny of a tortoise-shell male cat, together with a discussion of his gametic constitution. *J. Genet.* 26:117-128.

Bane, A. 1954. Studies on monozygous cattle twins. *Acta Agric. Scand.* 4:95-208.

Barker, J. S. F. 1971. Ecological differences and competitive interaction between *Drosophila melanogaster* and *Drosophila simulans* in small laboratory populations. *Oecologia* 8:139-156.

Barr, M. L. 1959. Sex chromatin and phenotype in man. *Science* 130:679-685.

Barrows, E. M., W. J. Bell, and C. D. Michener. 1975. Individual odor differences and their social functions in insects. *Proc. Natl. Acad. Sci. U.S.A.* 72:2824-2828.

Barrows, S. L. 1945. The Inheritance of the Ability to Taste Brucine. Unpublished master's thesis, Stanford University.

Bartholomew, G. A., Jr., and J. B. Birdsell. 1953. Ecology and the protohominids. *Am. Anthropol.* 55:481-498.

Bastock, M. 1956. A gene mutation which changes a behavior pattern. *Evolution* 10:421-439.

Bastock, M. 1967. *Courtship: An Ethological Study.* Chicago: Aldine.

Bateman, A. J. 1948. Intra-sexual selection in *Drosophila. Heredity* 2:349-368.

Bateman, A. J. 1949. Analysis of data on sexual isolation. *Evolution* 3:174-177.

Beach, F. A. 1950. The snark was a boojum. *Am. Psychol.* 5:115-124.

Beardmore, J. A. 1970. Ecological Factors and the Variability of Gene-Pools in Drosophila. In *Essays in Evolution and Genetics in Honor of Theodosius Dobzhansky.* M. K. Hecht and W. C. Steere (eds.). New York: Appleton, pp. 299-314.

Becker, H. J. 1970. The genetics of chemotaxis in *Drosophila melanogaster:* Selection for repellent insensitivity. *Mol. Gen. Genet.* 107:194-200.

Beckett, P., and T. Bleakley (eds.). 1968. *A Teaching Program in Psychiatry,* Vol. I. Detroit: Wayne State University Press, p. 73.

Beckman, L. 1962. Assortative matings in man. *Eugenics Rev.* 54:63-67.

Bentley, D. R. 1971. Genetic control of an insect network. *Science* 174:1139-1141.

Bentley, D. R., and R. R. Hoy. 1972. Genetic control of the neuronal network generating cricket *(Teleogryllus gryllus)* song patterns. *Anim. Behav.* 20:478-492.

Bentley, D. R., and W. Kutsch. 1966. The neuromuscular mechanism of stridulation in crickets (Orthoptera:Gryllidae). *J. Exp. Biol.* 45:151-164.

Benzer, S. 1971. From the gene to behavior. *J. Am. Med. Assoc.* 218:1015-1026.

Benzer, S. 1973. Genetic dissection of behavior. *Sci. Am. 229*:24-37.

Bergsma, D. 1973a. *Birth Defects*. Baltimore: Williams & Wilkins.

Bergsma, D. (ed.). 1973b. *Birth Defects: Atlas and Compendium*. Baltimore: Williams & Wilkins.

Berry, R. J. 1963. Epigenetic polymorphism in wild populations of *Mus musculus*. *Genet. Res. 4*:193-220.

Berry, R. J. 1970. The natural history of the house mouse. *Field Stud. 3*:219-262.

Bigelow, R. S. 1960. Interspecific hybrids and speciation in the genus *Acheta* (Orthoptera: Gryllidae). *Can. J. Zool. 38*:509-524.

Bignami, G. 1965. Selection for high rates and low rates of avoidance conditioning in the rat. *Anim. Behav. 13*:221-227.

Birdsell, J. B. 1950. Some implications of the genetical concept of race in terms of spatial analysis. *Cold Spring Harbor Symp. Quant. Biol. 15*:259-314.

Birdsell, J. B. 1973. A basic demographic unit. *Curr. Anthropol. 14*:337-356

Blewett, D. B. 1954. An experimental study of the inheritance of intelligence. *J. Ment. Sci. 100*:922-933.

Blinkov, S. M. and I. I. Glezar. 1968. *The Human Brain in Figures and Tables*. New York: Plenum Press.

Bliss, E. (ed.). 1968. *Roots of Behavior*. New York: Hafner.

Blizard, D. A. 1971. Autonomic reactivity in the rat: Effects of genetic selection for emotionality. *J. Comp. Physiol. Psychol. 76*:282-289.

Bodmer, W. F. and L. Cavalli-Sforza. 1970. Intelligence and race. *Sci. Am. 223*(4):19-29.

Böök, J. A. 1953. A genetic and neuropsychiatric investigation. *Acta Genet. 4*:3-100.

Borisov, A. I. 1970. Disturbances of panmixia in natural populations of *Drosophila funebri* polymorphous by the inversion II-1. *Genetika 6*:61-67.

Bösiger, E. 1957. Sur l'activité sexuelle des males de plusieurs souches de *Drosophila melangaster*. *C. R. Acad. Sci. (D) (Paris) 244*:1419-1422.

Bösiger, E. 1967. La signification evolutive de la selection sexuelle chez les animaux. *Scientia 102*:207-223.

Bray, P. F. 1972. Inheritance of Focal and Petit Mal Seizures. In *The Epidemiology of Epilepsy: A Workshop*. M. Alter and W. Hauser (eds.). Pub. No. (NIH) 73-390. Washington, D.C.: U.S. Dept. of Health, Education and Welfare, pp. 109-112.

Brenner, S. 1973. The genetics of behavior. *Br. Med. Bull. 29*:269-271.

Bridges, C. B., and K. Brehme. 1944. *The Mutants of Drosophila melanogaster*. *Carnegie Institution Pub.* No. 552. Washington, D.C.: Carnegie Institution.

Brncic, D. 1966. Ecological and cytogenetic studies of *Drosophila flavopilosa*, a neotropical species living in *Cestrum* flowers. *Evolution 20*:16-29.

Brncic, D. 1968. The effects of temperature on chromosomal polymorphism of *Drosophila flavopilosa* larvae. *Genetics 59*:427-432.

Brncic, D., and S. Koref-Santibañez. 1963. Life cycle and mating activity as criteria of heterosis in heterokaryotypes in *Drosophila pavani*. *Genetics Today I*:157-158.

Brncic, D., and S. Koref-Santibañez. 1964. Mating activity of homo- and heterokaryotypes in *Drosophila pavani*. *Genetics 49*:585-591.

Broadhurst, P. L. 1960. Experiments in Psychogenetics: Applications of Biometrical Genetics to the Inheritance of Behavior. In *Experiments in Personality, Vol. 1, Psychogenetics and Psychopharmacology.* H. J. Eysenck (ed.). London: Routledge, pp. 1–102.

Broadhurst, P. L. 1967a. The biometrical analysis of behavioural inheritance. *Sci. Progr.* 55:123–129.

Broadhurst, P. L. 1967b. An Introduction to the Diallel Cross. In *Behavior-Genetic Analysis.* J. Hirsch (ed.). New York: McGraw-Hill, pp. 287–304.

Broadhurst, P. L., and J. L. Jinks. 1961. Biometrical genetics and behaviour: Reanalysis of published data. *Psychol. Bull.* 58:337–362.

Broadhurst, P. L., and J. L. Jinks. 1963. The inheritance of mammalian behavior re-examined. *J. Hered.* 54:170–176.

Bruell, J. H. 1964a. Heterotic inheritance of wheelrunning in mice. *J. Comp. Physiol. Psychol.* 58:159–163.

Bruell, J. H. 1964b. Inheritance of behavioral and physiological characters of mice and the problem of heterosis. *Am. Zool.* 4:125–138.

Bruere, A., R. Marshall, and D. Ward. 1969. Testicular hypoplasia and XXY sex chromosome complement in two rams: The ovine counterpart of Klinefelter's syndrome in man. *J. Reprod. Fertil.* 19:103–108.

Burks, B. S. 1928. The Relative Influence of Nature and Nurture upon Mental Development: A Comparative Study of Foster Parent-Foster Child Resemblance and True Parent-True Child Resemblances. *27th Yearbook of the National Society for the Study of Education.* Bloomington, Ill.: Public School.

Burns, M., and M. Fraser. 1966. *Genetics of the Dog,* 2nd ed. Edinburgh: Oliver & Boyd.

Burt, C. 1943. Ability and income. *Br. J. Educ. Psychol.* 13:83–98.

Burt, C. 1961. Intelligence and social mobility. *Br. J. Statist. Psychol.* 14:3–24.

Burt, C. 1966. The genetic determination of differences in intelligence: A study of monozygotic twins reared together and apart. *Br. J. Psychol.* 57:137–153.

Burt, C., and M. Howard. 1956. The multifactorial theory of inheritance and its application to intelligence. *Br. J. Statist. Psychol.* 8:95–131.

Calhoun, J. 1962. Population density and social pathology. *Sci. Am.* 206 (2):139–148.

Campbell, M., S. Wolman, H. Breurer, F. Miller and B. Perlman. 1972. Klinefelter's syndrome in a three-year-old severely disturbed child. *J. Autism Child Schizo.* 2:34–48.

Cancro, R. 1971. *Intelligence: Genetic and Environmental Influences.* New York: Grune & Stratton.

Carpenter, F. W. 1905. The reactions of the pomace fly (*Drosophila melanogaster* Loew) to light, gravity and mechanical stimulation. *Am. Nat.* 39:151–171.

Carson, H. L. 1951. Breeding sites of *Drosophila pseudoobscura* and *Drosophila persimilis* in the transition zone of the Sierra Nevada. *Evolution* 5:91–96.

Carson, H. L., D. E. Hardy, H. T. Spieth, and W. S. Stone. 1970. The Evolutionary Biology of the Hawaiian Drosophilidae. In *Essays in Evolution and Genetics in Honor of Theodosius Dobzhansky.* M. K. Hecht and W. C. Steere (eds.). New York: Appleton, pp. 437–453.

Carter, C. O. 1965. The inheritance of common congenital malformations. *Progr. Med. Genet.* 4:59–84.

Carter, J. E. L. 1970. The somatotypes of athletes: A review. *Hum. Biol.* 42:535–569.

Caspari, E. 1951. On the biological basis of adaptedness. *Am. Sci.* 39:441–451.

Caspari, E. 1955. On the formation in the testis sheath of *Rt* and *rt Ephestia kuhniella*. *Zell. Biol. Z.* 74:585–602.

Caspari, E. 1958. Genetic Basis of Behavior. In *Behavior and Evolution*. A. Roe and G. G. Simpson (eds.). New Haven: Yale University Press, pp. 103–127.

Caspari, E. 1967. Introduction to Part I and Remarks on Evolutionary Aspects of Behavior. In *Behavior-Genetic Analysis*. J. Hirsch (ed.). New York: McGraw-Hill, pp. 3–9.

Caspari, E. and F. J. Gottlieb. 1959. On a modifier of the gene *a* in *Ephestia kuhniella*. *Vererbungslehre* 90:263–272.

Cattell, R. B. 1953. Research designs in psychological genetics with special reference to the multiple variance analysis method. *Am. J. Hum. Genet.* 5:76–93.

Cattell, R. B. 1960. The multiple abstract variance analysis equations and solutions for nature-nurture research on continuous variables. *Psychol. Rev.* 67:353–372.

Cattell, R. B. 1965. Methodological and Conceptual Advances in Evaluating Herditary and Environmental Influences and their Interaction. In *Methods and Goals in Human Behavior Genetics*. S. G. Vandenberg (ed.). New York: Academic Press, pp. 95–139.

Cattell, R. B. 1971. The Structure of Intelligence in Relation to the Nature-Nurture Controversy. In *Intelligence: Genetic and Environmental Influences*. R. Cancro (ed.). New York: Grune & Stratton.

Cattell, R. B., H. Beloff, and R. N. Coan. 1958. *Handbook for the IPAT High School Personality Questionnaire*. Champaign, Ill.: Institute Personality Ability Testing.

Cavalli-Sforza, L. L., and W. F. Bodmer. 1971. *The Genetics of Human Populations*. San Francisco: Freeman.

Cavalli-Sforza, L. L., and M. W. Feldman, 1973. Cultural versus biological inheritance: Phenotypic transmission from parent to children. *Am. J. Hum. Genet.* 25:618–637.

Chang, S., and C. Kung, 1973. Temperature-sensitive pawns: Conditional behavioral mutants of *Paramecium aurelia*. *Science* 18:1197–1199.

Chang, S., J. van Houten, L. Robles, S. S. Lui, and C. Kung. 1974. An extensive behavioural and genetic analysis of the pawn mutants of *Paramecium aurelia*. *Genet. Res.* 23:165–173.

Chiarelli, B. 1963. Sensitivity to PTC (phenyl-thio-carbamide) in primates. *Folia Primatol.* 1:103–107.

Child, I. L. 1950. The relation of somatotype to self-ratings on Sheldon's temperamental traits. *J. Pers.* 18:440–453.

Chitty, D. 1967. The natural selection of self-regulatory behaviour in animal populations. *Proc. Ecol. Soc. Aust.* 2:51–78.

Clark, E., L. R. Aronson, and M. Gordon. 1954. Mating behavior patterns in two sympatric species of xiphophorin fishes: Their inheritance and significance in sexual isolation. *Bull. Am. Mus. Nat. Hist.* 103:135–226.

Cohen, J. 1959. The factorial structure of the WISC at ages 7-6, 10-6 and 13-6. *J. Consult. Psychol.* 23:285–299.

Cole, B. L. 1970. The colour blind driver. *Aust. J. Optometry* 53:261–269.

Cole, B. L. 1972. The handicap of abnormal colour vision. *Aust. J. Optometry* 58:304–310.

Coleman D. L. 1960. Phenylalanine hydroxylase activity in dilute and nondilute strains of mice. *Arch. Biochem. Biophys.* 91:300–306.

Coleman, J. S., E. Q. Campbell, C. J. Hobson, J. McPartland, A. M. Mood, F. D. Weinfeld, and R. L. York. 1966. *Equality of Educational Opportunity.* Washington, D.C.: GPO.

Collins, R. L. 1964. Inheritance of avoidance conditioning in mice: A diallel study. *Science* 143:1188–1190.

Connolly, K. 1966. Locomotor activity in Drosophila. 2. Selection for active and inactive strains. *Anim. Behav.* 14:444–449.

Connolly, K. 1968. The social facilitation of preening behavior in *Drosophila melanogaster. Anim. Behav.* 16:385–391.

Connolly, K., B. Burnet, and D. Sewell. 1969. Selective mating and eye pigmentation: An analysis of the visual component in the courtship behavior of *Drosophila melanogaster. Evolution* 238:548–559.

Conterio, F., and L. L. Cavalli-Sforza. 1959. Evolution of the human constitutional phenotype: An analysis of mortality effects. *Ricerca Sci.* (Suppl.) 29:3–14.

Cook, W. T., P. Siegel, and K. Hinkelmann. 1972. Genetic analyses of male mating behavior in chickens. 2. Crosses among selected and control lines. *Behav. Genet.* 2:289–300.

Cooke, F., G. H. Finney, and R. F. Rockwell. 1976. Assortative mating in lesser snow geese. *Behav. Genet.* 6:(2). In press.

Cooke, F., and C. M. McNally. 1975. Mate selection and colour preferences in lesser snow geese. *Behavior.* 53:151–170.

Cooke, F., and P. J. Mirsky. 1972. A genetic analysis of lesser snow goose families. *Auk* 89:863–871.

Cooper, R. M., and J. P. Zubek. 1958. Effects of enriched and restricted early environments on the learning ability of bright and dull rats. *Can J. Psychol.* 12:159–164.

Cornsweet, T. N. 1971. *Visual Perception.* New York: Academic Press.

Cotter, W. B. 1951. The Genetic and Physiological Analyses of the Silk-Spinning Behavior of *Ephestia kuhniella.* Unpublished masters thesis, Wesleyan University.

Cowan, B. D., and W. M. Rogoff. 1968. Variation and heritability of responsiveness of individual male house flies, *Musca domestica,* to the female sex pheromone. *Ann. Entomol. Soc. Am.* 61:1215–1218.

Craig, G. 1965. The role of genetics in mosquito control. *Proc. and Papers of the 33rd Annual Conf. of the Calif. Mosquito Control Assoc.* 33:80–82.

Craig, G., and R. Vandehey. 1962. Genetic variability in *Aedes aegypti.* 1. Mutations affecting color pattern. *Ann. Entomol. Soc. Am.* 55:58–67.

Craig, J. V. and R. A. Baruth. 1965. Inbreeding in social dominance ability in chickens. *Anim. Behav.* 13:109-113.

Craig, J. W., L. L. Ortman, and A. M. Guhl. 1965. Genetic selection for social dominance ability in chickens. *Animal. Behav.* 13:114-131.

Crow, J. F., and J. Felsenstein. 1968. The effect of assortative mating on the genetic composition of a population. *Eugen. Q.* 15:85-97.

Crowcroft, P. 1966. *Mice All Over.* Chester Springs, Pa.: Dufour.

Dahlstrom, W. G., and G. S. Welsh (eds.). 1960. *An MMPI Handbook.* Minneapolis: University of Minnesota Press.

Darwin, C. 1859. *The Origin of Species by Means of Natural Selection.* London: Murray.

Darwin, C. 1871. *The Descent of Man and Selection in Relation to Sex.* London: Murray.

Davis, P. C. 1956. A factor analysis of the Wechsler Bellevue Scale. *Educ. Psychol. Measurement* 16:127-146.

Dawson, W. M. 1932. Inheritance of wildness and tameness in mice. *Genetics* 17:296-326.

Day, R. H. 1969. *Human Perception.* New York: Wiley.

DeFries, J. C. 1964. Prenatal maternal stress in mice: Differential effects on behavior. *J. Hered.* 55:289-295.

DeFries, J. C., and J. P. Hegmann. 1970. Genetic Analysis of Open-Field Behavior. In *Contributions to Behavior-Genetic Analysis: The Mouse as a Prototype.* G. Lindzey and D. D. Thiessen (eds.). New York: Appleton, pp. 23-56.

DeFries, J. C., J. P. Hegmann, and M. W. Weir. 1966. Open-field behavior in mice: Evidence for a major gene effect mediated by the visual system. *Science* 154:1577-1579.

DeFries, J. C., and G. E. McClearn. 1972. Behavioral genetics and the fine structure of mouse populations: A study in microevolution. *Evol. Biol.* 5:279-291.

DeFries, J. C., E. A. Thomas, J. P. Hegmann, and M. W. Weir. 1967. Open-field behavior in mice: Analysis of maternal effects by means of ovarian transplantation. *Psychon. Sci.* 8:207-208.

DeFries, J. C., J. Wilson, and G. McClearn. 1970. Open-field behavior in mice: Selection response and situational generality. *Behav. Genet.* 1:195-211.

De Souza, H. M. L., A. B. Da Cunha, and E. P. Dos Santos. 1970. Adaptive polymorphism of behavior evolved in laboratory populations of *Drosophila willistoni. Am. Nat.* 104:175-189.

De Souza, H. M. L., A. B. Da Cunha, and E. P. Dos Santos. 1972. Assortative mating in polymorphic laboratory populations of *Drosophila willistoni. Egypt. J. Genet. Cytol.* 1:225-230.

DeVore, I. (ed). 1965. *Primate Behavior: Field Studies of Monkeys and Apes.* New York: Holt.

Dilger, W. 1962a. Behavior and Genetics. In *Roots of Behavior.* E. Bliss (ed.). New York: Harper & Row.

Dilger, W. 1962b. The behavior of lovebirds. *Sci. Am.* 206:88-98.

Dingman, H. 1968. Psychological Test Patterns in Down's Syndrome. In *Progress in Human Behavior Genetics.* S. Vandenberg (ed.). Baltimore: John Hopkins University Press, pp. 19-25.

Dobzhansky, T. 1937. *Genetics and the Origin of Species.* New York: Columbia University Press.

Dobzhansky, T. 1940. Speciation as a stage in evolutionary divergence. *Am. Nat.* 74:312-321.

Dobzhansky, T. 1947. Adaptive changes induced by natural selection in wild populations of Drosophila. *Evolution* 1:1-16.

Dobzhansky, T. 1950. Genetics of natural populations. 19. Origin of heterosis through natural selection in populations of *Drosophila pseudoobscura. Genetics* 35:288-302.

Dobzhansky, T. 1951. *Genetics and the Origin of Species,* 3rd ed. New York: Columbia University Press.

Dobzhansky, T. 1955. A review of some fundamental concepts and problems of populational genetics. *Symp. Quant. Biol.* 20:1-15.

Dobzhansky, T. 1964. *Heredity and the Nature of Man.* New York: Harcourt.

Dobzhansky, T. 1968. On some fundamental concepts of Darwinian biology. *Evol. Biol.* 2:1-34.

Dobzhansky, T. 1970. *Genetics of the Evolutionary Process.* New York: Columbia University Press.

Dobzhansky, T., D. M. Cooper, H. J. Phaff, E. P. Knapp, and H. L. Carson. 1956. Studies on the ecology of Drosophila in the Yosemite region of California. 4. Differential attraction of species of Drosophila to different species of yeasts. *Ecology* 37:544-550.

Dobzhansky, T., and C. Pavan. 1950. Local and seasonal variations in relative frequencies of species of Drosophila in Brazil. *J. Anim. Ecol.* 19:1-14.

Dobzhansky, T., and O. Pavlovsky. 1962. A comparative study of the chromosomes in the incipient species of the *Drosophila paulistorum* complex. *Chromosoma* 13:196-218.

Dobzhansky, T., and B. Spassky. 1962. Genetic drift and natural selection in experimental populations of *Drosophila pseudoobscura. Proc. Natl. Acad. Sci. U.S.A.* 48:148-156.

Dobzhansky, T., and B. Spassky. 1969. Artificial and natural selection for two behavioral traits in *Drosophila pseudoobscura. Proc. Natl. Acad. Sci. U.S.A.* 62:75-80.

Dunham, H. W. 1965. *Community and Schizophrenia: An Epidemiological Analysis.* Detroit: Wayne State University Press.

Dunlop, K. 1943. Mental maladjustment and color vision. *Science* 98:470-472.

Dustman, R. E., and E. C. Beck. 1965. The visually evoked potential in twins. *Electroencephalogr. Clin. Neurophysiol.* 19:570-575.

Eaton, S. W., and R. S. Weil. 1955. *Culture and Mental Disorders.* Glencoe, Ill.: Free Press.

Eaves, L. J. 1973. Assortative mating and intelligence: An analysis of pedigree data. *Heredity* 30:199-210.

Eckert, R. 1972. Bioelectric control of ciliary activity. *Science* 176:473-481.

Eckland, B. K. 1972. Comments on School Effects, Gene-Environment Covariance, and the Heritability of Intelligence. In *Genetics, Environment and Behavior: Implications for Educational Policy*. L. Ehrman, G. S. Omenn, and E. Caspari (eds.). New York: Academic Press, pp. 297-306.

Edwards, J. H. 1958. Congenital malformations of the central nervous system in Scotland. *Br. J. Prev. Soc. Med. 12*:115-130.

Edwards, J. H. 1960. The simulation of mendelism. *Acta Genet. Statist. Med. 10*:63-70.

Edwards, J. H. 1971. The analysis of X-linkage. *Am. J. Hum. Genet. 341*:229-250.

Ehrlich, P. R. and A. H. Ehrlich. 1970. *Population Resources Environment: Issues in Human Ecology*. San Francisco: Freeman.

Ehrman, L. 1960a. The genetics of hybrid sterility in *Drosophila paulistorum*. *Evolution 14*:212-223.

Ehrman, L. 1960b. A genetic constitution frustrating the sexual drive in *Drosophila paulistorum*. *Science 131*:1381-1382.

Ehrman, L. 1961. The genetics of sexual isolation in *Drosophila paulistorum*. *Genetics 46*:1025-1038.

Ehrman, L. 1964. Courtship and mating behavior as a reproductive isolating mechanism in Drosophila. *Am. Zool. 4*:147-153.

Ehrman, L. 1965. Direct observation of sexual isolation between allopatric and between sympatric strains of the different *Drosophila paulistorum* races. *Evolution 19*:459-464.

Ehrman, L. 1966. Mating success and genotype frequency in Drosophila. *Anim. Behav. 14*:332-339.

Ehrman, L. 1967. Further studies on genotype frequency and mating success. *Am. Nat. 101*:415-424.

Ehrman, L. 1968. Frequency dependence of mating success in *Drosophila pseudoobscura*. *Gent. Res. 11*:135-140.

Ehrman, L. 1969. Genetic divergence in M. Vetukhiv's experimental populations of *Drosophila pseudoobscura*. 5: A further study of rudiments of sexual isolation. *Am. Midland Nat. 82*:272-276.

Ehrman, L. 1970a. The mating advantage of rare males in Drosophila. *Proc. Natl. Acad. Sci. U.S.A. 515*:345-348.

Ehrman, L. 1970b. A release experiment testing the mating advantage of rare Drosophila males. *Behav. Sci. 15*:363-365.

Ehrman, L. 1972. Genetics and Sexual Selection. In *Sexual Selection and the Descent of Man*. B. Campbell (ed.). Chicago: Aldine, pp. 105-135.

Ehrman, L., G. S. Omenn, and E. Caspari (eds.). 1972. *Genetics, Environment, and Behavior: Implications for Educational Policy*. New York: Academic Press.

Ehrman, L., B. Spassky, O. Pavlovsky, and T. Dobzhansky. 1965. Sexual selection, geotaxis and chromosomal polymorphism in experimental populations of *Drosophila pseudoobscura*. *Evolution 19*:337-346.

Ehrman, L., and M. Strickberger. 1960. Flies mating: A pictorial record. *Nat. Hist. 69*:28-33.

Eiduson, S., E. Geller, A. Yuwiller, and B. T. Eiduson. 1964. *Biochemistry and Behavior*. Princeton: Van Nostrand.

Eisenberg, L. 1973. Psychiatric intervention. *Sci. Am. 229*:116.

Elens, A. A. 1965. Studies of selective mating using the melanistic mutants of *Drosophila melanogaster*. *Experientia 21*:1-5.

Elens, A. A. and J. M. Wattiaux. 1964. Direct observation of sexual isolation. *Drosophila Inf. Serv.* 39:118-119.

El-Helw, M. R., and A. M. M. Ali. 1970. Competition between *Drosophila melanogaster* and *D. simulans* on media supplemented with *Saccharomyces* and *Schizosaccharomyces*. *Evolution* 24:531-537.

Emlen, J. M. 1973. *Ecology: An Evolutionary Approach*. Reading, Mass.: Addison-Wesley.

Epps, S., and R. W. Parnell. 1952. Physique and temperament of women deliquents compared with women undergraduates. *Br. J. Med. Psychol.* 25:249-255.

Erlenmeyer-Kimling, L. 1968. Mortality rates in the offspring of schizophrenic parents and a physiological advantage hypothesis. *Nature* 220:798-800.

Erlenmeyer-Kimling, L. 1972. Gene-Environment Interactions and the Variability of Behavior. In *Genetics, Environment, and Behavior: Implications for Educational Policy*. L. Ehrman, G. S. Omenn, and E. Caspari (eds.). New York: Academic Press, pp. 181-208.

Erlenmeyer-Kimling, L., and L. F. Jarvik. 1963. Genetics and intelligence: A review. *Science* 142:1477-1479.

Erlenmeyer-Kimling, L., and W. Paradowski. 1966. Selection and schizophrenia. *Am. Nat.* 100:651-665.

Erway, L., L. S. Hurley, and A. Fraser, 1966. Neurological defect: Manganese in phenocopy and prevention of a genetic abnormality of inner ear. *Science* 152:1766-1768.

Ewing, A. W. 1963. Attempts to select for spontaneous activity in *Drosophila melanogaster*. *Anim. Behav.* 11:369-377.

Ewing, A. W. 1969. The genetic basis of sound production in *Drosophila pseudoobscura* and *D. persimilis*. *Anim. Behav.* 17:555-560.

Ewing, A. W., and A. Manning. 1967. The evolution and genetics of insect behavior. *Annu. Rev. Entom.* 12:471-494.

Eysenck, H. J. 1956. The inheritance of extraversion. *Acta Psychol.* 12:95-110.

Eysenck, H. J. 1964. *Crime and Personality*. London: Routledge.

Eysenck, H. J. 1967. Intelligence assessment: A theoretical and experimental approach. *Br. J. Educ. Psychol.* 37:81-98.

Eysenck, H. J., and P. L. Broadhurst. 1964. Experiments with Animals: Introduction. In *Experiments in Motivation*. H. J. Eysenck (ed.). Elmsford, N.Y.: Pergamon Press.

Eysenck, H. J., and D. B. Prell. 1951. The inheritance of neuroticism: An experimental study. *J. Ment. Sci.* 97:441-465.

Falconer, D. S. 1960. *Introduction to Quantitative Genetics*. Edinburg: Oliver & Boyd.

Falconer, D. S. 1965. The inheritance of liability to certain diseases, estimated from the incidence among relatives. *Ann. Hum. Genet.* 29:51-71.

Falconer, D. S. 1967. The inheritance of liability to diseases with variable age of onset, with particular reference to diabetes mellitus. *Ann. Hum. Genet.* 31:1-20.

Falek, A., and S. Britton. 1974. Phases in coping: The hypothesis and its implications. *Soc. Bio.* 21:1-7.

Falk, C., and L. Ehrman. 1975. Random mating revisited. *Behav. Genet. 3*:91–95.

Fischer, R., and F. Griffin. 1960. Factors involved in the mechanism of "taste-blindness." *J. Hered. 51*:182–183.

Fisher, R. A. 1930. *The Genetical Theory of Natural Selection.* Oxford: Clarendon Press.

Fisher, R. A. 1958. Cancer and smoking. *Nature 182*:596.

Fisher, R. A., E. B. Ford, and J. H. Huxley. 1939. Taste testing the anthropoid apes. *Nature 144*:750.

Fisher, R., F. Griffin, S. England, and S. M. Garen. 1961. Taste thresholds and food dislikes. *Nature 191*:1328.

Fisher, R. A., and F. Yates. 1967. *Statistical Tables for Biological, Agricultural and Medical Research,* 6th ed. Edinburgh: Oliver & Boyd.

Ford, E. B. 1964. *Ecological Genetics.* London: Methuen.

Ford, E. B. 1971. *Ecological Genetics,* 3rd ed. London: Chapman & Hall.

Franck, D. 1970. Verhaltengenetische Untersuchungen an Artbastarden der Gattung *Xiphophorus.* (Pisces). *Zeitschrift für Tierpsychologie 27*:1–34.

French, J. W., R. B. Ekstrom, and L. A. Price. 1963. *Kit of Reference Tests for Cognitive Factors,* rev. ed. Princeton: Educational Testing Service.

Friedlaender, J. S., L. A. Sgaramella-Zonta, K. K. Kidd, L. Y. C. Lai, P. Clark, and R. J. Walsh. 1971. Biological divergences in south-central Bougainville: An analysis of blood polymorphism gene frequencies and anthropometric measurements utilizing tree models and a comparison of these variables with linguistic, geographic, and migrational "distances." *Am. J. Hum. Genet. 23*:253–270.

Fuhrmann, W., and F. Vogel. 1969. *Genetic Counseling.* New York: Springer Verlag.

Fulker, D. W. 1966. Mating speed in male *Drosophila melanogaster:* A psychogenetic analysis. *Science 153*:203–205.

Fulker, D. W. 1970. Maternal buffering of rodent genotypic responses to stress: A complex genotype-environment interaction. *Behav. Genet. 1*:119–124.

Fulker, D. W. 1972. Applications of a simplified triple-test cross. *Behav. Genet. 2*:185–198.

Fulker, D. W., J. Wilcock, and P. L. Broadhurst. 1972. Studies in genotype-environment interaction. 1. Methodology and preliminary multivariate analysis of a diallel cross of eight strains of rats. *Behav. Genet. 2*:261–287.

Fuller, J. L., and W. R. Thompson. 1960. *Behavior Genetics.* New York: Wiley.

Galton, F. 1869. *Hereditary Genius.* London: Macmillan.

Galton, F. 1874. *English Men of Science: Their Nature and Nurture.* London: Macmillan.

Galton, F. 1883. *Inquiry into Human Faculty.* London: Macmillan.

Gardner, E. J. 1972. *Principles of Genetics,* 4th ed. New York: Wiley.

Gardner, L., and R. Neu. 1972. Evidence linking an extra Y chromosome to sociopathic behavior. *Arch. Gen. Psychiatry 26*:220–222.

Garn, S. M. (ed.). 1961. *Human Races.* Springfield, Ill.: Thomas.

Garside, R. F., and D. W. K. Kay. 1964. The Genetics of Stuttering. In *The Syndrome of Stuttering.* G. Andrews and M. M. Harris (eds.). London: Heinemann.

Gayral, L., M. Barraud, J. Carrie, and C. Candebat. 1960. Pseudoherma-phrodisme à type de "testicule feminisant": 11 cas. *Toulouse Med.* 9:637-647.

Giblett, E. R. 1969. *Genetic Markers in Human Blood.* Oxford: Blackwell.

Gibson, J. B., and C. G. N. Mascie-Taylor. 1973. Biological aspects of a high socio-economic group. 2. I.Q. components and social mobility. *J. Biosoc. Sci.* 5:17-30.

Gibson J. B., and J. M. Thoday. 1962. Effects of disruptive selection. 9. Low selection intensity. *Heredity* 19:125-130.

Gill, K. S. 1963. A mutation causing abnormal mating behavior. *Drosophila Inf. Serv.* 38:33.

Ginsburg, B. E. 1967. Genetic Parameters in Behavioral Research. In *Behavior Genetic Analysis.* J. Hirsch (ed.). New York: McGraw-Hill.

Glueck, S., and E. Glueck. 1956. *Physique and Delinquency.* New York: Harper & Row.

Gottesman, I. I. 1965. Personality and Natural Selection. In *Methods and Goals in Human Behavior Genetics.* S. G. Vandenberg (ed.). New York: Academic Press, pp. 63-80.

Gottesman, I. 1969. Differential Inheritance of the Psychoneuroses. In *Behavioral Genetics: Method and Research.* M. Manosevitz, G. Lindzey, and D. Thiessen (eds.). New York: Appleton, pp. 657-662.

Gottesman, I., and L. Heston. 1972. Human Behavioral Adaptations: Specula-tions on Their Genesis. In *Genetics, Environment, and Behavior: Implications for Educational Policy.* L. Ehrman, G. S. Omenn, and E. Caspari (eds.). New York: Academic Press.

Gottesman, I., and J. Shields. 1966. Schizophrenia in twins: 16 years' consecu-tive admissions to a psychiatric clinic. *Dis. Nerv. Syst.* 27:11-19.

Gottesman, I., and J. Shields. 1971. Schizophrenia: Geneticism and environ-mentalism. *Hum. Hered.* 21:517-522.

Gottesman, I., and J. Shields. 1972. *Schizophrenia and Genetics: A Twin Study Vantage Point.* New York: Academic Press.

Gottesman, I., and J. Shields. 1973. Genetic theorizing and schizophrenia. *Br. J. Psychiatry* 122:15-30.

Goy, R. W. and J. S. Jakway. 1959. The inheritance of patterns of sexual behavior in female guinea pigs. *Anim. Behav.* 7:142-149.

Gramberg-Danielson, B. 1962. Investigation of traffic accident frequency of persons with defective color sense. *Klin. Monatsbl. Augenheilkd.* 139:677-682.

Grant, B., G. A. Snyder, and D. L. Glessner. 1974. Frequency-dependent mate selection in *Mormoniella vitripennis. Evolution* 28:259-264.

Grant, V., and K. A. Grant. 1965. *Flower Pollination in the Phlox Family.* New York: Columbia Univ. Press.

Green, E. L. 1966. Breeding Systems. In *Biology of the Laboratory Mouse,* 2nd ed. E. L. Green (ed.). New York: McGraw-Hill.

Gregory, R. L. 1966. *Eye and Brain: The Psychology of Seeing.* London: Weidenfeld & Nicolson.

Griffing, B. 1956. Concept of general and specific combining ability in relation to diallel crossing systems. *Aust. J. Biol. Sci.* 9:463-493.

Griffiths, D. R. 1970. Assessment of Personality. In *Psychological Assessment of Mental and Physical Handicaps.* P. Mittler (ed.). London: Methuen.

Grossfield, J. 1971. Geographic distribution and light-dependent behavior in *Drosophila. Proc. Natl. Acad. Sci. U.S.A. 68*:2669-2673.

Grossfield, J. 1972. The use of behavioral mutants in biological control. *Behav. Genet. 2*:311-319.

Grossfield, J. 1975. Behavioral Mutants of *Drosophila.* In *A Handbook of Genetics,* Vol. 3, *Invertebrates of Genetic Interest.* R. C. King (ed.). New York: Plenum, pp. 679-701.

Grüneberg, H. 1963. *The Pathology of Development: A Study of Inherited Skeletal Disorders in Animals.* Oxford: Blackwell.

Guillery, R. W. 1974. Visual pathways in albinos. *Sci. Am. 230*:44-54.

Guillery, R. W., and J. H. Kass. 1973. Genetic abnormality of the visual pathways in a "white" tiger. *Science 180*:1287-1289.

Gwadz, R. 1970. Monofactorial inheritance of early sexual receptivity in the mosquito, *Aedes Atropalpus. Anim. Behav. 18*:358-361.

Hadler, N. M. 1964. Heritability and phototaxis in *Drosophila melanogaster. Genetics 50*:1269-1277.

Hafez, E. S. 1968. *Reproduction in Farm Animals,* 2nd ed. Philadelphia: Lea & Febiger.

Hafez, E. S. E. (ed.). 1969. *The Behavior of Domestic Animals,* 2nd ed. London: Bailliere.

Haldane, J. B. S. 1946. The interaction of nature and nurture. *Ann. Eugen. 13*:197-205.

Hall, C. S. 1951. The Genetics of Behavior. In *Handbook of Experimental Psychology.* S. S. Stevens (ed.). New York: Wiley, pp. 304-329.

Hall, C. S. and G. Lindzey. 1957. *Theories of Personality.* New York: Wiley.

Hamerton, J. 1971. *Human Cytogenetics.* New York: Academic Press.

Hamilton, W. D. 1964. The genetical evolution of social behavior. *J. Theor. Biol. 7*:1-52.

Hancock, J. 1954. *Studies in monozygotic cattle twins.* Pub. No. 63, Washington, D.C.: U.S. Department of Agriculture.

Harrell, R. F., E. R. Woodyard, and A. I. Gates. 1956. The influence of vitamin supplementation of the diets of pregnant and lactating women on the intelligence of their offspring. *Metabolism 5*:555-562.

Harris, H. 1959. *Human Biochemical Genetics.* London: Cambridge University Press.

Harris, V. T. 1952. An experimental study of habitat selection by prairie and forest races of the deermouse, *Peromyscus maniculatus. Contrib. Lab. Vertebrate Biol. Univ. Michigan 56*:1-53.

Harrison, G. A., R. J. Morton, and J. S. Weiner. 1959. The growth in weight and tail length of inbred and hybrid mice reared at two different temperatures. *Philos. Trans. R. Soc. Lond. (Biol. Sci.) 242*:479-516.

Harrison, G. A., J. S. Weiner, J. M. Tanner, and N. A. Barnicot. 1964. *Human Biology: An Introduction to Human Evolution, Variation and Growth.* Oxford: Clarendon Press.

Harvald, B., and M. Hauge. 1965. Hereditary Factors Elucidated by Twin Studies. In *Genetics and the Epidemiology of Chronic Diseases.* J. V. Neel, M. W.

Shaw, and W. J. Schull (eds.). Washington, D.C.: U.S. Department of Health, Education and Welfare, pp. 61-76.

Hay, D. A. 1972. Recognition by Drosophila melanogaster of individuals from other strains or cultures: Support for the role of olfactory cues in selective mating. *Evolution* 26:171-176.

Hay, D. A. 1975. Strain differences in the maze-learning abilities of D. melanogaster. *Nature* 257:44-46.

Hayman, B. I. 1958. The theory and analysis of diallel crosses. *Genetics* 43:63-85.

Hayman, R. H. 1964. Exercise of mating preference by a Merino ram. *Nature* 203:160-162.

Hazelbauer, G. L., and J. Adler. 1971. Role of the galactose binding protein in chemotaxis of Escherichia coli toward galactose. *Nature New Biol.* 230:101-104.

Heed, W. B., and H. W. Kircher. 1965. Unique sterol in the ecology and nutrition of Drosophila pachea. *Science* 149:758-761.

Hegmann, J. 1966. A Quantitative Genetic Analysis of Open-Field Behavior in Two Inbred Strains of Mice. Unpublished masters thesis, University of Illinois.

Henderson, N. D. 1968. The confounding effects of genetic variables in early experience research: Can we ignore them? *Dev. Psychobiol.* 1:146-152.

Henderson, N. D. 1970. Genetic influences on the behavior of mice can be obscured by laboratory rearing. *J. Comp. Physiol. Psychol.* 72:505-511.

Herrnstein, R. 1971. I.Q. *Atlantic* 228:43-64.

Herskowitz, I. 1973. *Principles of Genetics.* New York: Macmillan.

Heston, L. L. 1966. Psychiatric disorders in foster home reared children of schizophrenic mothers. *Br. J. Psychol.* 112:819-825.

Heston, L. L. 1970. The genetics of schizophrenic and schizoid disease. *Science* 167:249-256.

Heston, L. L. 1971. Discussion: Schizophrenia — Onset in infancy. *Soc. Biol.* 18:5114-5116.

Heston, L. L. 1972. Discussion. In *Genetics, Environment, and Behavior: Implications for Educational Policy.* L. Ehrman, G. S. Omenn, and E. Caspari (eds.). New York: Academic Press, pp. 99-102.

Hiernaux, J. 1956. Analyse de la variation des caractères physiques humains en une region de l'Afrique Centrale: Ruanda-Urundi et Kivu. *Anthropologie* 3:1-131.

Higgins, J. V., E. W. Reed, and S. C. Reed. 1962. Intelligence and family size: A paradox resolved. *Eugen. Q.* 9:84-90.

Hill, K. G., J. J. Loftus-Hills, and D. F. Gartside. 1972. Premating isolation between the Australian field crickets, Teleogryllus commodus and T. oceanicus. *Aust. J. Zool.* 20:153-163.

Himwich, H. E. 1971. *Biochemistry, Schizophrenia and Affective Illnesses.* Baltimore: William Wilkins.

Hirsch, J. 1963. Behavior genetics and individuality understood. *Science* 142:1436-1442.

Hirsch, J. 1964. Intellectual Functioning and the Dimensions of Human Variation. In *Genetic Diversity of Human Behavior.* J. N. Spuhler (ed.). Chicago: Aldine.

Hirsch, J. (ed.). 1967. *Behavior-Genetic Analysis*. New York: McGraw-Hill.

Hirsch, J., and J. Boudreau. 1958. Studies in experimental behavior genetics. The heritability of phototaxis in a population of *Drosophila melanogaster*. *J. Comp. Physiol. Psychol.* 51:647-651.

Hirsch, J., and L. Erlenmeyer-Kimling. 1962. Individual Differences in Behavior and Their Genetic Basis. In *Roots of Behavior*. E. L. Bliss (ed.). New York: Harper & Row, pp. 3-23.

Hoagland, H. 1966. Cybernetics of Population Control. In *Human Ecology: Collected Readings*. J. B. Bresler (ed.). Reading, Mass.: Addison-Wesley, pp. 351-359.

Hodgson, R. E. 1935. An eight generation experiment in inbreeding swine. *J. Hered.* 26:209-217.

Hollaender, A. (ed.). 1971. *Chemical Mutagens: A Method for Their Detection*. New York: Plenum Press.

Hölmberg, L. 1972. Genetic studies in a family with testicular feminization, hemophilia A and color blindness. *Clin. Genet.* 3:253-257.

Holzinger, K. J. 1929. The relative effect of nature and nurture influences on twin differences. *J. Educ. Psychol.* 20:241-248.

Honzik, M. P. 1957. Developmental studies of parent-child resemblance in intelligence. *Child Dev.* 28:215-228.

Hook, E. B. 1973. Behavioral implications of the human XYY genotype. *Science* 179:139-150.

Hotta, Y., and S. Benzer. 1969. Abnormal electroretinograms in visual mutants of Drosophila. *Nature (Lond.)* 222:354-356.

Hotta, Y., and S. Benzer. 1970. Genetic dissection of the Drosophila nervous system by means of mosaics. *Proc. Natl. Acad. Sci. U.S.A.* 67:1156-1163.

Hotta, Y., and S. Genzer. 1972. Mapping of behavior in Drosophila mosaics. *Nature (Lond.)* 240:527-535.

Hotta, Y., and S. Benzer. 1973. Mapping of Behavior in Drosophila Mosaics. In *Genetic Mechanisms of Development*. F. H. Ruddle (ed.). New York: Academic Press.

Howe, W. L., and P. A. Parsons. 1967. Genotype and environment in the determination of minor skeletal variants and body weight in mice. *J. Embryol. Exp. Morphol.* 17:283-292.

Howells, W. W. 1966. Population distances: Biological, linguistic geographical, and environmental. *Curr. Anthropol.* 7:531-540.

Hoy, R. R. 1974. Genetic control of acoustic behavior in crickets. *Am. Zool.* 14:1067.

Hoy, R. R., and R. L. Paul. 1973. Genetic control of song specificity in crickets. *Science 180*:82-83.

Hsu, T. C., and K. Benirschke. 1967. *An Atlas of Mammalian Chromosomes*. New York: Springer Verlag.

Hungerford, D. 1971. Chromosome structure and function in man. 1. Pachytene mapping in the male, improved methods and general discussion of initial results. *Cytogenetics 10*:23-32.

Hungerford, D., G. U. LaBadie, and G. B. Balaban. 1971. Chromosome structure and function in man. 2. Provisional maps of the two smallest autosomes (chromosomes 21 and 22) at pachytene in the male. *Cytogenetics 10*:33-37.

Hungerford, D., W. Mellman, G. Balaban, G. LaBadie, and L. Messatzzia. 1970. Chromosome structure and function in man. 3. Pachytene analysis and identification of the supernumerary chromosome in a case of Down's syndrome (mongolism). *Proc. Natl. Acad. Sci. U.S.A.* 67:221-224.

Huxley, J. 1942. *Evolution: The Modern Synthesis.* New York: Harper & Row.

Ikeda, K., and W. D. Kaplan. 1970a. Patterned neural activity of a mutant *Drosophila melanogaster. Proc. Natl. Acad. Sci. U.S.A.* 66:765-772.

Ikeda, K., and W. D. Kaplan. 1970b. Unilaterally patterned neural activity of gynandromorphs: Mosaic for a neurological mutant of *Drosophila melanogaster. Proc. Natl. Acad. Sci. U.S.A.* 67:1480-1487.

Jakway, J. S. 1959. Inheritance of patterns of mating behavior in the male guinea pig. *Anim. Behav.* 7:150-162.

Jay, B. 1974. Recent advances in ophthalmic genetics. *Br. J. Ophthalmol.* 58:427-437.

Jennings, H. S. 1906. *Behavior of the Lower Organisms.* New York: Columbia University Press.

Jensen, A. R. 1972. Discussion. In *Genetics, Environment, and Behavior: Implications for Educational Policy.* L. Ehrman, G. S. Omenn, and E. Caspari (eds.). New York: Academic Press, pp. 240-246.

Jensen, A. R. 1973. *Educability and Group Differences.* London: Methuen.

Jinks, J. L., and D. W. Fulker. 1970. Comparison of the biometrical genetical, MAVA, and classical approaches to the analysis of human behavior. *Psychol. Bull.* 73:311-349.

Judd, D. 1943. Color blindness and the detection of camouflage. *Science* 97:544-546.

Jude, A., and A. Searle. 1957. A fertile tortoiseshell tomcat. *Nature* 179:1087-1088.

Juel-Nielsen, N., and B. Harvald. 1958. The electroencephalogram in uniovular twins brought up apart. *Acta Genet.* 8:57-64.

Kalmus, H. 1965. *Diagnosis and Genetics of Defective Colour Vision.* Elmsford, N.Y.: Pergamon Press.

Kalmus, H. 1967. Sense Perception and Behavior. In *Genetic Diversity and Human Behavior.* J. N. Spuhler (ed.). Chicago: Aldine.

Kaplan, W. D., and W. E. Trout. 1969. The behavior of four neurological mutants of Drosophila. *Genetics* 61:399-409.

Kaul, D., and P. Parsons. 1965. The genotypic control of mating speed and duration of copulation in *Drosophila pseudoobscura. Heredity* 20:381-392.

Kaul, D., and P. Parsons. 1966. Competition between males in the determination of mating speed in *Drosophila pseudoobscura. Aust. J. Biol. Sci.* 19:945-947.

Kempthorne, O. 1969. *An Introduction to Genetic Statistics.* Ames: Iowa State Univ. Press.

Kennedy, W. A., V. Van De Riet, and J. C. White, Jr. 1963. A normative sample of intelligence and achievement of Negro elementary school children in the southeastern United States. *Mongr. Soc. Res. Child Dev.* 28(6):1-112.

Kessler, S. 1966. Selection for and against ethological isolation between *Drosophila pseudoobscura* and *Drosophila persimilis*. *Evolution* 20:634-645.

Kessler, S. 1968. The genetics of *Drosophila* mating behavior. *Anim. Behav.* 16:485-491.

Kessler, S. 1969. The genetics of *Drosophila* mating behavior. 2. The genetic architecture of mating speed in *Drosophila pseudoobscura*. *Genetics* 62:421-433.

Kety, S. S. 1967. Biochemical Aspects of Mental States. In *The Human Mind*. J. D. Roslansky (ed.). Amsterdam: North Holland, pp. 141-152.

Kidd, K., and L. L. Cavalli-Sforza. 1973. An analysis of the genetics of schizophrenia. *Soc. Biol.* 20:254-265.

Kidd, K., T. Reich, and S. Kessler. 1973. A genetic analysis of stuttering suggesting a single major locus. *Genetics* 74(2, Pt. 2):s137.

Kilgour, R. 1975. The open-field test as an assessment of the temperament of dairy cows. *Anim. Behav.* 23:615-624.

King, J. A. 1967. Behavioral Modification of the Gene Pool. In *Behavior-Genetic Analysis*. J. Hirsch (ed.). New York: McGraw-Hill, pp. 22-43.

King, J. A., D. Maas, and R. Weisman. 1964. Geographic variations in nest size among species of *Peromyscus*. *Evolution* 18:230-234.

Kirk, R. L. 1966. Population Genetic Studies in Australia and New Guinea. In *The Biology of Human Adaptability*, P. T. Baker and J. S. Weiner (eds.). Oxford: Clarendon Press, pp. 395-430.

Klawans, L., Jr., W. Paulson, P. Ringel, and A. Barbeau. 1972. Use of L-dopa in the detection of presymptomatic Huntington's chorea. *N. Engl. J. Med.* 286:1332-1334.

Knight, G., A. Robertson, and C. Waddington. 1956. Selection for sexual isolation within a species. *Evolution* 10:14-22.

Koch, R. 1967. Tagesperidik der Activität und der Orientierung nach Wald und Feld von *D. subobscura* und *D. obscura*. *Z. Vergl. Physiol.* 54:353-394.

Konopka, R. J., and S. Benzer. 1971. Clock mutants of *Drosophila melanogaster*. *Proc. Natl. Acad. Sci. U.S.A.* 68:2112-2116.

Koopman, K. R. 1950. Natural selection for reproductive isolation between *Drosophila pseudoobscura* and *Drosophila persimilis*. *Evolution* 4:135-148.

Kraepelin, E. 1896. *Psychiatrie*, 5th ed. Leipzig: Barth.

Krebs, C. J., M. S. Gaines, B. L. Keller, J. H. Myers, and R. H. Tamarin. 1973. Population cycles in small rodents. *Science* 179:35-41.

Kretchmer, N. 1972. Lactose and lactase. *Sci. Am.* 227:70-78.

Kruijt, J. P. 1964. Ontogeny of social behavior in Burmese red junglefowl (*Gallus gallus spaediceus*). *Behaviour* (Suppl.) 12:1-201.

Kruijt, J. P., and J. A. Hogan. 1967. Social behavior on the lek in black grouse, *Lyrurus tetrix tetrix* (L.) *Ardea* 55:203-240.

Kung, C., and Y. Naitoh. 1973. Calcium-induced ciliary reversal in the extracted models of "pawn," a behavioral mutant of *Paramecium*. *Science* 179:195-196.

Landis, B., and E. S. Tauber (eds.). 1972. *In the Name of Life: Essays in Honor of Erich Fromm*. New York: Holt.

Leader, R. 1967. The kinship of animal and human diseases. *Sci. Am.* 216:110-116.

Leader, R., and I. Leader. 1971. *Dictionary of Comparative Pathology and Experimental Biology.* Philadelphia: Saunders.

Lee, B. T., and P. A. Parsons. 1968. Selection prediction and response. *Biol. Rev.* 43:139-174.

LeFrancois, G. R. 1972. *Psychological Theories and Human Learning: Kongor's Report.* Monterey, Cal.: Brooks/Cole.

Lennox, W. 1959. Epilepsy. In *A Textbook of Medicine.* K. Cecil and R. F. Loeb (eds.). Philadelphia: Saunders.

Leonard, J., L. Ehrman, and A. Pruzan. 1974. Pheromones as a means of genetic control of behavior. *Annu. Rev. Genet.* 8:179-193.

Leonard, J. L., L. Ehrman, and M. Schorsch. 1974. Bioassay of a Drosophila pheromone influencing sexual selection. *Nature* 250:261-262.

Lerner, I. M. 1968. *Heredity, Evolution and Society,* San Francisco: Freeman.

Leroy, Y. 1964. Transmission du paramètre fréquence dans le signal acoustique des hybrides F_1 et P × P_1, de deux Grillons: *Teleogryllus commodus* Walker et *F. oceanicus* Le Guillon (Orthoptères, ensifères). *C. R. Acad. Sci.* 259:892-895.

Leroy, Y. 1966. Signaux acoustiques, comportement, et systèmatique de quelques espèces de Gryllides (Orthopetères, ensifères). *Bull. Biol. Fr. Belg.* 100:63-134.

Levene, H., L. Ehrman, and R. C. Richmond. 1970. Theodosius Dobzhansky Up to Now. In *Essays in Evolution and Genetics in Honor of Theodosius Dobzhansky.* M. K. Hecht and W. Steere (eds.). New York: Appleton, pp. 1-41.

Levine, L. 1958. Studies on sexual selection in mice. *Am. Nat.* 92:21-26.

Levine, L. 1969. *Biology of the Gene.* St. Louis: Mosby.

Levins, R. 1969. Thermal acclimation and heat resistance in *Drosophila* species. *Am. Nat.* 103:483-499.

Levitan, M., and A. Montagu. 1971. *Textbook of Human Genetics.* New York: Oxford Univ. Press.

Lewontin, R. C. 1959. On the anomalous response of *Drosophila pseudoobscura* to light. *Am. Nat.* 93:321-328.

Lewontin, R. C., and L. C. Dunn. 1960. The evolutionary dynamics of a polymorphism in the house mouse. *Genetics* 45:705-722.

L'Héritier, P., and G. Teissier. 1933. Etude d'une population de drosophiles en équilibre. *C. R. Acad. Sci.* 197:1765-1767.

Li, C. 1955. *Population Genetics.* Chicago: Univ. of Chicago Press.

Li, C. 1961. *Human Genetics.* New York: McGraw-Hill.

Lidicker, W. Z., Jr. 1962. Emigration as a possible mechanism permitting the regulation of population density below carrying capacity. *Am. Nat.* 96:29-33.

Lindsley, D., and E. Grell. 1968. *Genetic Variations of Drosophila melanogaster.* Washington, D.C.: Carnegie Institution.

Lindzey, G. 1967. Behavior and Morphological Variation. In *Genetic Diversity and Human Behavior.* J. N. Spuhler (ed.). Chicago: Aldine, pp. 227-240.

Lindzey, G., J. Loehlin, M. Manosevitz, and D. Thiessen. 1971. Behavioral genetics. *Annu. Rev. Psychol.* 22:39-94.

Lindzey, G., and D. Thiessen. 1970. *Contributions to Behavior-Genetic Analysis: The Mouse as a Prototype.* New York: Appleton.

Livingstone, F. B. 1958. Anthropological implications of sickle cell gene distribution in West Africa. *Am. Anthropol.* 60:553-562.

Livingstone, F. B. 1963. Blood groups and ancestry: A test case from the New Guinea highlands. *Curr. Anthropol.* 4:541-542.

Livingstone, F. B. 1967. *Abnormal Hemoglobins in Human Populations.* Chicago: Aldine.

Loehlin, J. C. 1965a. A Heredity-Environment Analysis of Personality Inventory Data. In *Methods and Goals in Human Behavior Genetics.* S. G. Vandenberg (ed.). New York: Academic Press.

Loehlin, J. C. 1965b. Some methodological problems in Cattell's Multiple Abstract Variance Analysis. *Psychol. Rev.* 72:156-161.

Lyman, F. L. 1963. *Phenylketonuria.* Springfield, Ill.: Thomas.

Lynch, C. B., and J. P. Hegmann. 1972. Genetic differences influencing behavioral temperature regulation in small mammals. 1. Nesting by *Mus musculus. Behav. Genet.* 2:43-53.

Lynch, C. B., and J. P. Hegmann. 1973. Genetic differences influencing behavioral temperature regulation in small mammals. 2. Genotype-environment interactions. *Behav. Genet.* 3:145-154.

Lynch, H. T. 1969. *Dynamic Genetic Counseling for Clinicians.* Springfield, Ill.: Thomas.

Lyon, M. F. 1962. Sex chromatin and gene action in the mammalian X-chromosome. *Am. J. Hum. Genet.* 14:135-148.

MacBean, I. T., and P. A. Parsons. 1967. Directional selection for duration of copulation in *Drosophila melanogaster. Genetics* 56:233-239.

MacMahon, B. 1968. Gene-Environment Interaction in Human Disease. In *The Transmission of Schizophrenia.* Rosenthal & Kety (eds.). New York: Pergamon Press, pp. 393-402.

Malogolowkin-Cohen, C., A. S. Simmons, and H. Levene. 1965. A study of sexual isolation between certain strains of *Drosophila paulistorum. Evolution* 19:95-103.

Manning, A. 1959. The sexual behavior of two sibling Drosophila species. *Behaviour* 15:123-145.

Manning, A. 1961. The effects of artificial selection for mating speed in *Drosophila melanogaster. Anim. Behav.* 9:82-92.

Manning, A. 1962. A sperm factor affecting the receptivity of *Drosophila melanogaster. Nature* (Lond.) 194:252-253.

Manning, A. 1963. Selection for mating speed in *Drosophila melanogaster* based on the behavior of one sex. *Anim. Behav.* 11:116-120.

Manning, A. 1965. Drosophila and the Evolution of Behavior. In *Viewpoints in Biology.* J. Carthy and C. Duddington (eds.). London: Butterworth, pp. 125-169.

Manning, A. 1966. Corpus allatum and sexual receptivity in female *Drosophila melanogaster. Nature* (Lond.) 211:1321-1322.

Manning, A. 1967. The control of sexual receptivity in female Drosophila. *Anim. Behav.* 15:239-250.

Manning, A. 1968. The effects of artificial selection for slow mating in *D. simulans. Anim. Behav.* 16:108-113.

Mather, K. 1942. The balance of polygenic combinations. *J. Genet.* 43:309-336.

Mather, K. 1949. *Biometrical Genetics.* London: Methuen.

Mather, K. 1966. Variability and selection. *Proc. R. Soc. Lond. (Biol.) 164*:328–340.

Mather, K., and B. J. Harrison. 1949. The manifold effect of selection. *Heredity* 3:1–52, 131–162.

Mather, K., and J. L. Jinks. 1971. *Biometrical Genetics,* 2nd ed. London: Chapman & Hall.

Maxwell, J. 1969. Intelligence, education and fertility. A comparison between the 1932 and 1947 Scottish surveys. *J. Biosoc. Sci. 1*:217–247.

Maynard-Smith, S., L. Penrose, and C. Smith. 1961. *Mathematical Tables for Research Workers in Human Genetics.* London: Churchill.

Mayr, E. 1942. *Systematics and the Origin of Species.* New York: Columbia Univ. Press.

Mayr, E. 1963. *Animal Species and Evolution.* Cambridge, Mass.: Harvard Univ. Press.

Mayr, E. 1970. *Populations, Species, and Evolution.* Cambridge, Mass.: Harvard Univ. Press.

Mayr, E., and T. Dobzhansky. 1945. Experiments on sexual isolation in *Drosophila.* 4. Modification of the degree of isolation between *Drosophila pseudoobscura* and *Drosophila persimilis* and of sexual preferences in *Drosophila prosaltans. Proc. Natl. Acad. Sci. U.S.A. 31*:75–82.

McClearn, G. E. 1972. Genetic Determination of Behavior (Animal). In *Genetics, Environment, and Behavior: Implications for Educational Policy.* L. Ehrman, G. S. Omenn, and E. Caspari (eds.). New York: Academic Press, pp. 55–67.

McClure, H., K. Belden, W. Pieper. 1969. Autosomal trisomy in a chimpanzee: Resemblance to Down's syndrome. *Science 165*:1010–1011.

McCracken, R. 1971. Lactase deficiency: An example of dietary evolution. *Curr. Anthropol. 12*:479–517.

McDonald, J., and P. A. Parsons. 1973. Dispersal activities of the sibling species *Drosophila melanogaster* and *Drosophila simulans. Behav. Genet. 3*:293–301.

McGaugh, J. L. 1971. *Psychobiology.* New York: Academic Press.

McGaugh, J. L. (ed.). 1972. *The Chemistry of Mood-Motivation and Memory.* New York: Plenum Press.

McGill, T. E. 1970. Genetic Analysis of Male Sexual Behavior. In *Contributions to Behavior-Genetic Analysis: The Mouse as a Prototype.* G. Lindzey and D. D. Thiessen (eds.). New York: Appleton, pp. 57–88.

McKenzie, J. A. 1974. The distribution of vineyard populations of *Drosophila melanogaster* and *Drosophila simulans* during vintage and non-vintage periods. *Oecologia 15*:1–16.

McKenzie, J. A., and P. A. Parsons. 1971. Variations in mating propensities in strains of *Drosophila melanogaster* with different scutellar chaeta numbers. *Heredity* (Lond.) 26:313–322.

McKenzie, J. A., and P. A. Parsons. 1972. Alcohol tolerance: An ecological parameter in the relative success of *Drosophila melanogaster* and *Drosophila simulans. Oecologia 10*:373–388.

McKusick, V. 1971. *Mendelian Inheritance in Man.* Baltimore: Johns Hopkins Univ. Press.

McLaren, A., and D. Michie. 1956. Studies on the transfer of fertilized mouse eggs to uterine foster mothers. 1. Factors affecting the implantation and survival of native and transferred eggs. *J. Exp. Biol.* 33:394–416.

McLaren, A., and D. Michie. 1959. Studies on the transfer of fertilized mouse eggs to uterine foster mothers. 2. The effect of transferring large numbers of eggs. *J. Exp. Biol.* 36:40–50.

McNeil, E. B. 1970. *The Psychoses.* Englewood Cliffs, N.J.: Prentice-Hall.

Médioni, J. 1962. Contribution à l'étude psycho-physiologique et génétique du phototrophisme d'un insecte *Drosophila melanogaster* Meigen. *Bull. Psychol. (Paris)* 16(2):8

Mednick, S. A., E. Mura, F. Schulsinger, and B. Mednick. 1971. Perinatal conditions and infant development in children with schizophrenic parents. *Soc. Biol.* 18:103–113.

Melnyk, J., F. Vanasek, H. Thompson, and A. Rucci. 1969. Failure of transmission of supernumerary Y chromosomes in man. *Abstr. Am. Soc. Hum. Genet.*, San Francisco. (Oct. 1–4):42.

Mendel, G. 1865. Versuche über pflanzen-hybriden. *Verh. Naturf. Verein Brünn* 4:3–47. Translation reprinted by Harvard Univ. Press.

Mendlewicz, S., J. Fleiss, and R. Fieve. 1972. Evidence for X-linkage in the transmission of manic-depressive illness. *J. Am. Med. Assoc.* 222:1624–1627.

Merrell, D. J. 1953. Selective mating as a cause of gene frequency changes in laboratory populations of *Drosophila melanogaster. Evolution* 7:287–296.

Mesibov, R., and J. Adler. 1972. Chemotaxis toward amino acids in *Escherichia coli. J. Bacteriol.* 112:315–326.

Metrakos, J. D., and K. M. Metrakos. 1972. Genetic Factors in the Epilepsies. In *The Epidemiology of Epilepsy: A Workshop.* M. Alter and W. Hauser (eds.). Pub. No. (NIH) 73-390. Washington, D.C.: U.S. Dept. of Health, Education, and Welfare, pp. 97–102.

Mitchell, T. B. 1929. Sex anomalies in the genus *Megachile* with descriptions of new species (Hymenoptera: Megachilidae). *Trans. Am. Entomol. Soc.* 54:333.

Mittler, P. 1971. *The Study of Twins.* Baltimore: Penguin.

Money, J. 1970. Behavior genetics: Principles, methods and examples for XO, XXY, and XYY syndromes. *Sem. Psychiatry* 2:11–29.

Money, J., and S. Mittenthal. 1970. Lack of personality pathology in Turner's syndrome: Relation to cytogenetics, hormones, and physique. *Behav. Genet.* 1:43–56.

Moor, L. 1967. Niveau intellectuel et polygonosomie: Confrontation du caryotype et du niveau mental de 374 malades dont le caryotype comporte un exces de chromosomes X ou Y. *Rev. Neurophsychiatr. Infant.* 15:325–348.

Moore, J. A. 1952. Competition between *Drosophila melanogaster* and *Drosophila simulans.* 1. Population cage experiments. *Evolution* 6:407–420.

Morgan, T. H., and C. B. Bridges. 1919. Contributions to the Genetics of *Drosophila melanogaster.* In *The Origin of Gynandromorphs.* Carnegie Institution Pub. No. 278. Washington, D.C.: Carnegie Institution, pp. 1–122.

Morris, J. M. 1953. Testicular feminization. *Am. J. Obstet. Gynecol.* 65:1192–1211.

Morton, N. E. 1972. Human Behavioral Genetics. In *Genetics, Environment, and Behavior: Implications for Educational Policy.* L. Ehrman, G. S. Omenn, and E. Caspari (eds.). New York: Academic Press.

Morton, N. E., C. S. Chung, and M. P. Mi. 1967. *Genetics of Interracial Crosses in Hawaii*. New York: Karger.

Mourant, A. 1954. *The Distribution of the Human Blood Groups*. Springfield, Ill.: Thomas.

Muller, H. 1942. Isolating mechanisms, evolution, and temperature. *Biol. Symp.* 6:71–125.

Myers, J. H., and C. J. Krebs. 1971. Genetic, behavioral and reproductive attributes of dispersing field voles *Microtus pennsylvanicus* and *Microtus ochrogaster*. *Ecol. Monogr.* 41:53–78.

Myrianthropoulos, N. 1972. Maternal Factors Inflencing Risk of Epilepsy. In *The Epidemiology of Epilepsy: A Workship*. M. Alter and W. Hauser (eds.). Pub. No. (NIH) 73–390. Washington, D.C.: U.S. Dept. of Health, Education and Welfare, pp. 103–107.

Nachman, M., C. Larue, and J. LeMagnen. 1971. The role of olfactory and orosensory factors in the alchohol preference of inbred strains of mice. *Physiol. Behav.* 6:53–59.

Nagle, J. 1974. *Heredity and Human Affairs*. St. Louis: The C. V. Mosby Co.

Neel, J., S. Fajans, J. Conn, and R. Davison. 1965. Diabetes Mellitus. In *Genetics and Epidemiology of Chronic Diseases*. Pub. No. 1163. Washington D.C.: U.S. Dept. of Health, Education, and Welfare, pp. 105–132.

Neel, J. V., and R. H. Post. 1963. Transitory "positive" selection for color blindness. *Eugen. Q.* 10:33–35.

Neel, J. V., and W. J. Schull. 1968. On some trends in understanding the genetics of men. *Perspect. Biol. Med.* 11:565–602.

Nei, M., K. Kojima, and H. Schaffer. 1967. Frequency changes of new iversions in populations under mutation-selection equilibria. *Genetics* 57:741–750.

Newman, H. H., F. N. Freeman, and K. J. Holzinger. 1937. *Twins: A Study of Heredity and Environment*. Chicago: University of Chicago Press.

Nicholls, J. R., and S. Hsiao. 1967. Addiction liability of albino rats: Breeding for quantitative differences in morphine drinking. *Science* 157:561–563.

Nielson, J., and F. Henriksen. 1972. The incidence of chromosome aberrations among males in a Danish youth prison. *Acta Psychiatr. Scand.* 48:87–102.

Ödegaard, Ö. 1963. The psychiatric disease entitles in the light of genetic investigation. *Acta Psychiatr. Scand.* 39:s169–194.

Ogilvie, D. M., and R. H. Stinson. 1966. Temperature selection in *Peromyscus* and laboratory mice, *Mus musculus*. *J. Mammal.* 47:655–660.

Olsen, H. H., and W. E. Peterson. 1951. Uniformity of semen production and behavior in monozygous triplet bulls. *J. Dairy Sci.* 34:489–490.

Olsen, H. H., and W. E. Peterson. 1952. Uniformity and Nutritional Studies with Monozygotic Bulls. Presented at 47th annual meeting, American Dairy Science Assoc., Davis, Cal.

O'Meara, G. 1972. Polygenic regulation of fecundity in autogenous *Aedes atropalpus*. *Entomol. Exp. Appl.* 15:81–89.

Omenn, G., E. Caspari, and L. Ehrman. 1972. Epilogue: Behavior Genetics and Educational Policy. In *Genetics, Environment, and Behavior: Implications for*

Educational Policy. L. Ehrman, G. Omenn, and E. Caspari (eds.). New York: Academic Press, pp. 307-310.

Omenn, G. S., and A. G. Motulsky. 1972. Biochemical Genetics and the Evolution of Human Behavior. In *Genetics, Environment, and Behavior: Implications for Educational Policy*. L. Ehrman, G. S. Omenn, and E. Caspari (eds.). New York: Academic Press, pp. 129-171.

Osborne, R. H. (ed.). 1971. *The Biological and Social Meaning of Race*. San Francisco: Freeman.

Parnell, R. W. 1958. *Behavior and Physique: An Introduction to Practical and Applied Somatometry*. London: Arnold.

Parsons, P. A. 1964. A diallel cross for mating speeds in *Drosophila melanogaster*. *Genetica* 35:141-151.

Parsons, P. A. 1965. Assortative mating for a metrical characteristic in *Drosophila*. *Heredity* 20:161-167.

Parsons, P. A. 1967a. Behaviour and random mating. *Experientia* 23:161-167.

Parsons, P. A. 1967b. *The Genetic Analysis of Behaviour*. London: Methuen.

Parsons, P. A. 1971. Extreme-environment heterosis and genetic loads. *Heredity* 26:579-583.

Parsons, P. A. 1972a. Genetic Determination of Behavior (Mice and Men). In *Genetics, Environment, and Behavior: Implications for Educational Policy*. L. Ehrman, G. S. Omenn, and E. Caspari (eds.). New York: Academic Press, pp. 75-98.

Parsons, P. A. 1972b. Variations between strains of *Drosophila melanogaster* and *D. simulans* in giving offspring in interspecific crosses. *Can. J. Genet. Cytol.* 14:77-80

Parsons, P. A. 1973. *Behavioural and Ecological Genetics: A Study in Drosophila*. Oxford: Clarendon Press.

Parsons, P. A. 1974. The behavioral phenotype in mice. *Am. Nat.* 108:377-385.

Parsons, P. A. 1975. The comparative evolutionary biology of the sibling species *Drosophila melanogaster* and *D. simulans*. *Q. Rev. Biol.* 50:151-169.

Parsons, P. A., and M. Green. 1959. Pleiotropy and competition at the vermilion locus in *Drosophila melanogaster*. *Proc. Natl. Acad. Sci. U.S.A.* 45:993-996.

Parsons, P. A., S. M. Hosgood and B. T. Lee. 1967. Polygenes and polymorphism. *Mol. Gen. Genet.* 99:165-170.

Parsons, P. A., and D. Kaul. 1966. Mating speed and duration of copulation in *Drosophila pseudoobscura*. *Heredity* 21:219-225.

Patterson, D., and J. Slack. 1972. Lipid abnormalities in male and female survivors of myocardial infarction and their first degree relations. *Lancet* 1:392-399.

Patterson, J. 1942. Isolating mechanisms in the genus *Drosophila*. *Biol. Symp.* 6:271-287.

Payne, W. J. A., and J. Hancock. 1957. The direct effect of tropical climate on the performance of European-type cattle. *Emp. J. Exp. Agric.* 25:321-338.

Pearson, K., and A. Lee. 1903. On the laws of inheritance in man. 1. Inheritance of physical characters. *Biometrika* 2:357-462.

Penrose, L. S. 1954. Distance, size and shape. *Ann. of Eugen.* 18:337-343.

Penrose, L. S. 1961. Genetics of Growth and Development of the Fetus. In *Recent Advances in Human Genetics*. L. S. Penrose and H. L. Brown (eds.). London: Churchill, pp. 56–75.

Penrose. L. S. 1963. *The Biology of Mental Defect*, 3rd ed. London: Sidgwick & Jackson.

Pérez-Miravete, A. (ed.). 1973. *Behavior of Micro-organisms*. New York: Plenum Press.

Perris, L. 1971. Abnormality on paternal and maternal sides: Observations in bipolar (manic-depressive) and unipolar depressive psychoses. *Br. J. Psychiatry 118*:207–210.

Perttunen, V. 1963. Effect of desiccation on the light reactions of some terrestrial arthropods. *Ergeb. Biol. 26*:90–97.

Petit, C. 1958. Le determinisme génétique et psycho-psysiologique de la competition sexuelle chez *Drosophila melanogaster*. *Bull. Biol. Fr. Belg. 92*:1–329.

Petit, C., and L. Ehrman. 1969. Sexual selection in Drosophila. *Evol. Biol. 3*:177–223.

Petras, M. L. 1967. Studies of natural populations of *Mus*. 1. Biochemical polymorphisms and their bearing on breeding structure. *Evolution 21*:259–274.

Pettigrew, T. F. 1971. Race, Mental Illness and Intelligence: A Social Psychological View. In *The Biological and Social Meaning of Race*. R. H. Osborne (ed.). San Francisco: Freeman, pp. 87–124.

Pickford, R. W. 1972. Colour-defective art students in four art schools. *Br. J. Physiol. Opt. 27*:102–114.

Pittendrigh, C. S. 1958. Adaptation, Natural Selection, and Behavior. In *Behavior and Evolution*. A. Roe and G. Simpson (eds.). New Haven: Yale Univ. Press, pp. 390–416.

Pollitzer, W. S. 1972. Discussion. In *Genetics, Environment, and Behavior: Implications for Educational Policy*. L. Ehrman, G. S. Omenn, and E. Caspari (eds.). New York: Academic Press.

Porter, I. H. 1968. *Heredity and Disease*. New York: McGraw-Hill.

Porter, R. B., and R. B. Cattell. 1968. *The Children's Personality Questionnaire*. Champaign, Ill.: The University of Illinois Press.

Potegal, M. 1971. A note on spatial-motor deficits in patients with Huntington's disease: A test of a hypothesis. *Neuropsychologia 9*:233–235.

Pratt, R. T. C. 1967. *The Genetics of Neurological Disorders*. London: Oxford Univ. Press.

Price, W., and P. Whatmore. 1967. Criminal Behavior and the XYY male. *Nature 213*:815–816.

Prout, T. 1971a. The relation between fitness components and population prediction in Drosophila. 1. The estimation of fitness components. *Genetics 68*:127–149.

Prout, T. 1971b. The relation between fitness components and population prediction in Drosophila. 2. Population prediction. *Genetics 68*:151–167.

Pruzan, A. 1975. Effect of age, rearing and mating experiences on frequency dependent sexual selection in *Drosophila pseudoobscura*. *Evolution 29* (in press).

Pruzan, A., and L. Ehrman. 1974. Age, experience, and rare-male mating advantages in Drosophila pseudoobscura. Behav. Genet. 4:159–164.

Quinn, W. G., W. A. Harris, and S. Benzer. 1974. Conditioned behavior in Drosophila melanogaster. Proc. Natl. Acad. Sci. U.S.A. 71:708–712.

Race, R., and R. Sanger. 1968. Blood Groups in Man. Philadelphia: Davis.

Reed, E. W., and S. C. Reed. 1965. Mental Retardation: A Family Study. Philadelphia: Saunders.

Reed, S. C. 1964. Parenthood and Heredity, 2nd ed. New York: Science Editions.

Regen, J. 1914. Über die Anlockung des Weibchens von Gryllus campestris L. durch telephonisch übertrangene Stridulationslaute des Männchens. Pflüger's Arch. 155:193–200.

Reimer, J. D., and M. L. Petras. 1967. Breeding structure of the house mouse, Mus musculus, in a population cage. J. Mammal. 48:88–89.

Reiss, A. J. 1961. Occupations and Social Studies. New York: Free Press.

Richmond, R. C. 1969. Heritability of phototactic and geotactic responses in Drosophila pseudoobscura. Am. Nat. 103:315–322.

Riekhof, P. L., W. A. Horton, D. J. Harris, and R. N. Schimke. 1972. Monozygotic twins with the Turner syndrome. J. Obstet. Gynecol. 112:59–61.

Riesen, A. H., and E. F. Kinder. 1952. Postural Development of Infant Chimpanzees. New Haven: Yale Univ. Press. pp. 153–155.

Rife, D. C. 1938. Genetic studies of monzygotic twins. J. Hered. 29:83–90.

Rimoin, D. L., and R. N. Schimke, 1971. Genetic Disorders of the Endocrine Glands. St. Louis: Mosby.

Roberts, J. A. F. 1952. The genetics of metal deficiency. Eugen. Rev. 44:71–83.

Rockwell, R. F., F. Cooke, and R. Harmsen. 1975. Photobehavioral differentiation in natural population of D. pseudoobscura and D. persimilis. Behav. Genet. 5:189–202.

Rockwell, R. F., and M. B. Seiger. 1973a. Phototaxis is Drosophila: A critical evaluation. Am. Sci. 61:339–345.

Rockwell, R. F., and M. B. Seiger. 1973b. A comparative study of photoresponse in Drosophila pseudoobscura and Drosophila persimilis. Behav. Genet. 3:163–174.

Roderick, G. W. 1968. Man and Heredity. New York: Macmillan.

Rodgers, D. A., and G. E. McClearn. 1962. Mouse strain differences in preference for various concentrations of alcohol. Q. J. Stud. Alcohol 23:26–33.

Rogoff, W. M., G. H. Gretz, M. Jacobson, and M. Beroza. 1973. Confirmation of (Z)-9-tricosene as a sex pheromone of the house fly. Ann. Entomol. Soc. Am. 66:739–741.

Rose, A., and P. A. Parsons. 1970. Behavioural studies in different strains of mice and the problems of heterosis. Genetica 41:65–87.

Rosenthal, D. (ed.). 1963. The Genain Quadruplets. New York: Basic Books.

Rosenthal, D. 1970. Genetic Theory and Abnormal Behavior. New York: McGraw-Hill.

Rosenthal, D. 1971. Genetics of Psychopathology. New York: McGraw-Hill, p. 106.

Rothenbuhler, N. 1964. Behavior genetics of nest cleaning in honey bees. 4. Responses of F_1 and backcross generations to disease-killed brood. *Am. Zool.* 4:111–123.

Rundquist, E. A. 1933. Inheritance of spontaneous activity in rats. *J. Comp. Psychol.* 16:415–438.

Rusk, R. R. 1940. *39th Yearbook of the National Society for the Study of Education,* Part II. Bloomington, Ill.: Public School, p. 269.

Russell, E. 1972. Genetic Considerations in the Selection of Rodent Species and Strains for Research in Aging. In *Development of the Rodent as a Model System of Aging.* D. Gibson (ed.). Pub. No. (NIH)72-121. Washington, D.C.: U.S. Department of Health, Education and Welfare, pp. 33–53.

Russell, L. 1961. Genetics of mammalian sex chromosomes. *Science* 133:1795–1803.

Sank, D. 1963. Genetic Aspects of Early Total Deafness. In *Family and Mental Health Problems in a Deaf Population.* J. D. Rainer, K. Z. Altschuler, and F. J. Kallmann (eds.). New York: Columbia Univ. Press.

Satinder, K. P. 1971. Genotype-dependent effects of D-amphetamine sulphate and caffeine on escape-avoidance behavior of rats. *J. Comp. Physiol. Psychol.* 76:359–364.

Satow, Y., and C. Kung. 1974. Genetics dissection of the active electrogenesis in *Paramecium aurelia. Nature* 247:69–71.

Saunders, P. R. 1959. On the dimensionality of the WAIS battery for two groups of normal males. *Psychol. Rep.* 5:529–541.

Savage, T. F., and W. M. Collins. 1972. Inheritance of star-gazing in Japanese quail. *J. Hered.* 63:88.

Scarr, S. 1970. Effects of birth weight on later intelligence. *Soc. Biol.* 16:249–256.

Schaffer, J. 1962. A specific cognition deficit observed in gonadal aplasia (Turner's syndrome). *J. Clin. Psychol.* 18:403.

Schàrloo, W. 1971. Reproductive isolation by disruptive selection: Did it occur? *Am. Nat.* 105:83–96.

Schlesinger, K., R. C. Elston, and W. Boggan. 1966. The genetics of sound-induced seizure in inbred mice. *Genetics* 54:95–103.

Schlesinger, K., and B. J. Griek. 1970. The Genetics and Biochemistry of Audiogenic Seizures. In *Contributions to Behavior-Genetic Analysis: The Mouse as a Prototype.* G. Lindzey and D. D. Thiessen (eds.). New York: Appleton, pp. 219–257.

Schroder, J., P. Halkka, and M. Brumer-Korvenkontia. 1970. Meiotic aneuploidy in a mouse strain with latent ectromelia infection. *Hereditas* 65:297–300.

Scott, J. P. 1943. Effects of single genes on the behavior of Drosophila. *Am. Nat.* 77:184–190.

Scott, J. P. 1964. Genetics and the development of social behavior in dogs. *Am. Zool.* 4:161–168.

Scott, J. P., and J. L. Fuller. 1965. *Genetics and the Social Behavior of the Dog.* Chicago: Univ. of Chicago Press, p. 468.

Searles, H. F. 1971. Pathologic Symbiosis and Autism. In *In the Name of Life: Essays in Honor of Erich Fromm.* B. Landis and E. S. Tauber (eds.). New York: Holt, pp. 69–83.

374 THE GENETICS OF BEHAVIOR

Selander, R. K. 1970. Behavior and genetic variation in natural populations. *Am. Zool.* 10:53–66.

Selander, R. K. 1972. Sexual Selection and Dimorphism in Birds. In *Sexual Selection and the Descent of Man.* B. Campbell (ed.). Chicago: Aldine.

Selander, R. K., and D. W. Kaufman. 1973. Genic variability and strategies of adaptation in animals. *Proc. Natl. Acad. Sci. U.S.A.* 70:1875–1877.

Sharpe, R. S., and P. A. Johnsgard. 1966. Inheritance of behavioral characters in F$_2$ mallard × pintail (*Anas platyrhynchos* L. × *Anas acuta* L.) hybrids. *Behaviour* 27:259–272.

Sheldon, W. H., S. S. Stevens, and W. B. Tucker. 1940. *The Varieties of Human Physique: An Introduction to Constitutional Psychology.* New York: Harper & Row.

Sheldon, W. H., and S. S. Stevens. 1942. *The Varieties of Temperament: A Psychology of Constitutional Differences.* New York: Harper & Row.

Shepard, T. H. 1961. Increased incidence of nontasters of phenylthiocarbamine among congenital athyrotic cretins. *Science* 131:929.

Shields, J. 1962. *Monozygotic Twins Brought up Apart and Brought up Together.* London: Oxford Univ. Press.

Shorey, H., and R. J. Bartell. 1970. Role of a volatile female sex pheromone in stimulating male courtship behavior in *Drosophila melanogaster. Anim. Behav.* 18:159–164.

Sidman, R. L., S. H. Appel, and J. L. Fuller. 1965. Neurological mutants of the mouse. *Science* 150:513–516.

Sidman, R. L. and M. C. Green. 1965. Retinal degeneration in the mouse: Location of the *rd* locus in linkage group XVII. *J. Hered.* 56:23–29.

Siegel, I. M. 1967. Heritability and threshold determinations of the optomotor response in *Drosophila melanogaster. Anim. Behav.* 15:299–306.

Siegel, P. 1972. Genetic analysis of male mating behavior in chickens. 1. Artificial selection. *Anim. Behav.* 20:564–570.

Silcock, M., and P. A. Parsons. 1973. Temperature preference differences between strains of *Mus musculus,* associated variables, and ecological implications. *Oecologia* 12:147–160.

Simoons, F. 1970. Primary adult lactose intolerance and the milking habit: A problem in biologic and cultural interrelations. 2. A culture historical hypothesis. *Am. J. Digest. Dis.* 15:695–710.

Sinnock, P. 1969. Non-random mating in *Tribolium castaneum. Genetics* (Suppl.) 61:55.

Sinnock, P. 1970. Frequency dependence and mating behavior in *Tribolium castaneum. Am. Nat.* 104:469–476.

Skinner, B. F. 1971. *Beyond Freedom and Dignity.* New York: Knopf.

Skodak, M., and H. M. Skeels. 1949. A final follow-up study of one hundred adopted children. *J. Genet. Psychol.* 75:85–125.

Slater, E., J. Maxwell, and J. S. Price. 1971. Distribution of ancestral secondary cases in bipolar affective disorders. *Br. J. Psychiatry* 118:215–218.

Smith, C. 1970. Heritability of liability and concordance in monozygous twins. *Ann. Hum. Genet.* 34:85–91.

Snezhnevsky, A. V., and M. Vartanyan. 1971. The Forms of Schizophrenia and Their Biological Correlates. In *Biochemistry, Schizophrenia and Affective Illnesses*. H. E. Hinwich (ed.). Baltimore: Williams & Wilkins.

Snyder, L. H., and P. R. David. 1957. *The Principles of Heredity*, 5th ed. Boston: Heath.

Snyder, L., and D. F. Davidson. 1937. Studies of human inheritance. 18. The inheritance of taste deficiency to diphenylguanidine. *Eugen. News* 22:1-2.

Sonneborn, T. M. 1970. Methods in *Paramecium* research. In *Methods of Cell Physiology*. D. M. Prescott (ed.). New York: Academic Press, Vol. 4, pp. 241-339.

Southern, H. N., and E. M. O. Laurie. 1946. The house-mouse *(Mus musculus)* in corn ricks. *J. Anim. Ecol.* 15:134-149.

Spassky, B., and T. Dobzhansky, 1967. Responses of various strains of *Drosophila pseudoobscura* and *Drosophila persimilis* to light and gravity. *Am. Nat.* 101:59-63.

Sperlich, D. 1966. Equilibria for inversions induced by X-ray in isogenic strains of *Drosophila pseudoobscura*. *Genetics* 53:835-842.

Spiess, E. 1962. *Papers on Animal Population Genetics*. Boston: Little, Brown.

Spiess, E. B. 1968. Low frequency advantage in mating of *Drosophila pseudoobscura* karyotype. *Am. Nat.* 102:363-379.

Spiess, E., and B. Langer. 1961. Chromosomal adaptive polymorphism in *Drosophila persimilis*. 3. Mating propensity of homokaryotypes. *Evolution* 15:535-544.

Spiess, E., and B. Langer. 1964a. Mating speed control by gene arrangements in *Drosophila pseudoobscura* homokaryotypes. *Proc. Natl. Acad. Sci. U.S.A.* 51:1015-1018.

Spiess, E. and B. Langer. 1964b. Mating speed control by gene arrangements in *Drosophila persimilis*. *Evolution* 18:430-444.

Spiess, E., B. Langer, and L. Spiess. 1966. Mating control by gene arrangements in *Drosophila pseudoobscura*. *Genetics* 54:1139-1149.

Spiess, E., and L. Spiess. 1967. Mating propensity, chromosomal polymorphism, and dependent conditions in *Drosophila persimilis*. *Evolution* 21:672-678.

Spiess, L. D., and E. B. Spiess. 1969. Minority advantage in interpopulational matings of *Drosophila persimilis*. *Am. Nat.* 103:155-172.

Spieth, H. 1952. Mating behavior within the genus *Drosophila* (Diptera). *Bull. Am. Museum Nat. Hist.* 99:105-145.

Spieth, H. T. 1958. Behavior and Isolating Mechanisms. In *Behavior and Evolution*. A. Roe and G. G. Simpson (eds.). New Haven: Yale Univ. Press, pp. 363-389.

Sprott, R. L., and J. Staats. 1975. Behavioral studies using genetically defined mice: A bibliography. *Behav. Genet.* 5:27-82.

Spuhler, J. N. 1962. *Empirical Studies on Quantitative Human Genetics*. WHO seminar on the Use of Vital and Health Statistics for Genetic and Radiation Studies, pp. 241-252.

Spuhler, J. N. 1968. Assortative mating with respect to physical characteristics. *Eugen. Q.* 15:128-140.

Spuhler, J. N., and G. Lindzey. 1967. Racial Differences in Behavior. In *Behavior-Genetic Analysis.* J. Hirsch (ed.). New York: McGraw-Hill, pp. 366–414.

Staats, J. 1966. The Laboratory Mouse. In *Biology of the Laboratory Mouse.* E. L. Green (ed.). New York: McGraw-Hill, pp. 1–9.

Stalker, H. D. 1942. Sexual isolation studies in the species complex *Drosophila virilis. Genetics* 27:238–257.

Stebbins, G. L. 1950. *Variation and Evolution in Plants.* New York: Columbia Univ. Press.

Stein, L., C. D. Wise, and B. D. Berger. 1972. Noradrenergic Reward Mechanisms, Recovery of Function, and Schizophrenia. In *The Chemistry of Mood, Motivation, and Memory.* J. L. McGaugh (ed.). New York: Plenum Press.

Stern, C. 1960. *Principles of Human Genetics.* San Francisco: Freeman.

Stern, C. 1973. *Principles of Human Genetics,* 3rd ed. San Francisco: Freeman.

Stevenson, A. C., B. C. C. Davidson, and M. W. Oakes. 1970. *Genetic Counseling.* Philadelphia: Lippincott.

Strickberger, M. W. 1968. *Genetics.* New York: Macmillan.

Sturtevant, A. H. 1915. Experiments on sex recognition and the problem of sexual selection in Drosophila. *J. Anim. Behav.* 5:351–366.

Sturtevant, A. H. 1929. *The Genetics of Drosophila simulans.* Carnegie Institution Pub. No. 399. Washington, D.C.: Carnegie Institution.

Sved, J., T. Reed, and W. Bodmer. 1967. The number of balanced polymorphisms that can be maintained in a natural population. *Genetics* 55:469–481.

Szebenyi, A. 1969. Cleaning behavior in *Drosophila melanogaster. Anim. Behav.* 17:641–651.

Tan, C. C. 1946. Genetics of sexual isolation between *Drosophila pseudoobscura* and *Drosophila persimilis. Genetics* 31:558–573.

Taylor, B. 1972. Genetic relationships between inbred strains of mice. *J. Hered.* 63:83–86.

Taylor, W. O. 1971. Effects on employment of defects in colour vision. *Br. J. Ophthalmol.* 55:753–760.

Thiessen, D. D., K. Owen, and M. Whitsett. 1970. Chromosome Mapping of Behavioral Activities. In *Contributions to Behavior Genetic Analysis: The Mouse as a Prototype.* G. Lindzey and D. D. Thiessen (eds.). New York: Appleton, pp. 161–204

Thoday, J. M. 1961. Location of polygenes. *Nature* 191:368–370.

Thoday, J. M., and J. B. Gibson. 1970. Environmental and genetical contributions to class difference: A model experiment. *Science* 167:990–992.

Thompson, J., and M. Thompson. 1973. *Genetics in Medicine,* 2nd ed. Philadelphia: Saunders.

Thompson, W. R. 1953. The inheritance of behaviour: Behavioural differences in fifteen mouse strains. *Can. J. Psychol.* 7:145–155.

Thompson, W. R. 1956. The inheritance of behavior: Activity differences in five inbred mouse strains. *J. Hered.* 47:147–148.

Thuline, H., and D. Norby. 1961. Spontaneous occurrence of chromosomal abnormality in cats. *Science* 134:554–555.

Thurstone, L. L., and T. G. Thurstone. 1941. *The Primary Mental Abilities Tests.* Chicago: Science Research.

Thurstone, T. G., L. L. Thurstone, and H. H. Strandskov. 1955. *A Psychological Study of Twins.* Psychometric Laboratory Report No. 12. Chapel Hill, N.C.: Univ. of North Carolina Press.

Tindale, N. B. 1953. Tribal and intertribal marriage among the Australian aborigines. *Hum. Biol. 25*:169–190.

Tobach, E. 1972. The Meaning of Cryptanthroparian. In *Genetics, Environment, and Behavior: Implications for Educational Policy.* L. Ehrman, G. Omenn, and E. Caspari (eds.). New York: Academic Press.

Tolman, E. C. 1924. The inheritance of maze-learning ability in rats. *J. Comp. Psychol. 4*:1–18.

Tryon, R. C. 1942. Individual differences. In *Comparative Psychology,* 2nd ed. F. A. Moss (ed.). Englewood Cliffs, N.J.: Prentice-Hall.

Tshudy, D. P. 1973. Acute Intermittent Porphyria. In *Birth Defects Atlas and Compendium.* D. Bergsma (ed.). Baltimore: Williams & Wilkins, pp. 146–147.

Turpin, R., and J. Lejeune. 1969. *Human Afflictions and Chromosomal Aberrations.* Elmsford, N.Y.: Pergamon Press.

Vale, J. R. 1973. Role of behavior genetics in psychology. *Am. Psychol. 28*:871–882.

Vale, J. R., C. A. Vale, and J. P. Harley, 1971. Interaction of genotype and population number with regard to aggressive behavior, social grooming, and adrenal and gonadal weight in male mice. *Commun. Behav. Biol. 6*:209–221.

Valenstein, E. S., W. Riss, and W. C. Young. 1955. Experiential and genetic factors in the organization of sexual behavior in male guinea pigs. *J. Comp. Physiol. Psychol. 48*:397–403.

Vandenberg, S. 1962. Hereditary abilities study: Hereditary components in a psychological test battery. *Am. J. Human Genet. 14*:220–237.

Vandenberg, S. 1964. The value of twin studies: A review of methods and new ideas. *Am. J. Physical Anthropol. 22*:355–358.

Vandenberg, S. 1967. Hereditary Factors in Psychological Variables in Man, with Special Emphasis on Cognition. In *Genetic Diversity and Human Behavior.* J. N. Spuhler (ed.). Chicago: Aldine.

Vandenberg, S. 1968. *Progress in Human Behavior Genetics.* Baltimore: Johns Hopkins Univ. Press.

Vandenberg, S. 1971. What Do We Know Today about the Inheritance of Intelligence and How Do We Know It? In *Intelligence: Genetic and Environmental Influences.* R. Cancro (ed.). New York: Grune & Stratton.

Vandenberg, S. 1972. The Future of Human Behavior Genetics. In *Genetics, Environment, and Behavior: Implications for Educational Policy.* L. Ehrman, G. Omenn, and E. Caspari (eds.). New York: Academic Press.

Van Riper, C. 1971. *The Nature of Stuttering.* Englewood Cliffs, N.J.: Prentice-Hall.

Van Valen, L., L. Levine, and J. A. Beardmore. 1962. Temperature sensitivity of chromosomal polymorphism in *Drosophila pseudoobscura*. *Genetica* 33:113-127.

Vessie, P. R. 1932. On the transmission of Huntington's chorea for 300 years: The Bures family group. *J. Nerv. Ment. Dis.* 76:553-573.

Voaden, D. J., M. Jacobson, W. M. Rogoff, and G. H. Gretz. 1972. Chemical investigation of the sex pheromone of the house fly. *J. Econ. Entomol.* 65:385-359.

Vogel, F., and G. Köhrborn. 1970. *Chemical Mutagenesis in Mammals and Man.* New York: Springer Verlag.

Waddington, C. H. 1957. *The Strategy of the Genes.* London: Allen & Unwin.

Wahlund, S. 1928. Zusammensetzung von populationen un Koerelationserscheinungen von Standpunkt der Vererbungslehre aus betrachet. *Hereditas* 11:65-106.

Walker, R. N. 1962. Body build and behavior in young children. 1. Body build and nursery school teachers' ratings. *Monogr. Soc. Res. Child Dev.* 27(3):1-94.

Wallace, B. 1954. Genetic divergence of isolated populations of *Drosophila melanogaster*. *Caryologia 6* (suppl.):761-764.

Wallace, B. 1968. *Topics in Population Genetics.* New York: Norton.

Waller, J. H. 1971. Achievement and social mobility: Relationships among IQ score, education, and occupation in two generations. *Soc. Biol.* 18:252-259.

Ward, S. 1973. Chemotaxis by the nematode *Caenorhabditis elegans:* Identification of attractants and analysis of the response by use of mutants. *Proc. Natl. Acad. Sci. U.S.A* 70:817-821.

Washburn, S. L., and P. C. Jay. 1968. *Perspectives on Human Evolution.* New York: Holt.

Watson, J. B. 1926. *Behaviorism.* New York: Norton.

Watson, J. D., and F. H. Crick. 1953. Molecular structure of nucleic acids: A structure for deoxyribose nucleic acid. *Nature* 171:737-738.

Wechsler, D. 1971. Intelligence: Definition, Theory, and the IQ. In *Intelligence: Genetic and Environmental Influences.* R. Cancro (ed.). New York: Grune & Stratton.

Wecker, S. C. 1964. Habitat selection. *Sci. Am.* 211(4):109-116.

Weisz, P. 1969. *Elements of Biology.* New York: McGraw-Hill, p. 441.

White, M. J. D. 1973. *The Chromosomes,* 6th ed. New York: Halsted Press.

White, N. G., and P. A. Parsons. 1973. Genetic and socio-cultural differentiation in the aborigines of Arnhem Land, Australia. *Am. J. Phys. Anthropol.* 38:5-14.

Whiting, P. W. 1932. Reproductive reactions of sex mosaics of a parasitic wasp. *J. Comp. Psychol.* 14:345-363.

Whiting, P. W. 1934. Eye colors in the parasite wasp *Habrobracon* and their behavior in multiple recessives and mosaics. *J. Genet.* 29:99-107.

Whiting, P. W. 1939. Mutant body colors in the parasitic wasp *Habrobracon juglandis* and their behavior in multiple recessives and mosaics. *Proc. Am. Phil. Soc.* 80:65-85.

Whittier, J. R., A. Heimler, and C. Korenyi. 1972. The psychiatrist and Huntington's disease (chorea). *Am. J. Psychiatry* 128:96-100.

Wienckowski, L. A. 1972. Schizophrenia: Is There An Answer? Washington, D.C.: U.S. Department of Health, Education, and Welfare.

Wiersma, L. (ed.). 1967. *Invertebrate Nervous Systems: Their Significance for Mammalian Neurophysiology.* (Conference on Invertebrate Nervous Systems. California Institute of Technology.) Chicago: University of Chicago Press.

Wilcock, J. 1969. Gene action and behavior: An evaluation of major gene pleiotropism. *Psychol. Bull. 72*:1-29.

Wilson, E. O. 1973. Group selection and its significance for ecology. *Biol. Sci. 23*:631-638.

Winokur, G. 1973. Genetic Aspects of Depression. In *Separation and Depression.* J. P. Scott and E. C. Senay (eds.). Washington, D.C.: American Association for the Advancement of Science.

Wolken, J. J., A. D. Mellom, and G. Contis. 1957. Photoreceptor structure. 2. *Drosophila melanogaster. J. Exp. Zool. 134*:383-406.

Wood-Gush, D. G. M. 1954. Observations on the nesting habits of brown leghorn hens. *Tenth World's Poult. Congr.* pp. 187-192.

Wood-Gush, D. G. M. 1963. The control of nesting behavior in the domestic hen. 1. The role of the oviduct. *Anim. Behav. 11*:293-299.

Wood-Gush, D. G. M. 1969. Laying in battery cages. *Wild's Poultry Sci. J. 25*:145.

Wood-Gush, D. G. M. 1972. Strain differences in response to sub-optimal stimuli in the fowl. *Anim. Behav. 20*:72-76.

Woolf, C. M. 1971. Congenital cleft lip: A genetic study of 496 propositi. *J. Med. Genet. 8*:65-84.

Wright, J., and R. Pal. (eds.). 1967. *Genetics of Insect Vectors of Disease.* New York: Elsevier.

Wright, S. 1934. The results of crosses between inbred strains of guinea pig, differing in number of digits. *Genetics 19*:537-551.

Wright, S. 1955. Population genetics: The nature and cause of genetic variability in populations. *Cold Spring Harbor Symp. Quant. Biol. 7*:16-24.

Wright, S., and T. Dobzhansky. 1946. Genetics of natural populations. 12. Experimental reproduction of some of the changes caused by natural selection in certain populations of *Drosophila pseudoobscura. Genetics 31*:125-156.

Wright, T., and M. Ashburner. 1976. *Genetics and Biology of Drosophila.* New York: Academic Press.

Wung-Wai, T., and J. Adler. 1974. Negative chemotaxis in *Escherichia coli. J. Bacteriol. 118*:560-576.

Wynne-Edwards, V. C. 1962. *Amimal Dispersion in Relation to Social Behaviour.* Edinburg: Oliver & Boyd.

Zerbin-Rudin, E. 1969. Zur Genetik der Depressiven Erkrankungen. In *Das Depressive Syndrom.* H. Hippius and H. Selbach (eds.). Munich: Urban & Schwarzenberg.

Index

ABOUT THE BOOK

The text of this book is set in V.I.P. Palatino, a contemporary typeface designed by Herman Zapf, based on Renaissance type styles. The book was edited by Dorothy Obre, designed by Jeanne Juster, set by DEKR Corporation, printed and bound by The Murray Printing Company.